Dense Phase Carbon Dioxide

Dense Phase Carbon Dioxide

Food and Pharmaceutical Applications

Edited by

Murat O. Balaban
Department of Chemical and Materials Engineering
University of Auckland
New Zealand

Giovanna Ferrentino
Department of Materials Engineering and
Industrial Technologies
University of Trento
Italy

WILEY-BLACKWELL
A John Wiley & Sons, Ltd., Publication

Blackwell Publishing was acquired by John Wiley & Sons in February 2007. Blackwell's publishing program has been merged with Wiley's global Scientific, Technical and Medical business to form Wiley-Blackwell.

Editorial Offices
2121 State Avenue, Ames, Iowa 50014-8300, USA
The Atrium, Southern Gate, Chichester, West Sussex, PO19 8SQ, UK
9600 Garsington Road, Oxford, OX4 2DQ, UK

For details of our global editorial offices, for customer services and for information about how to apply for permission to reuse the copyright material in this book please see our website at www.wiley.com/wiley-blackwell.

Library of Congress Cataloging-in-Publication Data

Dense phase carbon dioxide: food and pharmaceutical
applications / edited by Murat O. Balaban, Giovanna Ferrentino.
p. cm.
Includes bibliographical references and index.
ISBN 978-0-8138-0649-5 (hardback)
1. Chemical preservation. 2. Dense phase carbon dioxide pasteurization.
3. Food–Pasteurization. 4. Liquid carbon dioxide–Industrial applications.
5. Dense nonaqueous phase liquids–Industrial applications. I. Balaban, Murat O.
II. Ferrentino, Giovanna.
TP371.46.D46 2012
664'.0286–dc23

2011049103

A catalogue record for this book is available from the British Library.

Wiley also publishes its books in a variety of electronic formats. Some content that appears in print may not be available in electronic books.

Cover image: © Sean Gladwell / Fotolia.com
Cover design by Meaden Creative

Set in 9.5/12.5pt Palatino by SPi Publisher Services, Pondicherry, India
Printed and bound in Malaysia by Vivar Printing Sdn Bhd

1 2012

Contents

Preface

There are social evolutionary theories that propose that a critical factor in the progress and advancement of human societies was, is, and will be the production of, and access to, adequate food supplies. Despite the accelerating increase in the internationalization of food trade, the supply chain is far from perfect. Improvement in the production, processing, storage, and distribution of food is still an unfinished business.

Yet consumers worldwide are becoming more sophisticated and knowledgeable. Foods that used to be local or regional are now coming to be known worldwide. Consumers demand more, and expect more. Safe food is the primary expectation, but still people all over the world experience food poisoning with unacceptable human and economic consequences. Consumers also demand and expect "better" foods. Quality expectations in terms of nutritional value and sensory attributes are getting more stringent. At the same time, there is fierce competition in the marketplace with thousands of new products being introduced every year, and thousands failing and being replaced. This creates serious economic pressures to reduce cost, to increase production, processing, and distribution efficiency, and to optimize their integration. On top of all this, there is a growing concern about the effect of human activities, including agriculture and food production, on the environment, with its consequences on the sustainability of these endeavors.

This is a complex scene where the players (the food industry, regulatory agencies, scientists, and consumers) have sometimes cooperating and sometimes adversarial roles. Yet, our overall objectives are the same: to provide a safe, excellent-quality (in terms of nutritional value and sensory properties), affordable, convenient, optimally distributed, readily available, and sustainable food supply.

In this complex, sometimes conflicting, and ever-changing environment, the introduction of new technologies, new methods, and new products is becoming even more challenging. The food industry, with its dependence on volume sales and its small profit margin, is conservative and reluctant to change. Yet the changing needs of the marketplace, and increasing and conflicting pressures from consumers, are felt best by the industry. The increasing adoption of nonthermal technologies is therefore an indication that they fulfill a need, a niche in the marketplace.

Dense phase carbon dioxide (DPCD) is a nonthermal method of food processing. Its application is quite different from, for example, supercritical extraction with CO_2. In extraction, the typical solubility of materials in CO_2

is in the order of 1%. This requires large volumes of CO_2 to be used. In contrast, processing with DPCD requires much less CO_2. The typical solubility of CO_2 in many types of liquid foods is in the order of 5%. Even if more CO_2 was added, the excess would not dissolve and therefore would not be effective in, for example, microbial reduction or enzyme inactivation. The pressures used are at least one order of magnitude less than those typically used in ultra-high pressure (UHP) processing. There is no noticeable temperature increase with DPCD due to pressurization. The small weight percentage of CO_2 used assures this. The typical process temperatures are less than 40°C.

DPCD temporarily reduces the pH of the liquid foods with effects on microorganisms and enzymes. Because oxygen is removed from the environment, and because temperature is not high during the short process time (typically about 5 min in continuous systems), nutrients, antioxidant activity, and vitamins are preserved much better than in the case of thermal treatments. The economics of the operation have been studied, and look promising. Although the capital and operating costs are still higher than those of thermal treatments, they are much lower than those of other nonthermal technologies (e.g., UHP operations).

With all its potential for safe foods without nutrient or quality loss, DPCD still has not achieved commercial operation status. A significant number of scientific publications demonstrate its effectiveness against microorganisms and enzymes, and the rate of increase of publications is accelerating. Some of those recent publications are excellent reviews of the technology and its applications. Yet there has been no book that brought together this increasing body of knowledge. This book benefits from the experience and knowledge of individual and groups of international scientists and members of the industry. The editors are indebted to the contributors: the book could not have become a reality without their expertise, experience, and willingness to contribute to this effort.

On behalf of all authors, the editors hope that this book would benefit researchers working in the area of DPCD, and in other nonthermal or traditional technologies with the potential to be used in conjunction with DPCD. We also hope that the industry, the regulatory agencies, and, most importantly, consumers would find interesting and useful information in the book. We believe that DPCD will find its niche in the safe and minimal processing of foods, and will take its place among other commercial nonthermal food- and pharmaceutical-processing operations.

Murat O. Balaban
Giovanna Ferrentino

Contributors

Murat O. Balaban
Department of Chemical and Materials Engineering
University of Auckland
Auckland, New Zealand

Patricia Ballestra
Department of Bioengineering
IUT Périgueux – Bordeaux IV
Périgueux, France

Thelma Calix
Department of Food Science
Zamorano University
Francisco Morazán, Honduras

Maria G. Corradini
Instituto de Tecnología
Facultad de Ingeniería y Ciencias Exactas
Universidad Argentina de la Empresa
Ciudad de Buenos Aires, Argentina

Luu Thai Danh
School of Chemical Engineering
University of New South Wales
Sydney, Australia

Frank Devlieghere
Department of Food Safety and Food Quality
Laboratory of Food Microbiology and Food Preservation
Food2Know
Ghent University
Ghent, Belgium

Kathy Elst
Business Unit Separation and Conversion Technology
Flemish Institute for Technological Research (VITO)
Mol, Belgium

Osman Erkmen
Department of Food Engineering
University of Gaziantep
Gaziantep, Turkey

Giovanna Ferrari
Chemical and Food Engineering Department
ProdAl S.c.ar.l. – Centro Regionale di Competenza sulle Produzioni Agroalimentari
University of Salerno
Fisciano, Salerno, Italy

Giovanna Ferrentino
Department of Materials Engineering and Industrial Technologies
University of Trento
Trento

Neil R. Foster
School of Chemical Engineering
University of New South Wales
Sydney, Australia

Linsey Garcia-Gonzalez
Business Unit Separation and Conversion Technology
Flemish Institute for Technological Research (VITO)
Mol, Belgium

Ireneo Kikic
Department of Materials and Natural Resources
University of Trieste
Trieste, Italy

Xiaojun Liao
National Engineering Research Center for Fruit
and Vegetable Processing
College of Food Science and Nutritional Engineering
China Agricultural University
Beijing, China

Raffaella Mammucari
School of Chemical Engineering
University of New South Wales
Sydney, Australia

John S. Novak
Food Lab
City of New York Public Health Laboratory
New York, NY, USA

Micha Peleg
Department of Food Science
University of Massachusetts
Amherst, MA USA

Massimo Poletto
Department of Chemical and Food Engineering
University of Salerno
Fisciano (SA), Italy

Sara Spilimbergo
Department of Materials Engineering and Industrial Technologies
University of Trento
Trento, Italy

Wen Hui Teoh
Department of Chemical Engineering
Faculty of Engineering
University of Malaya, Malaysia
and
School of Chemical Engineering
University of New South Wales
Sydney, Australia

Luc Van Ginneken
Business Unit Separation and Conversion Technology
Flemish Institute for Technological Research (VITO)
Mol, Belgium

James T.C. Yuan
Global Beverage R&D
PepsiCo
Valhalla, NY, USA

1 Introduction to Dense Phase Carbon Dioxide Technology

Giovanna Ferrentino and Murat O. Balaban

Abstract: The introduction aims to bring together accumulated knowledge in the areas of supercritical and dense phase CO_2 technology. A summary is given of the areas covered by the book's chapters. Recognized experts in their fields discuss the following topics: solubility of CO_2 in liquids; the effects of supercritical and dense phase CO_2 on microorganisms, including vegetative cells and spores; the application of supercritical and dense phase CO_2 on juices, beverages, and dairy products; progress in the use of pressurized CO_2 in pharmaceuticals. Finally an outlook regarding the future of the technology is presented.

Keywords: overview; dense phase; carbon dioxide; nonthermal processing.

Nonthermal technologies have gained increasing importance in recent years as potentially valuable processes to replace or complement the traditional technologies, currently used for preserving foods and other biological materials. Traditionally, many foods are thermally processed by subjecting them to a temperature range from 60°C to more than 100°C for few seconds to several minutes (Jay 1992). During thermal treatments, heat transferred to the food kills vegetative cells of microorganisms, yeast, and molds, and also inactivates spores depending on the severity of the applied conditions. This process also inactivates many undesirable enzymes in foods that cause quality loss. However, thermal treatment may cause unwanted reactions leading to undesirable changes or formation of by-products in the food.

Thermally processed foods can undergo organoleptic changes and a cooked flavor accompanied by a loss of vitamins, essential nutrients, and flavors.

Increased consumer demand for safe, nutritious, fresh-like food products with a high organoleptic quality and an extended shelf life resulted in the concept of preserving foods using nonthermal methods.

Dense Phase Carbon Dioxide: Food and Pharmaceutical Applications, First Edition. Edited by Murat O. Balaban and Giovanna Ferrentino.

During nonthermal processing, the temperature of the food is held below temperatures normally used in thermal processing. Therefore, the quality degradation expected from high temperatures is reduced and some vitamins, essential nutrients, and flavors are expected to undergo minimal or no changes (Barbosa-Cánovas 1998). High hydrostatic pressure (HHP), dense phase carbon dioxide (DPCD), oscillating magnetic fields, high-intensity pulsed electric fields, intense light pulses, irradiation, cold plasma, chemicals, biochemicals, and hurdle technology are some of the possibilities in the area of nonthermal technologies. Compared to traditional techniques, these avoid drawbacks such as loss of flavors and nutrients, production of undesirable side reactions, as well as changes in physical, mechanical, and optical properties of the food treated.

The use of dense phase carbon dioxide (DPCD) has been proposed as an alternative nonthermal pasteurization technique for foods (Spilimbergo *et al.* 2002), for the first time in the 1950s by Fraser (1951) and Foster *et al.* (1962) who reported the disruption of bacterial cells by the rapid decompression of carbon dioxide (CO_2) from a pressure of 500 lbf/in^2 (about 3.45 MPa) to ambient pressure. DPCD involves mostly the supercritical state of CO_2, but may also involve subcritical gases and sometimes liquids under pressure. In the DPCD technique, food is contacted with (pressurized) sub- or supercritical CO_2 for a certain amount of time in batch, semibatch, or continuous equipment. The DPCD technique presents some advantages over HHP due to its milder process conditions. The pressures applied are much lower (generally <30 MPa) compared to those used in HHP (300–1200 MPa). In addition, capital expenditure is considerably lower. In 1969 Swift & Co. (Chicago, IL) obtained the first US patent for food product pasteurization with CO_2 at "super atmospheric" pressure. Since 1980, others demonstrated the bacteriostatic action and inhibitory effect of CO_2 on the growth and metabolism of some microorganisms. *Pseudomonas* was found to be very sensitive while other types, such as *Lactobacillus* and *Clostridium*, were less sensitive. However, it was with the work published by Kamihira *et al.* (1987) that the inhibitory effect of CO_2 under pressure started to be addressed systematically. These authors tested the sterilizing effect of CO_2 in supercritical, liquid and gaseous phase on wet and dry *Escherichia coli, Staphilococcus aureus* and conidia of *Aspergillus niger* by using a supercritical fluid extraction apparatus. Since then, many studies investigated the effects of CO_2 on pathogenic and spoilage organisms, vegetative cells and spores, yeasts and molds, and enzymes. It has been proven that this technique can be considered a cold pasteurization method that affects microorganisms and enzymes, using CO_2 under pressures below 50 MPa without exposing foods to the adverse effects of heat. Thus foods retain their physical, nutritional and sensory qualities.

DPCD pasteurization of liquid foods is operational and almost ready to be employed on a commercial scale. Most of the commercialization efforts have been performed so far by Praxair Inc. (Burr Ridge, IL, US). Based on the technology, licensed from the University of Florida (Balaban *et al.* 1995; Balaban 2004a, b), Praxair developed a continuous process system which utilizes the DPCD as a nonthermal process alternative to thermal pasteurization (Connery *et al.* 2005). This system has been registered under the trademark "Better Than Fresh (BTF)." Praxair constructed four mobile BTF units for processing about 1.5 L/min of liquid foods for demonstration purposes. In addition, a commercial-scale unit of 150 L/min has also been constructed (Connery *et al.* 2005) and tested in an orange juice–processing plant in Florida. For the continuous treatment of liquid foods, pilot-scale equipment was also manufactured by Mitsubishi Kakoki Co. (Tokyo, Japan) on behalf of and according to the patents owned by Shimadzu Co. (Kyoto, Japan) (Osajima *et al.* 1997, 1999a, b). This equipment consisted of a vessel of 5.8 L through which CO_2 and liquid foodstuff were simultaneously pumped at maximum flow rates of 3.0 kg/h and 20 kg/h, respectively. In 2003, the apparatus was made available only to research laboratories in Japan (private communication, Shimadzu Belgium). At the moment, we have no information available on further commercialization efforts of Shimadzu in the field of DPCD processing for liquid foods. On the basis of their own patent (Sims 2000), PoroCrit LLC (Berkeley, CA, US) also developed a membrane contactor consisting of several hollow-fiber membrane modules for the continuous DPCD pasteurization of liquid foods, mainly beverages, juices, milk, and wine.

As for all non-thermal technologies, the most important issue involved in the commercialization of DPCD process is the regulatory approval. Foods processed thermally or nonthermally must comply with the safety regulations set forth by the US Food and Drug Administration prior to being marketed or consumed. For example, the regulations for thermally processed low-acid canned foods are contained in Title 21, Part 113 of the US Code of Federal Regulations, entitled "Thermally Processed Low-Acid Foods Packaged in Hermetically Sealed Containers." The regulations in Title 21 were established to evaluate (1) the adequacy of the equipment and procedures to perform safe processing operations, (2) the adequacy of record keeping proving safe operation, (3) justification of the adequacy of process time and temperature used, and (d) the qualifications of supervisory staff responsible for thermal-processing and container closure operations (Teixeira 1992). However, the validation of DPCD as a nonthermal method and the determination of compliance regulations necessary for commercialization are complex and challenging. The progress in the validation needs to be encouraged to address the regulatory needs in the near future.

This volume attempts to bring together the accumulated knowledge in the area of DPCD. Experts in many areas have contributed to this book regarding the following topics:

- Solubility of CO_2 in liquids, both from a thermodynamics–theoretical perspective, including models, and from an experimental approach. This is critical because the accumulated evidence points to the critical contribution of dissolved (and saturated) CO_2 to microbial inactivation.
- Effects of DPCD on microorganisms, including vegetative cells and spores; kinetics of microbial inactivation with DPCD; and inactivation of certain enzymes with DPCD. This area is very important because, first and foremost, a new technology must prove the safety of the foods that are processed using it.
- Application of DPCD on juices, beverages, and dairy products. There is a growing body of work in this area, reporting on effects on microorganisms, and especially on quality attributes and nutritional contents. If DPCD applications are to be successful, the organoleptic and nutritional quality of foods processed using DPCD must be known.
- Progress in the use of pressurized CO_2 in pharmaceuticals. This is a well-developed area, with many applications. It also opens up the exciting field of treatment of solids and powders with DPCD.
- An overview of the current technology. This chapter is written by experts in the industry, who have been involved with the design, development, and commercialization efforts of DPCD. Their hands-on views are very valuable.
- An outlook regarding the future of DPCD technology.

It is sincerely hoped that the reader will find the book valuable in bringing information, research results, and most importantly an extensive bibliography in the nonthermal field of DPCD.

2 Thermodynamics of Solutions of CO_2 with Effects of Pressure and Temperature

Sara Spilimbergo and Ireneo Kikic

Abstract: Knowledge of the thermodynamics of CO_2 solutions under pressure is fundamental to the investigation of both the inactivation mechanism and the efficiency of the process as a function of the process parameters of temperature and pressure. In order to evaluate the properties of the solutions during DPCD treatment, it is essential to have reliable high-pressure experimental data and accurate thermodynamic models over a broad range of conditions. A large number of experimental and theoretical studies have been conducted on CO_2 solubility in pure water; however, the literature lacks studies concerning the phase behavior of CO_2–water solutions or CO_2–solid components in food applications. In the present chapter. a general survey of the published knowledge concerning the thermodynamics of CO_2–water phase systems under pressure is given focusing on both the theoretical aspects and the applications, for electrolyte and non-electrolyte models.

Keywords: thermodynamic model; liquid–vapor phase equilibria; equation of state.

2.1 Introduction

The phase behaviour of the system carbon dioxide (CO_2) + water is of great industrial and scientific interest. For instance, in the petroleum industry many natural gases contain acidic gases which have to be removed, such as CO_2 and hydrogen sulphide. Another industrial application is the decontamination of wastewater streams containing dissolved acidic gases. The capture of CO_2, a greenhouse gas, has become a great concern: different storage techniques have been considered such as the storage of CO_2 in deep saline reservoirs. Also, in geochemical applications, especially in the analysis of CO_2-bearing fluid inclusions in minerals, the accurate description of solubility of CO_2 in pure-water solutions is required. Recently, the increasing attention to CO_2 processing in food applications, in particular in

Dense Phase Carbon Dioxide: Food and Pharmaceutical Applications, First Edition. Edited by Murat O. Balaban and Giovanna Ferrentino.

microbial inactivation of food products at low temperature compared to the traditional thermal treatments, makes the knowledge of a CO_2–water phase system under pressure fundamental to investigate both the inactivation mechanism and the efficiency of the process as a function of the operating parameters of temperature and pressure.

Therefore, it is essential to have reliable high-pressure experimental data and an accurate thermodynamic model over a broad range of conditions to evaluate the solubility in different conditions. For most of these applications, temperatures (*T*) up to 100°C and pressures (*P*) up to 100 MPa are particularly relevant.

A large number of experimental and theoretical studies have been conducted on CO_2 solubility in pure water. However, to the best of our knowledge, literature lacks studies concerning phase behaviour of CO_2–liquid solutions or CO_2–solid components in food applications. The objective of the present chapter is to give a general survey of the published knowledge concerning the thermodynamics of CO_2–water phase systems under pressure, focusing on the theoretical aspects and the applications, for both electrolyte and non-electrolyte models.

2.2 Thermodynamics of liquid–vapour phase equilibria

According to the second law of thermodynamics, the total Gibbs energy of a closed system at constant temperature and pressure is minimum at equilibrium. If this condition is combined with the condition that the total number of moles of component *i* is constant in a closed system,

$$\sum_{\alpha} n_i^{\alpha} = \text{constant} \tag{2.1}$$

Where n_i^{α} is the number of moles of component *i* in phase α, it can be derived that for a system of Π phases and *N* components, the equilibrium conditions expressed in terms of chemical potential (μ_i) are as follows (Smith *et al.* 2005):

$$\mu_i^{\alpha} = \mu_i^{\beta} = \ldots\ldots = \mu_i^{\Pi} \tag{2.2}$$

for $i = 1$ to *N*.

The chemical potential of component *i* in phase α is defined by

$$\mu_i^{\alpha} = \left(\frac{\partial \left(\sum_i n_i^{\alpha} g^{\alpha} \right)}{\partial n_i^{\alpha}} \right)_{P,T,n_{j \neq i}^{\alpha}} \tag{2.3}$$

where *g* is the molar Gibbs energy. Since μ_i^{α} is a function of *P*, *T* and (*N*–1) mole fractions (the additional condition $\sum_i x_i^{\alpha} = 1$ makes one of the mole fractions a dependent variable), Equation (2.2) represents $N(\Pi - 1)$ equations in $2 + \Pi(N - 1)$ variables. Therefore the number of degrees of freedom F is

$$F = 2 + \Pi\,(N - 1) - N(\Pi - 1) = 2 - \Pi + N \tag{2.4}$$

Equation (2.4) is the *phase rule* of Gibbs. According to this rule a state with Π phases in a system with N components is determined (all intensive thermodynamic properties can be calculated) if the values of F variables are fixed, provided that g values of all phases as functions of pressure, temperature and composition are known.

For practical applications (calculation of the composition of phases) the explicit dependence of chemical potentials on composition, temperature and pressure is needed. For this reason auxiliary functions such as the fugacity coefficient and the activity coefficient are used. These functions are closely related to the Gibbs energy and assume the ideal gas and the ideal solution behaviour in the reference state.

The fugacity of component i in a mixture $\hat{f}_i$ is defined by

$$d\mu_i \equiv RT \ln \hat{f}_i \tag{2.5a}$$

at constant T (where the symbol ^ indicates the value of the quantity in a mixture) with

$$\lim_{P \to 0} \frac{\hat{f}_i}{P_i} = 1 \tag{2.5b}$$

According to this definition, $\hat{f}_i$ is equal to the partial pressure P_i in the case of an ideal gas. The fugacity coefficient $\hat{\phi}_i$ is defined by

$$\hat{\phi}_i = \frac{\hat{f}_i}{P_i} \tag{2.6}$$

and is a measure for the deviation from ideal gas behaviour.

The fugacity coefficient ($\hat{\phi}_i$ can be calculated from an equation of state by one of the following expressions (Prausnitz *et al.* 1999):

$$\ln \hat{\phi}_i = \frac{1}{RT} \int_0^P \left(\left(\overline{v}_i - \frac{RT}{P} \right) \right) dP \tag{2.7}$$

$$RT \ln \hat{\phi}_i = \frac{1}{RT} \int_{V=\infty}^{V} \left[\frac{RT}{V} - \left(\frac{\partial P}{\partial n_i} \right)_{T,V,n_{j\neq 1}} \right] dV + RT \ln \left(\frac{\sum_i n_i RT}{PV} \right) \tag{2.8}$$

According to Equation (2.5a), the equilibrium relation Equation (2.2) can be replaced by

$$\hat{f}_i^{\alpha} = \hat{f}_i^{\beta} = ... = \hat{f}_i^{\pi} \tag{2.9}$$

for $i = 1$ to N.

This approach for the calculation of the fugacities in mixtures can be used for gaseous and condensed phases. This approach followed for all the phases at equilibrium is named the $\hat{\phi} - \hat{\phi}$ method since the fugacities of both phases are calculated with Equation (2.7) or (2.8) using an equation of state.

However, the calculation of the integral in Equation (2.8) needs an equation of state for the description of volumetric properties of the phase considered from low density (ideal gas) to the actual density of the phase. For this reason an alternative method for the calculation of the fugacities in condensed-phase mixtures is proposed. In this case, the reference is not the behaviour of an ideal gas mixture (that normally is far away from that of a solid or liquid mixture) but that of an 'ideal mixture'. Excess functions are defined to describe the departure of the properties of a real mixture from the ideal behaviour.

It is very useful from a practical point of view to define as 'ideal' the behaviour of a mixture when it is possible to predict it from the knowledge of the properties of the pure components involved at the same temperature and pressure.

The activity a_i is defined as the ratio of $\hat{f}_i$ and the fugacity of component i in the standard state at the same P and T:

$$a_i = \frac{\hat{f}_i(T,P,x)}{f_i^0(T,P,x^0)} \tag{2.10}$$

In the ideal solution:

$$a_i^{id} \equiv x_i \tag{2.11}$$

The activity coefficient of component i, γ_i, measures the deviation from ideal solution behaviour:

$$\gamma_i = \frac{a_i}{a_i^{id}} \tag{2.12}$$

so the fugacity of a solid or liquid solution can be written as

$$\hat{f}_i = x_i \gamma_i f_i^0 \tag{2.13}$$

The activity coefficient γ_i can be calculated from the molar excess Gibbs energy g^E:

$$RT \ln \gamma_i = \left(\frac{\partial \left(\sum_i n_i g^E \right)}{\partial n_i} \right)_{P,T,n_{j \neq i}} \tag{2.14}$$

The standard-state fugacity of the liquid or solid component f_i^0 is usually the fugacity of the pure solid or liquid component at the same temperature and pressure and is related to the sublimation pressure P_i^{sub} or vapour pressure P_i^{sat}, respectively.

On the sublimation curve of a pure component, we have

$$f_i^S(P_i^{sub},T) = f_i^V(P_i^{sub},T) = \phi_i^V(P_i^{sub},T)P_i^{sub} \tag{2.15}$$

where superscripts *s* and *V* indicate a solid or vapour phase, respectively.

$$f_i^S(P,T) = f_i^S(P_i^{sub},T)\exp\left(\int_{P_i^{sub}}^{P}\frac{d\mu_i^S}{RT}\right) = f_i^S(P_i^{sub},T)\exp\left(\int_{P_i^{sub}}^{P}\frac{v_i^S}{RT}dP\right) \tag{2.16}$$

where v_i^S is the molar volume of pure solid *i*.

Combining Equations (2.15) and (2.16), we get

$$f_i^S(P,T) = \phi_i^V(P_i^{sub},T)P_i^{sub}\exp\left(\int_{P_i^{sub}}^{P}\frac{v_i^S}{RT}dP\right) \tag{2.17}$$

A similar derivation is possible for a liquid:

$$f_i^L(P,T) = \phi_i^V(P_i^{sat},T)P_i^{sat}\exp\left(\int_{P_i^{sat}}^{P}\frac{v_i^L}{RT}dP\right) \tag{2.18}$$

At low pressure the fugacity coefficients and the exponential terms are close to 1, so

$$f_i^S \approx P_i^{sub} \quad \text{and} \quad f_i^L \approx P_i^{sat} \tag{2.19}$$

From this assumption for the standard state fugacity, it follows that the activity coefficient of component i is equal to unity when the composition $x_i = 1$ (i.e. for the pure component *i*).

This assumption is convenient if the component, at the temperature and pressure conditions of the mixture, exists in the same physical state. This is neither realistic nor practical when considering liquid mixtures if one of the components is supercritical. In this case the more convenient reference state is not that of pure component but that of the component infinitely diluted. The fugacity of the component i in an ideal mixture when $x_i \to 0$ is expressed by

$$\hat{f}_i^L(T,P,x_i) = x_i H_i(T,P) \tag{2.20}$$

where H_i is the Henry constant and is a function of temperature, pressure and the solvent–solute pair in the system. At high concentration there is no linear proportionality between $f_i^L(T,P,x_i)$ and the molar fraction x_i. Equation (2.20) can also be used for high concentrations considering a new activity coefficient γ_i^*:

$$\hat{f}_i^L(T,P,x_i) = x_i H_i(T,P)\gamma_i^*(T,P,x_i) \tag{2.21}$$

This new coefficient is different from γ_i: when $x_i \rightarrow 0$, $\gamma_i^* \rightarrow 1$ while when $x_i \rightarrow 1$, $\gamma_i \rightarrow 1$. It is possible to write

$$\hat{f}_i^L(T,P,x_i) = x_i H_i(T,P)\gamma_i^*(T,P,x_i) = x_i \gamma_i(T,P,x_i) f_i^0(T,P) \quad (2.22)$$

If the fugacities of the phases at equilibrium are calculated using the equation of state approach through Equation (2.8) for one phase and the activity coefficient approach (Equation (2.22) for the other phase, the method is named the γ–φ or activity–fugacity coefficient approach.

2.2.1 Calculation of γ

In typical mixtures, the ideal (or Raoult law) approximation provides no more than a rough approximation and it is valid only when the components are similar. The activity coefficient, therefore, plays a key role in the calculation of vapour–liquid equilibria.

Classical thermodynamics gives information on the effect on the activity coefficients of pressure (related to the partial molar volume) and of temperature (related to the partial molar enthalpy).

The Gibbs–Duhem equation, however, is a useful tool for correlating and extending limited experimental data due to the fact that, in a mixture, the activity coefficients of the individual components are not independent of one another but are related by a differential equation. For a binary mixture, the equation is

$$x_1\left(\frac{\partial \ln \gamma_1}{\partial x_1}\right)_{T,P} = x_2\left(\frac{\partial \ln \gamma_2}{\partial x_2}\right)_{T,P} \quad (2.23)$$

This equation means that in a binary mixture, activity coefficient data for one component can be used to predict the activity coefficient of the other component. Alternatively, with extensive experimental data for both activity coefficients as a function of composition, it is possible to test the data for thermodynamic consistency by determining whether or not the data obey Equation (2.23). In the case of limited data, the integral form of the Gibbs–Duhem equation provides an equation to extend the information. To do so, it is necessary to assume a mathematical expression of the excess energy as a function of composition. The numerical values of the constants are calculated from the fitting of the limited data. Normally these constants, independent from composition, depend on temperature. Once the values of the constants are known, the activity coefficients can be calculated by differentiation.

Different expressions are proposed for the evaluation of the excess energy (Prausnitz *et al.* 1999): usually for moderately non-ideal systems, all equations containing two or more parameters give good results.

The older expressions are those proposed by Van Laar and by Margules (see the list following in this section). These expressions are mathematically easier to handle than the modern and new models (the Wilson, Nonrandom

Two Liquids (NRTL) and Universal Quasi-Chemical (UNIQUAC) equations), but their use is often limited to the correlation of data for binary mixtures alone (for this reason, in the list for these models, the equations are reported for binary systems).

The Wilson equation, with two parameters, gives good results for strongly non-ideal binary mixtures and it is often used for vapour–liquid calculations: it is simpler than the UNIQUAC equation and contains only two parameters per binary mixtures, whereas the NRTL equation requires the knowledge of three parameters. Also the dilute region is represented with reasonable accuracy. The main deficiency of the Wilson equation is the impossibility to represent mixtures which exhibit a miscibility gap. For the systems that have incomplete miscibility and as a consequence are very non-ideal, the NRTL and UNIQUAC equations are very useful. The NRTL equation contains three parameters but very often the numerical value of the nonrandomness parameter α_{ji} is fixed (at the value 0.2 or 0.3). The UNIQUAC equation has the advantage that it uses only two parameters (with often a lesser dependence on temperature) and since the primary concentration variable is a surface fraction (rather than mole fraction), it is applicable to solutions containing small or large molecules.

Sometimes there is a total absence of experimental data to fit for the evaluation of the parameters used in the models for the activity coefficients. Then, a completely predictive model must be used. These models are based on the groups' contributions concept. Each molecule in the mixture is considered as formed by functional groups. The behaviour of the mixture can be predicted by knowing the interaction between the functional groups in the mixture. The interaction between functional groups, called a_{mn}, is evaluated by studying the experimental data of known mixtures and it is assumed to be the same in every mixture in which the functional groups are considered. This allows predicting the equilibrium for systems with no experimental data. The UNIFAC (Fredenslund *et al.* 1977) and ASOG (Kojima and Tochigi 1979) are the most important models based on the functional groups method for mixtures. The most used is the UNIFAC model because it can be applied to a great number of compounds. The UNIFAC is the version based on the contribution of the functional groups of the UNIQUAC model, while the ASOG uses the Wilson model. To use the UNIFAC model it is necessary to identify the functional groups of the molecules in the system and evaluate the values of the parameters from its tables. The equations describing this model are very similar to that one of the UNIQUAC model:

Van Laar equation

$$\frac{g^E}{RT} = \frac{A_{12}A_{21}}{A_{12}x_1 + A_{21}x_2} x_1 x_2 \tag{2.24a}$$

$$\ln \gamma_1 = A_{12}\left(\frac{A_{12}x_1}{A_{12}x_1 + A_{21}x_2}\right)^2 \tag{2.24b}$$

Margules equation

$$\frac{g^E}{RT} = \left[A_{12} + \left(A_{21} - A_{12}\right)x_1\right]x_1x_2 \tag{2.25a}$$

$$\ln\gamma_1 = \left[A_{12} + 2\left(A_{21} - A_{12}\right)x_1\right]x_2^2 \tag{2.25b}$$

Wilson equation

$$\frac{g^E}{RT} = -\sum_i x_i \ln\left(\sum_j x_j\Lambda_{ij}\right) \tag{2.26a}$$

$$\ln\gamma_i = 1 - \ln\sum_j x_j\Lambda_{ij} - \sum_j \frac{x_j\Lambda_{ji}}{\sum_k x_k\Lambda_{jk}} \tag{2.26b}$$

Parameters

$$\Lambda_{ij} = \frac{v_j}{v_i}\exp - \frac{\lambda_{ij} - \lambda_{ii}}{RT} \tag{2.26c}$$

NRTL equation

$$\frac{g^E}{RT} = \sum_i x_i \frac{\sum_j \tau_{ji}G_{ji}x_j}{\sum_k G_{ki}x_k} \tag{2.27a}$$

$$\ln\gamma_i = \frac{\sum_j \tau_{ji}G_{ji}x_j}{\sum_k G_{ki}x_k} + \sum_j \frac{x_jG_{ij}}{\sum_k G_{kj}x_k}\left[\tau_{ij} - \frac{\sum_k x_k\tau_{kj}G_{kj}}{\sum_m G_{mj}x_m}\right] \tag{2.27b}$$

Parameters

$$\tau_{ji} = \frac{g_{ji} - g_{ii}}{RT} \tag{2.27c}$$

$$G_{ji} = \exp(-\alpha_{ji}\tau_{ji}) \tag{2.27d}$$

UNIQUAC equation

$$\frac{g^E}{RT} = \sum_i x_i \ln\frac{\varphi_i}{x_i} + \frac{z}{2}\sum_i q_ix_i \ln\frac{\vartheta_i}{\varphi_i} - \sum_i q_ix_i \ln\left[\sum_j \vartheta_j\tau_{ji}\right] \tag{2.28}$$

$$\ln\gamma_i = 1 - \frac{\varphi_1}{x_i} + \ln\frac{\varphi_i}{x_i} + \frac{z}{2}q_i\left(1 - \frac{\vartheta_i}{\varphi_i} + \ln\frac{\vartheta_i}{\varphi_i}\right) + q_i\left[1 - \ln\left(\sum_j \vartheta_j\tau_{ji}\right)\right] - \sum_j \vartheta_j\frac{\tau_{ij}}{\sum_k \vartheta_k\tau_{kj}} \tag{2.29a}$$

Parameters

$$\tau_{ji} = \exp -\frac{u_{ji} - u_{ii}}{RT} \tag{2.29b}$$

$$\varphi_i = \frac{x_i r_i}{\sum_j x_j r_j} \qquad \vartheta_i = \frac{x_i q_i}{\sum_j x_j q_j} \tag{2.29c}$$

UNIFAC equation

$$\ln y_i = \ln y_i^C \text{ (combinatorial)} + \ln y_i^R \text{ (residual)}$$

$$\ln \gamma_i \left(combinatorial \right) = \left(\ln \frac{\varphi_i}{x_i} + 1 - \frac{\varphi_i}{x_i} - \frac{\varphi_i}{x_i} \right) - \frac{1}{2} z q_i \left(\ln \frac{\varphi_i}{\vartheta_i} + 1 - \frac{\varphi_i}{\vartheta_i} \right) \tag{2.30a}$$

where φ_i (volume fraction) and ϑ_i (surface fraction) are expressed as in the UNIQUAC equation, but the values of r_i and q_i of the components are calculated by the addition of the corresponding contributions (R_k and Q_k) of the constituent groups:

$$r_i = \sum_k \upsilon_k^{(i)} R_k \qquad q_i = \sum_k \upsilon_k^{(i)} Q_k \tag{2.30b}$$

and $v_k^{(i)}$ is the number of the k functional group in the i species.

$$\ln \gamma_i (residual) = \sum_k \upsilon_k^{(i)} \left[\ln \Gamma_k - \ln \Gamma_k^{(i)} \right] \tag{2.30c}$$

$$\ln \Gamma_k = Q_k \left[1 - in \left(\sum_m \theta_m \Psi_{mk} \right) - \sum_m \frac{\theta_m \Psi_{mk}}{\Sigma_n \theta_n \Psi_{mn}} \right] \tag{2.30d}$$

$$\Psi_{mn} = \frac{\exp(-a_{mn})}{T} \tag{2.30e}$$

$$\theta_m = \frac{Q_m}{\Sigma_n Q_n X_n} \tag{2.30f}$$

where a_{nm} is the energy interaction between the m and n functional groups. $\Gamma_k^{(i)}$ is the residual activity coefficient of the k functional group if the solution contains only i species. This means that for the i pure species, the residual term is equal to zero.

2.2.2 Calculation of ϕ

The equation of state (EOS) is an analytical relationship between pressure P, temperature T and molar volume V:

$$f(P, T, v) = 0 \tag{2.31}$$

Starting from this equation, the calculation of volumetric and thermodynamic properties of a pure component or of a mixture is possible. Substituting this equation in Equation (2.7) or (2.8), the fugacity coefficients and as a consequence equilibrium between different phases are also calculated.

The equation of state models for the calculation of fugacities can be divided into classes on the basis of different criteria. One is based on the degree of the polynomial used in developing the equation of state in terms of volume. In this case it is possible to divide the EOS in cubic and non-cubic models.

An interesting observation, that can be useful for the classification, is that a common feature of the equations of state is that it is possible to recognize separate contributions resulting from repulsive and attractive interactions.

Following these criteria, the EOS can be separated into three families:

(1) Family of virial equation of state.
(2) Family of Van der Waals–type EOS where the contribution of repulsive and attractive forces is present.
(3) Molecular-based equation of state.

In the virial equation of state, the compressibility factor z is given as the power expansion of the density ρ:

$$z = \frac{Pv}{RT} = 1 + B\rho + C\rho^2 + ... \tag{2.32}$$

where B and C are the second and third virial coefficients that are a function of temperature for a pure fluid or a function of temperature and composition for a mixture.

This theoretical equation was empirically modified by different authors often introducing a large number of constants. These modifications can be useful for the evaluation of pure-component properties, but their extension to mixtures is generally questionable.

For a mixture, the composition dependence is expressed as

$$B = \sum_i \sum_j x_i x_j B_{ij}(T) \tag{2.33}$$

$$C = \sum_i \sum_j \sum_k x_i x_j x_k C_{ijk}(T) \tag{2.34}$$

where B_{ii} is the second virial coefficient for the pure component i and the terms with different indices (ij) are calculated from the average of the second virial coefficients of component i and j. Sometimes, an empirical correction to this average, based on the introduction of a binary interaction parameter k_{ij}, is used. This empirical modification improves the performance of the equation of state but its value must be determined by fitting experimental data.

It is important to emphasize that virial equation of states cannot be applied for the calculation of vapour–liquid equilibrium using the φ-φ approach.

The van der Waals family EOS are derived by modification of the classical van der Waals equation that represents the first attempt to describe the coexistence of a liquid and vapour phase:

$$P = \frac{RT}{v - b} - \frac{a}{v^2} \tag{2.35}$$

or, in term of the compressibility factor z,

$$z = \frac{v}{v - b} - \frac{a}{RTv} \tag{2.36}$$

In these equations the pressure (or the compressibility factor z) is given as a sum of two different contributions: the first, containing the co-volume b, represents the effect of repulsive forces, and the second, containing the 'a' parameter, takes into account the influence of attractive forces. The numerical values of the parameters can be calculated from the critical coordinates, from fixing the critical constraints or from vapour pressures and liquid or vapour densities. The modification of the Van der Waals equation by Redlich and Kwong (1949), who introduced a different temperature dependence and a slightly different volume dependency in the attractive term, is very important since it opened the way to a better description of the temperature-dependent properties like virial coefficients:

$$P = \frac{RT}{v - b} - \frac{a(T)}{v(v + b)} \tag{2.37}$$

where $a(T) = a/T^{1/2}$.

The Redlich-Kwong equation gives a somewhat better critical compressibility ($Z_c = 0.333$ instead of 0.375 from the van der Waals equation), but is still not very accurate for the prediction of vapour pressures and liquid densities.

Soave's modification (Soave 1972) of the temperature dependence of the a parameter, which resulted in accurate vapour pressure predictions (especially above 1 bar) for light hydrocarbons, led to cubic equations of state becoming important tools for the prediction of vapour–liquid equilibria at moderate and high pressures for nonpolar fluids.

$$a(T) = 0.4274\frac{R^2T_c^2}{P_c}\left[1 + (0.48 + 1.57\omega - 0.176\omega^2)\left(1 - \sqrt{T_r}\right)\right]^2 \tag{2.38}$$

where ω is the acentric factor defined as

$$\omega = -\log\frac{P(T_r = 0.7)}{P_c} - 1.0 \tag{2.39}$$

Table 2.1 Attractive terms used in cubic equations of state.

Equation	$-Z^{att}$
van der Waals	$\frac{a}{RTV}$
Redlich-Kwong (1949)	$\frac{a}{(V+b)RT^{1.5}}$
Soave (1972)	$\frac{a(T)}{(V+b)RT}$
Peng-Robinson (1976)	$\frac{Va(T)}{[V(V+b)+c(V-b)]RT}$
Patel-Teja (1982)	$\frac{Va(T)}{[V(V+b)+c(V-b)]RT}$
Trebble and Bishhnoi (1987)	$\frac{va(T)}{[v^2+(b+c)v-(bc+d^2)]}$

Peng and Robinson (1976) used a different volume dependency of the attractive term, which results in slightly improved liquid volumes and changed slightly the temperature dependence of a to give accurate vapour pressure predictions for hydrocarbons in the 6- to 10-carbon-number range.

$$P = \frac{RT}{V-b} - \frac{a(T)}{V(V+b)+b(V-b)} \tag{2.40}$$

The Peng–Robinson (PR) and the Soave–Redlich–Kwong (SRK) equations are widely used since they require little input (only critical properties and an acentric factor to calculate the generalized parameters) and require little computing power.

All these modifications and those proposed by different authors address the attractive part of the cubic equation of state. Some of these are reported in Table 2.1. Other expressions, based on some physical justifications, for the repulsive part, were also proposed, but no cubic equation of state can arise from the combination of some of these expressions.

The greatest use of cubic equation of state is for phase equilibrium calculations involving mixtures. This is most commonly done using the van der Waals one-fluid mixing rules,

$$a = \sum_i \sum_j a_{ij} z_i z_j \tag{2.41}$$

$$b = \sum \sum b_{ij} z_i z_j \tag{2.42}$$

In addition, combining rules are needed for the parameters a_{ij} and b_{ij}. The usual combining rules are

$$a_{ij} = \sqrt{a_{ii}a_{jj}}(1 - k_{ij}) \tag{2.43}$$

and

$$b_{ij} = \frac{(b_{ii} + b_{jj})}{2}(1 - l_{ij}) \tag{2.44}$$

where k_{ij} and l_{ij} are the binary interaction parameters obtained by fitting experimental vapour–liquid equilibrium (VLE) data or VLE and density data. Generally, l_{ij} is set equal to zero, in which case we have the linear mixing rule for the b-parameter $b = \Sigma x_i b_{ii}$.

A shortcoming of the van der Waals classical mixing rules is that they are not applicable to the so-called asymmetric mixtures and to mixtures containing polar compounds. For that reason, different mixing rules have been proposed for the a-parameter, involving essentially a concentration dependence of the k_{ij} (Adachi and Sugie 1986; Panagiotopoulos and Reid 1986).

Since many mixtures of interest in the chemical industry exhibit much greater degrees of non-ideality, and have been traditionally described by activity coefficient (Gibbs energy) models, Huron and Vidal (1979) suggested a method to use excess Gibbs energy models to represent the mixing rule for the a-parameter of the equation of state.

The basic assumptions of the Huron–Vidal method are as follows:

The excess Gibbs energy G^E calculated from a liquid-phase activity coefficient model, and the excess Gibbs energy G^E calculated from the equation of state, are equal at infinite pressure.

The co-volume b is equal to the molar volume V at infinite pressure.

The excess volume at infinite pressure is zero.

By using the linear mixing rule for the volume parameter b, the expression for the parameter a is

$$a = b\left[\sum_i z_i \frac{a_i}{b_i} - \frac{g_\infty^E}{\Lambda}\right] \tag{2.45}$$

where Λ is a constant depending on the equation of state used (Λ is ln 2 for the Redlich–Kwong equation) and g_∞^E is the value of the molar excess Gibbs energy at infinite pressure. For g_∞^E it is possible to choose between the different models proposed in the literature (Prausnitz *et al.* 1999): the Wilson, NRTL or UNIQUAC equations. Huron and Vidal (1979) suggested using the NRTL model:

$$g_\infty^E = \sum_i z_i \frac{\sum_j RTz_j G_{ji}\tau_{ji}}{\sum_k z_k G_{ki}} \tag{2.46}$$

In this equation $\alpha_{ij}, \tau_{ij}, \tau_{ji}$ are adjustable parameters. When $\alpha_{ij} = 0$, the van der Waals mixing rules are obtained.

This mixing rule, when combined with the Wilson or NRTL models, gives excellent results in describing the VLE of some highly non-ideal systems. However, the Huron–Vidal mixing rule has some theoretical and computational difficulties. The mixing rule may not be successful in describing nonpolar hydrocarbon mixtures, and this is a problem when a multicomponent mixture contains both polar and nonpolar components since all species must be represented by the same mixing rule. Furthermore, it is necessary to draw attention to the difficulties of this mixing rule in correlating low-pressure vapour–liquid equilibrium data.

Some efforts have been directed towards relaxing the infinite pressure limit in the Huron–Vidal model (Mollerup 1986; Michelsen 1990; Dahl and Michelsen 1990). The most successful of these is the so-called modified Huron–Vidal first order (MHV1) mixing rule, proposed by Michelsen (1990). In developing the new mixing rule, the Soave–Redlich–Kwong equation of state and the Huron–Vidal approach are used but with the equation of state and excess Gibbs energy models matched at liquid density and zero pressure at the temperature of interest:

$$\frac{b}{RT} = \sum_i z_i \frac{a_i}{RTb_i} + \frac{1}{q_1}\left[\frac{g_0^E}{RT} - \sum_i z_i \, Ln\frac{b_i}{b}\right] \tag{2.47}$$

for q_1, a recommended value of −0.593 is suggested.

In addition, an alternative mixing rule (referred as the second-order modified Huron–Vidal mixing rule, or MHV2) was also derived (Dahl and Michelsen 1990):

$$q_1\left(\frac{b}{RT} - \sum_i z_i \frac{a_i}{RTb_i}\right) + q_2\left(\frac{b^2}{R^2T^2} - \sum_i z_i \frac{a_i^2}{R^2T^2b_i^2}\right) = \frac{g_0^E}{RT} + \sum_i z_i \, Ln\frac{b}{b_i} \tag{2.48}$$

with $q_1 = -0.478$, and $q_2 = -0.0047$

These new mixing rules have the advantage that they allow the use of numerical parameters for the excess-Gibbs-energy models which were obtained by fitting low-pressure vapour–liquid equilibrium data. In particular, the MHV2 mixing rule was used in combination with the UNIFAC group contribution model with excellent results (Dahl *et al.* 1992).

However, these new mixing rules (based on both infinite- and zero-pressure limits) give, for the composition dependence on the second virial coefficient, results that are inconsistent with those obtained from statistical mechanics.

Wong and Sandler (1992; Orbey and Sandler 1995) used the Helmholtz excess energy to develop the following mixing rule that satisfies the second virial restriction:

$$\frac{a}{b} = \sum_i z_i \frac{a_i}{b_i} + \frac{a_\infty^E}{\Lambda} \tag{2.49}$$

where Λ is a constant dependent on the equation of state selected (Λ is equal to ln 2 for the Redlich–Kwong equation), and k_{ij} is a binary interaction parameter.

$$b = \frac{\sum_i \sum_j \left[\frac{(b_i + b_j)}{2} - \frac{(1 - k_{ij})\sqrt{a_i a_j}}{RT} \right] z_i z_j}{1 - \left[\sum_i z_i \frac{a_{ii}}{RTb_i} + \frac{a_\infty^E}{\Lambda RT} \right]} \tag{2.50}$$

These mixing rules were applied to perform critical-point calculations and the critical behaviour of some highly non-ideal systems (Castier and Sandler 1997a, b).

Statistical mechanics and computer simulations have contributed to the development of new generations of equations of state, the so-called molecular-based EOS which, in contrast with those discussed so far, have a sounder theoretical basis.

The original van der Waals idea was that the pressure in a fluid is the result of both repulsive forces or excluded volume effects, which increase as the molar volume decreases, and attractive forces which reduce the pressure. These assumptions can be justified since, having the molecules have a finite size, there would be a limiting molar volume, b, which could be achieved only at infinite pressure. At large separations, on the basis of the London dispersion theory, the attractive forces increase as r^{-6}, where r is the intermolecular distance. Since volume is proportional to r^3, this provides some explanations for the attractive term in the van der Waals equation of state. Nevertheless, modern statistical mechanics has shown that neither the repulsive nor the attractive term in the equation is correct.

The noncubic equations of state are characterized by the use of a repulsive term that is based on the Carnahan–Starling expression (Carnahan and Starling 1969, 1972) derived assuming that molecules behave as hard spheres:

$$z = \frac{1 + \eta + \eta^2 - \eta^3}{(1 - \eta)^3} \tag{2.51}$$

where $\eta = b/4V$.

The attractive part is generally based on that derived from the perturbed hard chain theory (PHCT) (Beret and Prausnitz 1975), or from the statistical associating fluid theory (SAFT) (Chapman *et al.* 1988, 1990). These theories consider that most molecules like polymers or high-molecular-weight compounds do not have a spherical but a more complex structure that is more similar to chains. The thermodynamic properties of the molecules are also more complex than those derived on the basis of hard-body considerations: the dependence on rotational and vibrational motions of the molecules must be taken in consideration.

These approaches were the precursors of many theoretical attractive terms and consequently of different equations of state. Different authors reported applications of PHCT models to polymer systems (Donohue and Prausnitz 1978; Liu and Prausnitz 1979, 1980) and to supercritical systems (Gregorowicz *et al.* 1991; Fermeglia and Kikic 1993). The SAFT equation of state was successively applied to asymmetric (Huang and Radosz 1991) and water–hydrocarbon systems (Economou and Tsonopoulos 1997).

These equations of state can be extended for the description of the properties of mixtures by introducing combination rules for the pure-component parameters. The main advantage is the reduced number of binary parameters required (normally one) that are unrelated to temperature. It is also interesting to emphasize that, in these equations, the binary parameter is a measure of interactions between the segments of the different molecules and it can be used for the evaluation of the properties of mixtures made with components belonging to the same families (Fermeglia and Kikic 1993).

2.2.3 Calculation of the liquid–vapour phase equilibria

The equilibrium equations describe the conditions of the thermodynamic stability of the system. The use of these equations is important to design a great number of unit operations based on the concept of the equilibrium stage and on the tendency of a given chemical species to be in a phase rather than in other one at equilibrium conditions. The measure of this tendency, for liquid–vapour equilibrium, is defined by the equilibrium ratio y_i/x_i:

$$K_i = \frac{y_i}{x_i} \tag{2.52}$$

where K_i is called *equilibrium constant* or *vapour–liquid distribution coefficient*. This parameter is a measure of the *volatility* of a species or its tendency to be in the vapour phase, if $K_i > 1$ the "i" species.

The Duhem theory (Sandler 1989) states that the equilibrium stage of a closed system can be described by pressure and temperature which are uniform in all the equilibrium phases. In this way it is possible to evaluate the equilibrium phase compositions by knowing the global compositions $z_1, z_2, z_3, \ldots, z_m$ of the m components of the system. This kind of calculation, applied to solve a liquid – vapour equilibrium problem, is defined as *flash calculation*. The algorithm to solve the problem is shown in Figure 2.1.

Conceptually the way to calculate the molar fraction of each species in both phases is simple. Comparing the fugacity it is possible to calculate the vapour and liquid fractions of each species i in the equilibrium conditions. It is easy to solve the equilibrium if a calculator is available and capable to solve N equations simultaneously. The algorithm can be considered as the one used for the calculation of the Bubble Point with the difference that

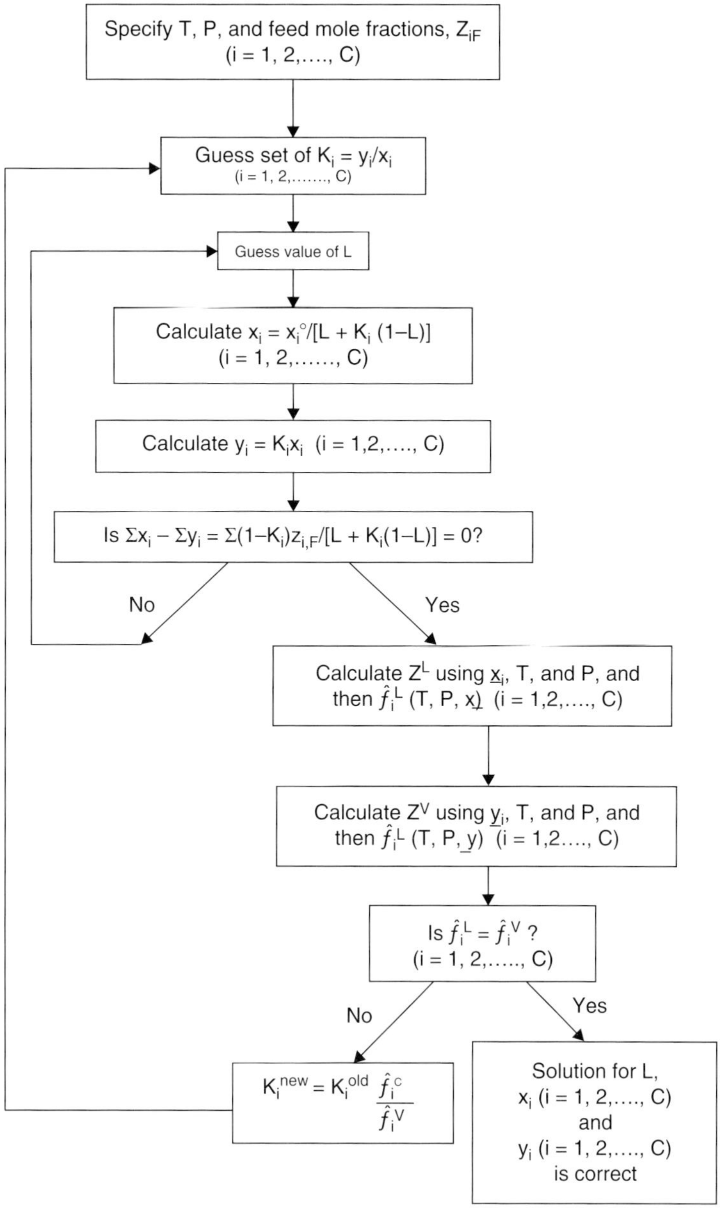

Figure 2.1 Algorithm to solve the flash equilibrium problem.

now it is necessary to consider a complete spectrum of equilibrium compositions ($x_1, x_2, x_3, \ldots\ldots, x_n$) and not a single equilibrium pressure. This means that the algorithm is more complex and it is necessary to have a numeric calculator to solve the equilibrium equations (Figure 2.2).

It is important to choose the thermodynamic model to solve the equilibrium equations. Usually it is necessary to choose different models for each

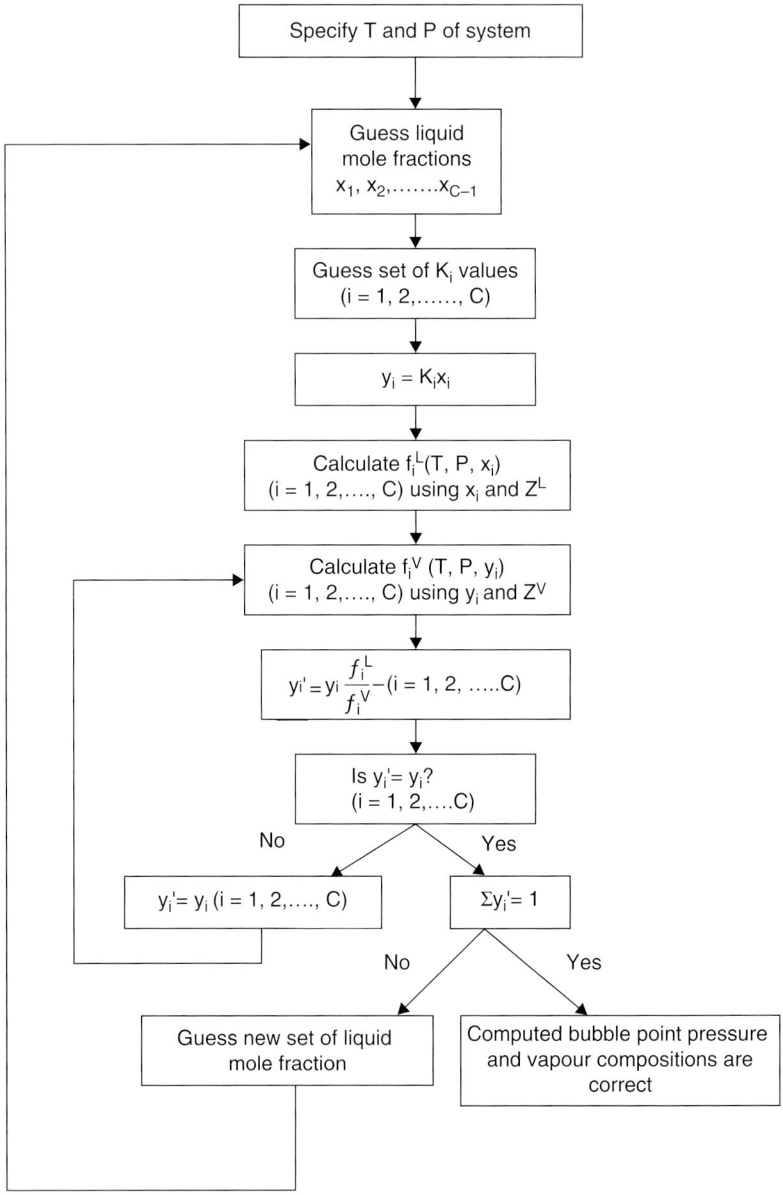

Figure 2.2 Algorithm to solve the phase equilibrium.

phase. To calculate the fugacity of the vapour phase as a function of pressure, temperature and composition an equation of state (EOS) must be used. For the liquid phase, the following can be considered:

An equation of state (EOS)
A method based on the activity coefficient

If an EOS is used also for the liquid phase, the method is named as $\phi - \phi$. If a method based on the activity coefficient is used, the method is named as $\phi - \gamma$.

When an EOS is used for the calculation of the vapour phase fugacity, it is necessary to solve the EOS and the expression of the phase volume is calculated. Between all the solutions, it is necessary to choose the one with the highest value. When an EOS is used for the calculation of the liquid phase fugacity, it is necessary to solve the EOS and the expression of the phase volume is calculated but in this case the solution with the low volume is chosen.

The calculated phase volume is used in these mathematical expressions to compute the fugacity:

$$\ln\left[\frac{\hat{f}_i^V(T,P,y_i)}{y_iP}\right] = \ln\hat{\phi}_i^v = \frac{1}{RT}\int_{V=\infty}^{V}\left[\frac{RT}{V^V} - \left(\frac{\partial P}{\partial n_i}\right)_{T,V,N_{j\neq 1}}\right]dV - \ln Z^V \quad (2.53)$$

$$\ln\left[\frac{\hat{f}_i^L(T,P,x_i)}{x_iP}\right] = \ln\hat{\phi}_i^L = \frac{1}{RT}\int_{V=\infty}^{V}\left[\frac{RT}{V^L} - \left(\frac{\partial P}{\partial n_i}\right)_{T,V,N_{j\neq 1}}\right]dV - \ln Z^L \quad (2.54)$$

where ϕ_i is the fugacity coefficient of the species i in the mixture and n_i is the number of moles of the species in the system. In this way, if an EOS is used to describe both phases, the equilibrium equations to consider are

$$K_i = \frac{\hat{\phi}_i^L(T,P,x_i)}{\hat{\phi}_i^V(T,P,Y_i)} \quad (2.55)$$

while the fugacity coefficients are evaluated with Equations (2.53) and (2.54).

Another method to solve the phase equilibrium considers an EOS to describe the vapour phase, and a model based on the activity coefficient to describe the liquid phase. The equilibrium equation is

$$\begin{aligned}\hat{f}_i^L(T,P,x_i) &= x_i\,\gamma_i(T,P,x_i)P^{sat}(T)\,\phi_i\left(T,P^{sat}(T)\right)\exp\left\{\frac{v_i^L\left[P-P^{sat}(T)\right]}{RT}\right\} \\ &= \hat{f}_i^V(T,P,y_i) = y_iP\hat{\phi}_i^V(T,P,y_i)\end{aligned} \quad (2.56)$$

and

$$K_i = \frac{\gamma_i(T,P,x_i)f_i(T,P)}{\hat{\phi}_i^V(T,P,y_i)P} \quad (2.57)$$

The two methods are different. In some process conditions it is better to use the $\phi - \phi$ method and in other conditions the $\phi - \gamma$ method. The method based on the activity coefficient or the excess of Gibbs free energy can be used to describe the behaviour of different liquid mixtures. This does not take into account the density, and it is not able to describe liquid mixtures near the critical conditions. In addition, if two phases are described with two different models, for example an EOS for the vapour phase and a model based on the activity coefficient for the liquid phase, then it could be that the properties of the phases are not equal.

The $\phi - \phi$ method can be used for all temperatures, pressures and densities and also in supercritical conditions. If the classical mixing rules are considered, this method is able to describe only inorganic gas and hydrocarbon mixtures. Other mixing rules such as the Wong and Sandler mixing rules are able to combine the model based on the activity coefficient with an EOS describing both the liquid and the vapour phases near the critical point.

2.3 Application to CO_2–H_2O system model

2.3.1 *Non-electrolyte models*

The solubility data of CO_2 in a non-electrolyte aqueous solution at high P have been extensively studied since 1883 (Wroblewski 1883): since then numerous experimental studies have been published, but important discrepancies remain in the database.

Several models have been proposed to describe the solubility of CO_2 in water.

Carroll *et al.* (1991) and Crovetto (1991) compiled and thermodynamically correlated the thermodynamic data in the range of pressures below 1 Mpa.

The models proposed by Duan and Sun (2003), Kiepe *et al.* (2004), Papaiconomou *et al.* (2003) and Spycher *et al.* (2003) for CO_2 + H_2O and CO_2 + H_2O + salt are examples of a $\gamma - \varphi$ approach, where an equation of state (EOS) is used to describe the non-ideality in the CO_2-rich phase and Henry's law or an activity model is used to describe the non-ideality in the H_2O-rich phase.

The EOS used may be Krichevsky–Kasarnovsky (KK) EOS, Peng–Robinson (PR) EOS and its modifications, Redlich–Kwong (RK) EOS and its modifications, Patel–Teja (PT) EOS (Patel and Teja 1982) and so on.

For example, Nicolas *et al.* (2003) calculated the mutual solubilities for the CO_2–H_2O system in a wide temperature and pressure range (285. 15–373.15 K, 0.1–60 MPa) by using a solubility formulation based on "true" equilibrium constants (K) and RK EOS. Values of K vary with temperature and pressure. The temperature and pressure dependence on K were taken into account in the formulation in the paper. The comparison with the experimental data shows that the calculation is reliable at temperatures lower than 323.15 K. However, the calculation of the water content has not been verified at higher temperatures (>348.15 K) and pressures (>5 MPa) because of the limited experimental data.

The inherent disadvantage of this approach is that it does not allow for estimating the density of the H_2O-rich phase. This is not an issue in a $\varphi-\varphi$ approach, where an EOS is used for both phases. The $\varphi-\varphi$ approach has been applied to the CO_2 + H_2O system. For example, Evelein *et al.* (1976) used the Redlich–Kwong EOS, Shyu *et al.* (1997) used the Peng–Robinson EOS, Valtz *et al.* (2004) used the SAFT-VR (Galindo *et al.* 1998; Gil-Villegas *et al.* 1997) EOS and Button and Gubbins (1999) used the SAFT EOS developed by Huang and Radosz (1991; Coquelet *et al.* 2004) (HR-SAFT). Button and Gubbins (1999) modelled CO_2 as a molecule with four association sites in order to represent the mole fraction of CO_2 in the liquid phase (H_2O-rich phase) but not the mole fraction of H_2O in the CO_2-rich phase.

Ji *et al.* (2005) used SAFT1-RPM to represent the phase behaviour and density of CO_2–H_2O. H_2O is modelled as a molecule with four association sites, two sites of type O and two sites of type H. The salt is modelled as a molecule composed of two charged, but non-associating, spherical segments, of which one represents the cation and one represents the anion. For the CO_2–H_2O system, only one type of cross-association is assigned (i.e. between the site of type O in CO_2 and the site of type H in H_2O). Using temperature-dependent parameters, SAFT1-RPM is found to represent the density and equilibrium data for the CO_2–H_2O system, including the minimum H_2O concentration in the CO_2-rich phase in the y–P diagram (Ji *et al.* 2005), from 285 to 473 K, 600 bar.

Valtz *et al.* (2004) used three different models to compare the $\gamma-\varphi$ *and* $\varphi-\varphi$ approaches. The first thermodynamic model is based on an asymmetric approach with the PR EOS including classical mixing rules for the vapour phase and a Henry's law treatment of the aqueous phase. The second and third thermodynamic models involve $\varphi-\varphi$ approaches: the second model uses the Peng–Robinson equation of state combined with the Wong–Sandler and Huron–Vidal mixing rules, and NRTL local composition model was used to obtain the excess molar Gibbs free energy; the third model the SAFT-VR equation of state based on a variable range square-well potential. The SAFT-VR theory, derived from statistical mechanics, is based on a representation of the interactions at the molecular level with well-defined intermolecular potentials (Gil-Villegas *et al.* 1997; Galindo *et al.* 1998).

All three approaches are in good agreement with the experimental data, and the solubility of CO_2 in water is particularly well represented. The second model gives the most accurate predictions of the composition of the vapour phase for all temperatures; however, the deviations on the CO_2 solubility are more important at high pressures. The third model (SAFT-VR) gives accuracy similar to the first model in predicting the CO_2 solubility. The UCEP was extrapolated by the SAFT-VR model. However, these results could be improved by using the SAFT-VR theory combined with a crossover approach.

A review (Diamond and Akinfiev 2003) about carbon dioxide solubility in water has been published, referring to a large number of experimental data, many of them tabulated in the 1996 IUPAC Solubility Series volume (Scharlin 1996), though without any discrimination of the conflicting measurements or attempts at correlation.

Diamond and Akinfiev (2003) assembled data from 25 literature studies and tested for their accuracy against simple thermodynamic criteria. Of the 520 data compiled, 158 data were excluded. The 362 measurements that satisfy the acceptance criteria have been correlated by a thermodynamic model based on Henry's law and on recent high-accuracy equations of state ($\gamma-\varphi$ approach). The assumption that the activity coefficients of aqueous CO_2 are equal to unity is found to be valid up to solubilities of approximately 2 mol%. At higher solubilities the activity coefficients showed a systematic trend from values greater than unity at low temperatures, to values progressively lower than unity at high temperatures. An empirical correction function that describes this trend is applied to the basic model. The resulting corrected model reproduces the accepted experimental solubilities with a precision of better than 2% over the entire $P–T–x$ range considered, whereas the data themselves scatter with a standard deviation of approximately 1.7%. The model is available as a computer code at http://www.geo.unibe.ch/diamond. In addition to calculating solubility, the code calculates the full set of partial molar properties of the CO_2-bearing aqueous phase, including activity coefficients, partial molar volumes and chemical potentials (Diamond and Akinfiev 2003).

2.3.2 Electrolyte models

The modelling of thermodynamic equilibrium properties of aqueous solutions containing electrolytes was studied for a long time but is still in a state of development. The reason is that there are some additional factors to be taken into account when dealing with these solutions since the dissolved electrolytes exist as ions and these ions can interact both with each other and with the solvent. The ions are charged particles, and therefore electrostatic interactions different from the dispersion and repulsive forces' interactions must be considered.

In considering electrolyte solution thermodynamics, care must be taken regarding the definition of the appropriate reference state and the choice of the concentration units. In literature different concentration units are often used as molarity (moles of solute per litre of solution) or molality (moles of solute per kg of solvent). As reference state for the evaluation of the non-ideality of solutions, as pointed out by Prausnitz *et al.* (1999), the right choice for a solute i is the hypothetical ideal solution of i in the solvent at system temperature and pressure and at unit concentration.

In addition, it is necessary to consider that in a solution of an electrolyte, the solute dissociates into cations and anions that are not independent components because of the requirement of electro-neutrality.

The thermodynamic properties that are normally measured are not the properties of the individual ionic species but of the neutral electrolytes formed by cations and anions.

Some useful quantities in the electrolyte solutions are the mean ionic activity coefficient $\gamma_{\pm}$:

$$\gamma_{\pm} = \left(\gamma_{+}^{\nu+}\gamma_{-}^{\nu+}\right)^{1/\nu} \tag{2.58}$$

And the osmotic coefficient Φ:

$$\ln \gamma_{\pm}^{m} = \Phi - 1 + 2\int_{0}^{m} \frac{\Phi - 1}{m^{1/2}}\, dm^{1/2} \tag{2.59}$$

where m is the molality of the solution.

Since at fixed concentration of ions, electrolytes containing ions with multiple charges have a stronger effect on the non-ideality of the solution than electrolytes with only singly charged ions, it is useful to introduce a parameter named ionic strength I:

$$I = \frac{1}{2}\sum_{i} m_i z_i^2 \tag{2.60}$$

where z_i is the charge of ion i and m_i is its concentration expressed in molality.

At low electrolyte concentrations, only long-range electrostatic forces are significant, while at high electrolyte concentrations, short-range attractive and repulsive forces become important as well. Another important effect is that the chemistry of the solution (types of species present in the solution) changes at different conditions: complete dissociation of ions, association of ions that behave as dipoles and so on.

For dilute solutions of electrolytes, the Debye–Hückel (DH) limiting law is generally used. To extend the applicability of the model to the description of concentrated electrolyte solutions, it is necessary to resort to semi-empirical models.

In the Pitzer model (Pitzer 1973), the deviations from the DH limiting law are attributed to physical interactions between ions due to excluded volume and van der Waals attraction effects.

Different models have been proposed to describe the thermodynamic properties of electrolyte solutions by introducing the local composition concept in order to account for nonrandomness. In these models it is assumed that the molar excess Gibbs energy of the electrolyte solutions is the sum of two contributions, one arising from the long-range (LR) coulombic forces (represented by the DH theory or its extension) and the other from short-range forces (SR):

$$\frac{g^{E*}}{RT} = \frac{g^{E*,LR}}{RT} + \frac{g^{E*,SR}}{RT} \tag{2.61}$$

This equation gives two contributions to the asymmetric mean ionic (mole fraction–based) activity coefficient:

$$\ln \gamma_{\pm}^{*} = \ln \gamma_{\pm}^{*,LR} + \ln \gamma_{\mp}^{*,SR} \tag{2.62}$$

To obtain $\ln \gamma_{\pm}^{*}$ from the excess Gibbs energy, it is necessary to calculate the contribution of the cations and anions and then use the equation:

$$\ln \gamma_{\pm}^{*} = \frac{1}{\nu}\left(\nu_{+} \ln \gamma_{+}^{*} + \nu_{-} \ln \gamma_{-}^{*}\right) \tag{2.63}$$

For the long-range contribution, different local composition models were proposed: Chen and Evans (1986) used an expression for the excess Gibbs energy that includes a contribution derived from NRTL equation, Zhao *et al.* (2000) have extended the Wilson model and Sander *et al.* (1986) and Macedo *et al.* (1990) have used the UNIQUAC equation. Kikic *et al.* (1991) used the UNIFAC equation using also for the short-range contribution the DH theory combined with the McMillan–Mayer solution theory as described by Cardoso and O'Connell (1987).

More recently a new segment-based NRTL was proposed by Chen (Chen *et al.* 2001; Chen and Song, 2005) in order to describe systems also containing large molecules.

Using perturbation theory, EOS models were developed to describe electrolyte solution properties (Jin and Donohue 1988a, b; Furst and Renon 1993; Wu and Prausnitz 1998). An equation of state covering a large range of conditions was developed using the Peng and Robinson (1976) expression for the short-range interactions (Myers *et al.* 2002).

When no complete dissociation is reached in the solution, then the dissociation chemical equilibrium, characterized by a chemical equilibrium constant, must be considered.

2.4 Thermodynamics of solid–vapour equilibria

Only the basic equations for the calculation of the solubility of a solid material (solute, component 2) in a dense gas will be reported. The equilibrium conditions for component 2 are

$$\hat{f}_2^{S} = \hat{f}_2^{SF} \tag{2.64}$$

where the superscript *S* indicates the solid phase and *SF* the supercritical phase. Since normally the solid phase is considered to be the pure solid solute, according to Equation (2.17), the left-hand side of this equation is given by

$$\hat{f}_2^{S} = f_2^{S} = P_2^{sub} \hat{\phi}_2^{V} \exp\left(\int_{P_2^{sub}}^{P} \frac{v_2^S dP}{RT} \right) \tag{2.65}$$

where P_2^{sub} is the sublimation pressure of the solid, ϕ_2^V is the fugacity coefficient of the pure component 2 at the pressure P_2^{sub} and temperature T in the gas phase, and v_2^S is the molar volume of the pure solid component, 2, at temperature T. The fugacity of component 2 in the supercritical phase is given by the relationship:

$$\hat{f}_2^{SF} = y_2 \hat{\phi}_2^{SF} P \tag{2.66}$$

And, finally, the solubility is given by the following equation (Prausnitz *et al.* 1999):

$$y_2 = \frac{P_2^{sub}}{P} \left[\frac{\phi_2^V \exp\left(\int_{P_2^S}^{P} \frac{v_2^S dP}{RT} \right)}{\hat{\phi}_2^{SF}} \right] \tag{2.67}$$

The quantity contained in square brackets represents the ratio of the real solubility to the ideal solubility (when the supercritical phase is supposed to obey the ideal gas law); it is always greater than 1 and it is also called 'enhancement factor' (E). E can have values of 10^{+3} or higher.

The enhancement factor contains three terms: $\hat{\phi}_2^{SF}$, the fugacity in the supercritical phase, ϕ_2^S, which takes into account the non-ideal behaviour of the pure component 2 in the vapour phase at the sublimation pressure, and the Poynting factor that describes the influence of the pressure on the fugacity of pure solid 2.

Since the sublimation pressure is normally very low, the fugacity coefficient ϕ_2^S is close to 1 and also the Poynting correction is not very different from 1 (usually less than 2). As a consequence, the most important term that contributes to a value of E is the fugacity coefficient in the supercritical phase.

The effect of pressure on $\hat{\phi}_2^{SF}$ is given by the equation:

$$\left(\frac{\partial \ln \hat{\phi}_2^{SF}}{\partial P} \right)_{T,y} = \frac{\overline{v}_2^{SF}}{RT} - \frac{1}{P} \tag{2.68}$$

where $\overline{v}_2^{SF}$ is the partial molar volume of component 2 in the supercritical phase. When the solubility y_2 is very low, and the temperature T and pressure P are close to the critical conditions of the component 1, $\overline{v}_2^{SF}$ assumes large and negative values.

As a consequence, a small variation of the pressure causes a large variation of the fugacity coefficient and the solubility.

It is necessary to observe, in particular, the effect of temperature on the solubility. At pressures below a value named 'crossover pressure', typical for each binary solute–solvent system, the solubility increases with decreasing temperature whereas at higher pressures the opposite effect is observed.

The temperature influences the sublimation pressure and the fugacity coefficient in the supercritical phase. The sublimation pressure increases with increasing temperature, and if this is the main effect, the solubility must always increase with temperature. On the contrary, at relatively low pressures (close to the critical pressure of the supercritical fluid) the fugacity coefficient in the supercritical phase plays the most important and preponderant role, and the temperature has an opposite effect.

For the evaluation of the solubility it is necessary to know the pure-component properties and to use an EOS model for the evaluation of the fugacity coefficients. Often the pure-component parameters required by the EOS for the heavy component are missing and the sublimation pressure of the solute is not known.

The critical properties, that are essential basic data if a cubic equation of state is used, must be evaluated using group contribution methods but the numerical values obtained depend on the method used. This represents a problem for multifunctional components that are generally involved in processing natural products and/or pharmaceuticals. As an example, depending on the prediction method used, a critical temperature ranging from 817.8 to 1254.0 K can be obtained for cholesterol (Neau *et al.* 1996).

Most of the compounds of interest as solutes in the supercritical fluids exhibit sublimation pressures of the order of 10^{-14} bar: these data cannot be determined experimentally and are estimated using empirical correlations. In the correlation of solubility data this problem can be solved empirically by considering the pure-component parameters as fitting parameters. Better results are obviously obtained (Alessi *et al.* 1997), but the physical significance is often missed. Some of these ambiguities can be partially solved using a simple approach (Garnier *et al.* 1999), in which the sublimation pressure of a solid can be estimated using experimental fusion properties and the vapourization enthalpy derived from the equation of state. Using the Clapeyron equation P_2^{sub} can be approximated by

$$\ln\left(\frac{P_2^{sub}}{P_t}\right) = -\frac{\Delta H_2^{sub}}{R}\left(\frac{1}{T} - \frac{1}{T_t}\right) \tag{2.69}$$

where T_t and P_t are the reference conditions chosen as the triple point of the pure component 2. The quantity ΔH_2^{sub} is the sublimation enthalpy in the reference state, and is calculated from the melting and vapourisation enthalpies $\Delta H_2^{sub} = \Delta H_2^{fus} + \Delta H_2^{vap}$.

Using the approximation that, for a solid, the triple-point temperature T_t is equal to the normal fusion temperature, T_2^{fus}, it is possible, from the same equation of state that is used for the calculation of $\hat{\phi}_2^{SF}$,

- to evaluate the reference pressure P_t at the temperature $T_t \approx T_2^{fus}$; and
- to calculate the vapourisation enthalpy ΔH_2^{vap} at the temperature T_t.

T_2^{fus} and ΔH_2^{fus} are normally easily measured.

The so-called subcooled liquid approach was also suggested in the literature in order to overcome the difficulties associated with the pure-component properties of the solid compound. This approach is commonly used for the calculation of the liquid–solid equilibria, and for solubility calculation of solids in supercritical fluids it was already suggested (Prausnitz *et al.* 1999) and subsequently extensively applied to different supercritical fluid processes (Kikic *et al.* 1997). The solubility y_2 is then expressed as

$$y_2 = \frac{P_0}{P}\frac{\phi_2^L}{\hat{\phi}_2^{SF}}\exp\left[\frac{v_2^s(P-P_0)}{RT}\right]\exp\left[-\frac{\Delta H_2^{fus}}{RT}\left(1-\frac{T}{T_2^{fus}}\right)\right] \tag{2.70}$$

where P_0 is the normal pressure and ϕ_2^L is the fugacity coefficient for the supercooled component 2, considered as a liquid at temperature T and pressure P_0.

Also with this approach the problem of the pure-component properties necessary for the determination of the parameters of the equation of state remains. The advantage lies in the use of experimentally accessible properties of heat-of-fusion and melting point, instead of the sublimation pressure.

2.5 List of symbols

a	attractive force's constant of the Van der Waals equation of state; J cm^3 mol^{-2}
$a(T)$	constant in the Redlich–Kwong equation of state; J cm^3 mol^{-2} K$^{-½}$
a_i	Equation (2.10) activity of component i in a mixture
$a_i = a_{ii}$	Equation (2.43) and (2.45) attractive force's constant for the pure component i in a cubic equation of state
a_i^{id}	activity of component i in an ideal mixture
a_{mn}	interaction parameter between group m and n in the U.N.I.F.A.C. model; K
a_∞^E	excess Helmholtz energy at infinite pressure; Joule mol^{-1}
A_{ij}	empirical constants of Van Laar (2.24) or Margules equation (2.25)
b	covolume; cm^3 mol^{-1}
B	second virial coefficient; cm^3 mol^{-1}
C	third virial coefficient; cm^6 mol^{-2}
B_{ii}	second virial coefficient of pure component i; cm^3 mol^{-1}
B_{ij}	second virial coefficient for the pair i–j; cm^3 mol^{-1}

C_{ijk}	third virial coefficient for the i–j–k interactions; $cm^6\ mol^{-2}$
$\hat{f}_i$	fugacity of component i in a mixture; bar
$\hat{f}_i^{\alpha}, \hat{f}_i^{\beta}$	fugacity of component i in the mixture α, β..; bar
$\hat{f}_i^{L}$	fugacity of component i in the liquid mixture; bar
$\hat{f}_i^{V}$	fugacity of component i in a vapour mixture; bar
$f_i^0\left(T.P,x^0\right)$	fugacity of a liquid in a reference state (T, P, x^0), bar
f_i^S	fugacity of the pure component i in the solid phase; bar
$\hat{f}_2^S$	fugacity of component 2 in a solid mixture; bar
$\hat{f}_2^{SF}$	fugacity of component 2 in the supercritical fluid; bar
f_i^V	fugacity of the pure component i in the vapour phase; bar
f_i^L	fugacity of the pure component i in the liquid phase; bar
$\hat{f}_i^L$	fugacity of the component i in the liquid phase; bar
$\hat{f}_i^V$	fugacity of the component i in the vapour phase; bar
F	degree of freedom
g^{α}	molar Gibbs energy of phase α; $J\ mol^{-1}$
g^E	molar excess Gibbs energy; $J\ mol^{-1}$
g^{E*}	molar excess Gibbs energy relative to unsymmetric reference state; $J\ mol^{-1}$
$g^{E*,LR}$	long range forces contribution to the molar excess Gibbs energy relative to unsymmetric reference state; $J\ mol^{-1}$
$g^{E*,SR}$	short range forces contribution to the molar excess Gibbs energy relative to unsymmetric reference state; $J\ mol^{-1}$
g_{ij}	energy of interaction between i and j molecules in the NRTL equation (2.27a); $J\ mol^{-1}$
g_{∞}^E	excess Gibbs energy at infinite pressure; $J\ mol^{-1}$
g_0^E	excess Gibbs energy at zero pressure; $J\ mol^{-1}$
G_{ij}	$= \exp\left(-\alpha_{ij}\tau_{ij}\right)$
H_i	Henry constant; bar
ΔH_2^{fus}	enthalpy of fusion of component 2; $J\ mol^{-1}$
ΔH_2^{sub}	enthalpy of sublimation of component 2; $J\ mol^{-1}$
I	ionic strength based on molality m
K_i	partition coefficient between vapour (y_i) and liquid phase (x_i)
k_{ij}	binary interaction parameter of equation of state
l_{ij}	binary interaction parameter of equation of state
m_i	molality of component i; $mol\ kg^{-1}$

n_i^{α}	number of moles of component i in the phase α; moles
n_i	number of moles of component i in the generic mixture; moles
N	number of components
N(t)	momentary number of cells per unit volume
N_0	initial number of cells per unit volume
P_t	triple point pressure; bar
p_i	partial pressure of component i in a mixture; bar
P_i^{sub}	sublimation pressure; bar
P_i^{sat}	vapour pressure of component i liquid; bar
P	pressure; bar
q_i	relative molecular surface area of component i
Q_k	relative molecular surface of group k (UNIFAC equation)
r_i	relative molecular volume of component i
R_k	relative molecular volume of group k (UNIFAC equation)
R	universal gas constant; J mol^{-1} K^{-1}
T	temperature; K
T_2^{fus}	fusion temperature of component 2; K
T_t	triple-point temperature; K
u_{ij}	energy of interaction between i and j molecules in the UNIQUAC equation (2.29b); J mol^{-1}
$\bar{v}_i$	partial molar volume of component i in the mixture;cm^3 mol^{-1}
$\bar{v}_i^S$	specific volume of the pure component i as a solid; cm^3 mol^{-1}
$\bar{v}_2^{SF}$	partial molar volume of component 2 in the supercritical fluid; cm^3 mol^{-1}
v_i	specific volume of liquid pure component i; cm^3 mol^{-1}
V	volume of the mixture; cm^3 mol^{-1}
V^L	volume of the liquid phase; cm^3 mol^{-1}
V^V	volume of the vapour phase; cm^3 mol^{-1}
z	compressibility factor
z_i	valency of the ion i
Z^L	compressibility factor of the liquid phase
Z^V	compressibility factor of the vapour phase
z_i	composition of a mixture in mole fraction
x_i	mole fraction of component i in the mixture
X_m	mole fraction of group m in the mixture

Greek symbols

α nonrandomness parameter in the NRTL equation

γ_i activity coefficient of component i in the mixture

γ_i^- activity coefficient of component i in the mixture according to the unsymmetric convention ($\gamma_i^- = 1$ for $x_i = 0$)

γ_+^- unsymmetrical cation activity coefficient

γ_-^- unsymmetrical anion activity coefficient

$\gamma_\mp$ mean ionic activity coefficient

$\gamma_\pm^-$ unsymmetrical mean ionic activity coefficient

$\gamma_\pm^{-,LR}$ long-range forces contribution to the unsymmetrical mean ionic activity coefficient

$\gamma_\pm^{-,SR}$ short-range forces contribution to the unsymmetrical mean ionic activity coefficient

$\gamma_\pm^m$ mean ionic activity coefficient on molal scale

Γ_k activity coefficient of the group k in the mixture

$\Gamma_k^{(i)}$ activity coefficient of the group k in the pure component i

η $= b/4v$ reduced volume in the equation (2.51)

ϑ_i area fraction of component i (2.29c)

θ_m area fraction of group m

Λ constant of equation (2.45)

Λ_{ij} Empirical constants of Wilson equation (2.26)

μ_i^α chemical potential of component i in the phase α, Joule mol^{-1}

μ_i^S chemical potential of component i in the solid phase S

$v_k^{(i)}$ number of groups of type *k* in a molecule of species i (2.30)

v sum of the electrolyte stoichiometric coefficients

v_+ electrolyte stoichiometric coefficient for the cation

v_- electrolyte stoichiometric coefficient for the anion

Π number of phases

ρ density; mol cm^{-3}

τ_{ij} dimensionless interaction between molecules i and j in the NRTL equation (2.27c) or in the UNIQUAC equation (2.29b)

φ_i volumetric fraction of component i (2.29c)

$\hat{\phi}_i$ fugacity coefficient of component i in a mixture

ϕ_i^V fugacity coefficient of pure component i in a vapour phase

$\hat{\phi}_i^L$ fugacity coefficient of component i in a liquid mixture

$\hat{\phi}_2^{SF}$	fugacity coefficient of component 2 in the supercritical fluid
$\hat{\phi}_i^{V}$	fugacity coefficient of component i in a vapour mixture
Φ	osmotic pressure
Ψ_{mn}	group interaction parameter
ω	acentric factor

3 Experimental Measurement of Carbon Dioxide Solubility

Giovanna Ferrentino, Thelma Calix, Massimo Poletto, Giovanna Ferrari, and Murat O. Balaban

Abstract: The application of dense phase carbon dioxide for the pasteurization of liquid foods has fostered research in the study of the phase behavior of systems in which CO_2 is in contact with liquid foods of different compositions in different processing conditions of pressure and temperature. CO_2 solubility in liquid phase is the key parameter for proper design of these processes. In this chapter, a review of the systems designed to experimentally measure CO_2 solubility and the resulting experimental data are reported starting from the results of CO_2 solubility in water and continuing with more complex aqueous solutions.

Keywords: CO_2 solubility; experimental systems; thermodynamic modeling.

3.1 Introduction

Solubility of gases in liquids has been studied since the nineteenth century (Battino and Clever 1965; King *et al.* 1992). The solubility of a solute is the analytical composition of a saturated solution, expressed in terms of the proportion of the designated solute in a designated solvent (Karukstis and Van Hecke 2000). Knowledge of carbon dioxide (CO_2) solubility and its phase behavior in different solutions is important considering the diverse applications in which CO_2 is involved.

The phase behavior of fluid mixtures can be determined experimentally with the design of suitable experimental systems, and theoretically with the application of thermodynamic models. The direct measurement of phase equilibrium data remains the most important source of reliable information despite the difficulty and expense of precise measurements since, with the exception of simple systems, theoretical and numerical models are not fully predictive. However, they provide a great help by significantly reducing the number of experimental data points needed to characterize the thermodynamic behavior of an unknown system. In

Dense Phase Carbon Dioxide: Food and Pharmaceutical Applications, First Edition. Edited by Murat O. Balaban and Giovanna Ferrentino.

general, the design of the experimental setup is critical in determining the reliability of the measurement procedures.

CO_2 solubility, whether obtained experimentally or by thermodynamic modeling, is needed in several scientific and technological applications, including (1) projects for CO_2 disposal on the sea floor or in sedimentary formations as a method to mitigate global warming; (2) CO_2-bearing fluid inclusion in minerals in geochemical applications; (3) the design of separation equipment in many applications such as the removal of CO_2 from gaseous streams by the hot carbonate process; (4) the production of soda from aqueous solutions containing ammonia, CO_2 and sodium chloride; and (5) cleaning gaseous effluents from power stations.

CO_2 is also used as supercritical fluid due to its physical properties such as high diffusivity, low viscosity, and high compressibility. However the use of supercritical CO_2 has its difficulties such as low solubility in strong polar substances and poor selectivity of some solutes. Some of these drawbacks can be overcome with the use of polar cosolvents, such as methanol or ethanol, or by mixed cosolvents such as acetic acid + acetonitrile or ethanol + acetonitrile, which can improve the solubility of a strong polar organic compound in a nonpolar supercritical fluid. All these applications require understanding and predictability of thermodynamic equilibria.

The application of dense phase carbon dioxide (DPCD) for the pasteurization of liquid foods has fostered the need to study the phase behavior of systems in which CO_2 is in contact with liquid foods of different composition in different processing conditions of pressure and temperature. CO_2 solubility in the liquid phase is a key parameter for a proper design of this process.

In this chapter a review of the equipment designed to measure CO_2 solubility and a description of CO_2 solubility results in water and in more complex aqueous solutions is reported.

3.2 Solubility of carbon dioxide in water

3.2.1 Definition and brief review of early studies

An accurate description of CO_2 solubility in aqueous solutions is critical for various scientific and technological applications.

The characterization of the thermodynamic equilibrium requires the evaluation of CO_2 solubility in the aqueous phase and the knowledge of its density at different pressures and temperatures. Most of the early experimental work on CO_2–H_2O mixtures focused on high temperatures and pressures applicable to the study of processes typically performed at several hundred Celsius degrees and up to several kilobars (Mader 1991). Published data in the two-phase region at temperatures below 100°C and at moderate pressures were initially limited. In the last two decades, more data became available in this P–T range. A summary of the experimental

Table 3.1 Summary of experimental data for the mutual solubilities of water and carbon dioxide, expressed in mol fractions. (Reproduced from Ji *et al.*, copyright 2007, with permission of Elsevier)

Temperature(K)	Pressure (MPa)	Number of experimental data	Type	References	Year
291.15–373.15	2.53–70.9	50	*x*	Wiebe and Gaddy	1939–1940
298.15–348.15	0.1–70.93	39	*y*	Wiebe and Gaddy	1941
283.15–303.15	0.101–2.027	15	*x*	Bartholome and Friz	1956
323.15–373.15	20–50	4 (*x*), 4 (*y*)	*x* and *y*	Toedheide and Frank	1963
383.15–473.15	10–70	7 (*x*), 7 (*y*)	*x* and *y*	Takenouchi and Kennedy	1964
303.15–353.15	0.993–3.891	13	*x*	Matous *et al.*	1969
298.15–373.15	1.733–5.147	22	*y*	Coan and King	1971
298.15–348.15	4.955	5	*x*	Malinin and Savelyeva	1972
298.15–373.15	4.955	5	*x*	Malinin and Kurovskaya	1975
323.15–473.15	0.154–5.389	33	*x*	Zawisza and Malesinska	1981
323.15	6.82–17.68	8 (*x*), 7 (*y*)	*x* and *y*	Briones *et al.*	1987
298.15	3.63	1	*y*	Nakayama *et al.*	1987
285.15–304.21	0.69–13.79	19	*y*	Song and Kobayashi	1987
323.15–348.15	10.13–15.20	4 (*x*), 4 (*y*)	*x* and *y*	D'Souza *et al.*	1988
373.1 5–473.1 5	0.325–8.0	49 (*x*), 49 (*y*)	*x* and *y*	Muller *et al.*	1988
3513.35–471.25	2.04–10.210	33	*x*	Nighswander *et al.*	1989
348.15	10.34–20.94	2 (*x*), 3 (*y*)	*x* and *y*	Sako *et al.*	1991
288.15–313.15	5.17–24.3	37 (*x*), 41 (*y*)	*x* and *y*	King *et al.*	1992
323.15	10.1–30.1	3 (*x*), 3 (*y*)	*x* and *y*	Dohm *et al.*	1993
323.15–348.15	34.48	2	*y*	Jackson *et al.*	1995
304.19–313.15	1.7 1–5 325	10	*x*	Gu	1998
298.31–298.57	2.7–5.33	7	*x*	Yang *et al.*	2000
323.15–353.15	4.05–14.11	29 (*x*), 29 (*y*)	*x* and *y*	Bamberger *et al.*	2000
274.15–288.15	0.07–2.179	54	*x*	Anderson	2002
313.2–393.17	0.0072–9.26	43	*x*	Kiepe *et al.*	2002
303.15–333.15	10–20	12	*x*	Bando *et al.*	2003
274.14–351.31	0.19–9.333	27	*x*	Chapoy *et al.*	2004
278.22–3 18.23	0.464–7.963	47 (*x*), 30 (*y*)	*x* and *y*	Valtz *et al.*	2004
332.15	3.34–19.89	6	*x*	Li *et al.*	2004

x: carbon dioxide solubility in water; *y*: water content in CO_2-rich phase.

data (Table 3.1) for the mutual solubilities of water and CO_2 has been reported by Yuanhui *et al.* (2007).

Some of the most extensive studies of CO_2 solubility in water were those of Wiebe and Gaddy (1939, 1940, 1941), Todheide and Franck (1963), and Takenouchi and Kennedy (1964, 1965). The data of Wiebe and Gaddy (1939, 1940, 1941) are the most comprehensive at temperatures below 373 K, while those of Todheide and Franck (1963) and Takenouchi and

Kennedy (1964, 1965) cover the widest temperature and pressure range from 323 to 623 K and from 10 to 350 MPa.

More recent work on the two-phase region below 373 K was performed by Gillepsie and Wilson (1982), Briones *et al.* (1987), Song and Kobayashi (1987), D'Souza and Teja (1988), Muller *et al.* (1988), Sako *et al.* (1991), King *et al.* (1992), Dohrn *et al.* (1993), and Bamberger *et al.* (2000). Other experimental studies providing additional data on CO_2 solubility in water (but no information on coexisting gas phase compositions) within our *P–T* range of interest include those of Teng and Yamasaki (1998), Jackson *et al.* (1995), and Rosenbauer *et al.* (2001) (Table 3.1).

A number of other experimental studies on CO_2 solubility in water have been conducted. They cover pressures mostly below 5 MPa and temperatures outside our range of interest, or they do not include data on the composition of the coexisting CO_2-rich phase. Most of these studies were reviewed by Crovetto (1991) and Carroll *et al.* (1991) to derive Henry's law constants for CO_2 in water. Data on CO_2 solubility are tabulated in the 1996 IUPAC Solubility Series volume (Scharlin 1996), without addressing the conflicting measurements or attempting to find some correlations.

Diamond and Akinfiev (2003) evaluated 520 experimental data from 25 references on CO_2 solubility in pure water from 271.65 to 373.15 K and from 0.1 to 100 MPa, and 158 data were discarded. Nicolas *et al.* (2003) reviewed the mutual solubilities of CO_2 and H_2O in the range of 285.15–373.15 K and 0–60 MPa. Chapoy *et al.* (2004) reviewed reliable experimental data for mutual solubility in the CO_2–H_2O system at 273.15–373.15 K and 0.07–70.9 MPa.

Table 3.1 shows that data regarding the mutual solubilities of CO_2 and H_2O at high temperature and pressure are scarce. This is due to the difficulties arising from the rigorous requirements for the design of experimental equipment at high temperatures and pressures and the high cost of personnel and resources needed to perform the experiments.

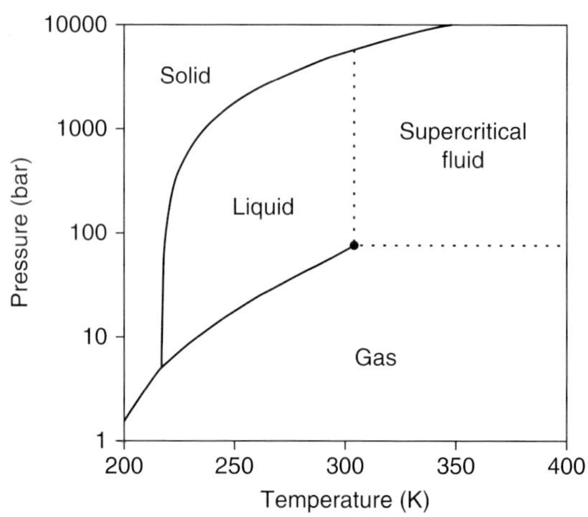

Figure 3.1 Carbon dioxide pressure – temperature phase diagram.

3.2.2 Physical properties associated with the phase diagram of carbon dioxide

Depending on temperature and pressure, CO_2 can be in liquid, gas, solid, or supercritical phase. Figure 3.1 shows the phase diagram of CO_2. The triple point occurs at a pressure of 5.2 atm (518 KPa) and a temperature of 216.6 K (–56.4°C). At a temperature of 197.5 K (–78.5°C), the vapor pressure of solid CO_2 is 1 atm. Therefore, at ambient pressure, the liquid phase is not stable, and the solid simply sublimates at a temperature of 197.5 K (–78.5°C). In these conditions, the solid CO_2 is called *dry ice*.

In the pressure–temperature phase diagram (Figure 3.1), the boiling line separates the gas and liquid regions and ends at the critical point, where the differences between the liquid and the gas phases disappear and the two phases merge in a single supercritical phase. This can be observed in the density–pressure phase diagram for CO (Figure 3.2). Well below the critical temperature (e.g. at 280 K), as the pressure increases, the gas compresses and eventually (at just over 40 bar) condenses into a much denser liquid, resulting in the discontinuity of the line (vertical dotted line). The system consists of two phases in equilibrium, a dense liquid and a low-density gas. As the critical temperature is approached (300 K), the density of the gas at the gas–liquid equilibrium increases, and that of the liquid decreases. At the critical point (304.1 K and 7.38 MPa) there is no phase transition, and above the critical temperature the gas cannot be liquefied by pressure. Slightly above the critical temperature (310 K), in the vicinity of the critical pressure, the isobaric line is very steep. At these conditions a small increase in pressure results in a large increase in the density of the supercritical phase. Not only the density but also many other physical properties of the fluid show large variations with the pressure near

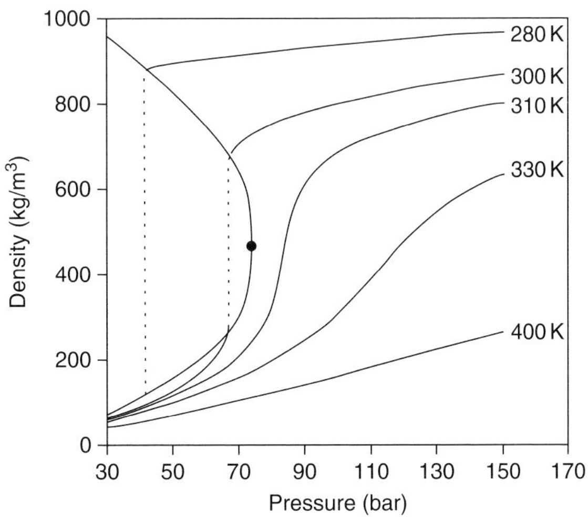

Figure 3.2 Carbon dioxide density – pressure phase diagram.

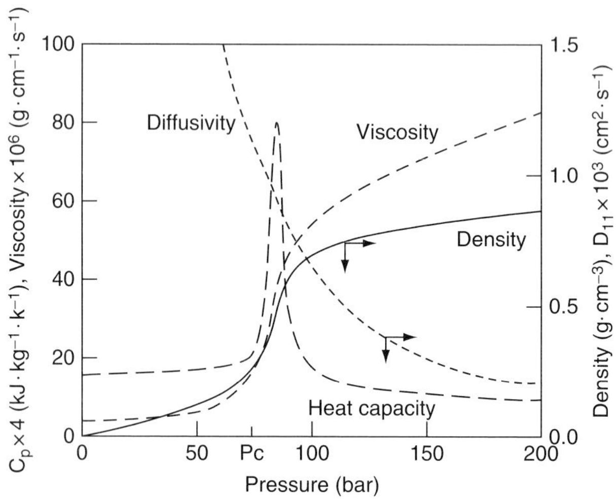

Figure 3.3 Variation of physicochemical properties of CO_2 at near-critical conditions. (Reproduced from Subramaniam *et al.*, copyright 2002, with permission of Elsevier)

Table 3.2 Physical properties associated with the phase diagram of CO_2 (Koutchma and Murakami 2005; Buhler *et al.* 1998)

States	Density (g/ cm^3)	Viscosity (cP)	Diffusion coefficient (cm^2/s)
Gas	0.002	0.014	0.01
Supercritical	0.467	0.02–0.12	0.0001
Liquid	1.0	1.0	0.00001

the critical point. Some of these are the viscosity, the relative permittivity, and the solvent strength, which are all closely related to the density. At higher temperatures, the fluid starts to behave like a gas (Figure 3.2). For CO_2 at 400 K, the density increases almost linearly with pressure.

As shown in Figure 3.3, not all the properties change at the same rate close to the critical pressure. For example, diffusivity changes more gradually than the other properties. Thus, CO_2 in the supercritical phase can diffuse through solids almost like a gas and can dissolve solutes almost like a liquid.

The density, the viscosity, and the diffusion coefficient for the different phases of CO_2 are shown in Table 3.2.

One of the advantages of working close to the critical point is the possibility of "tuning" the fluid properties with small changes in pressure or temperature.

3.2.3 *Effect of pressure and temperature on carbon dioxide solubility in water*

The solubility of a solute (solid, liquid, or gas) is the analytical composition of a saturated solution, expressed in terms of the proportion of the designated solute in a designated solvent (Karukstis and Van Hecke 2000).

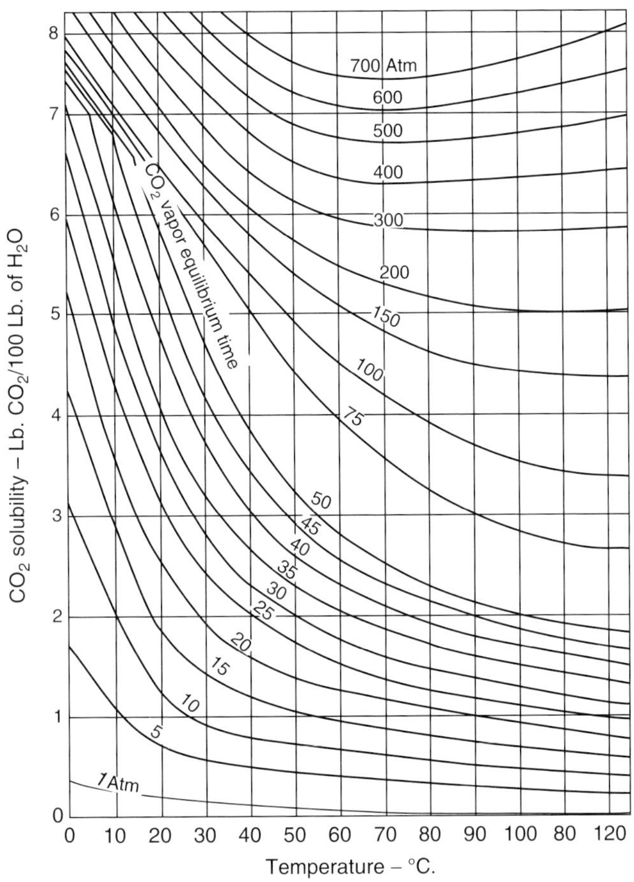

Figure 3.4 Solubility of carbon dioxide in water as a function of pressure and temperature. (Reproduced from Dodds *et al.*, copyright 1956, with permission of the American Chemical Society.)

Dodds *et al.* (1956) stated that CO_2 solubility in water depends on its pressure and temperature in equilibrium with the solution. The solubility of CO_2 decreases with rising temperature, but increases sharply with rising pressure. The effect of temperature and pressure on the solubility of CO_2 in pure water is summarized in Figure 3.4 (Dodds *et al.* 1956).

The presence of other substances may have a positive or negative effect on the solubility (Descoins *et al.* 2006; Meyssami *et al.* 1992). The dependence of CO_2 solubility on temperature follows the general trend of gases, whose solubility in water decreases with temperature, and has an opposite behavior with respect to the solubility of solids (e.g. sugar or salt) in water. Most solids dissolve more readily in hot water than in cold.

The complex phase behavior of the water–CO_2 system is summarized in Figure 3.5 (Spycher *et al.* 2003) where a projection of the CO_2–H_2O phase diagram is shown. The two- and three-phase coexistence curves and the critical points for this system are reported for pressures up to 300 bars. The

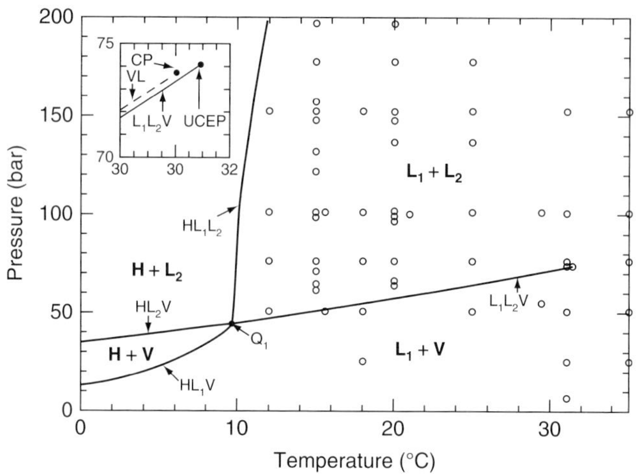

Figure 3.5 Enlarged *P–T* projection of the CO_2–H_2O phase diagram. (Reproduced from Spycher *et al.*, copyright 2003, with permission of Elsevier). The circles are literature data points. The solid curves are three-phase coexistence curves as labeled: V is the vapor phase, L_1 is the H_2O-rich liquid phase, and L_2 is the CO_2-rich liquid phase, and H is the hydrate phase. The dashed line in the inset is the pure-CO_2 liquid–vapor curve (VL), which almost coincides with the three-phase coexistence curve for the CO_2–H_2O system (L_1L_2V). CP is the critical point of pure carbon dioxide, UCEP is the upper critical end point for the CO_2–H_2O system, and Q_1 is a quadruple point for that system.

various phases considered include a solid hydrate (H), a water-rich liquid (L_1), a CO_2-rich liquid (L_2), and a vapor phase (V) consisting mostly of CO_2 in the *P–T* range considered. The L_1L_2V and HL_1V curves are based on the representations given by Wendland *et al.* (1999). The HL_1L_2 curve has been fitted to literature data (Ng and Robinson 1985; Fan and Guo 1999), evaluated using the functional form $P/P_q = 1 + 32.33(T/T_q - 1)^{1/2} + 91.169(T/T_q - 1)$, where $T_q = 9.77°C$ and $P_q = 44.60$ bar are the $P–T$ coordinates of the quadruple point Q_1 at which all four phases coexist. The data reported by Ng and Robinson (1985) for the hydrate formation curve HL_1L_2 go up to only 140 bar. For this reason, this representation is not reliable for extrapolation to pressures much above this value. The inset in Figure 3.5 shows that the vapor–liquid (VL) coexistence curve for pure CO_2 (Span and Wagner 1996) and the L_1L_2V curve almost coincide. The CO_2 critical point (31.06°C and 73.825 bar, from Angus *et al.* 1976; or 30.978±0.015°C and 73.773±0.003 bar, from Span and Wagner 1996) is very close to the upper critical end point (UCEP, 31.48°C and 74.11 bar; Wendland *et al.* 1999).

The three-phase ($L_1 + L_2 + V$) coexistence pressure range in the CO_2–H_2O system is also reported (Figure 3.6) in terms of a pressure–mole fraction cross-section of the CO_2–H_2O phase diagram at 25°C (Spycher *et al.* 2003).

Figure 3.6 shows a *P–X* cross-section of the CO_2–H_2O system obtained by drawing a perpendicular section to the plane of Figure 3.5 at 25°C. The inset of the figure was drawn using an approximate pressure value. From the

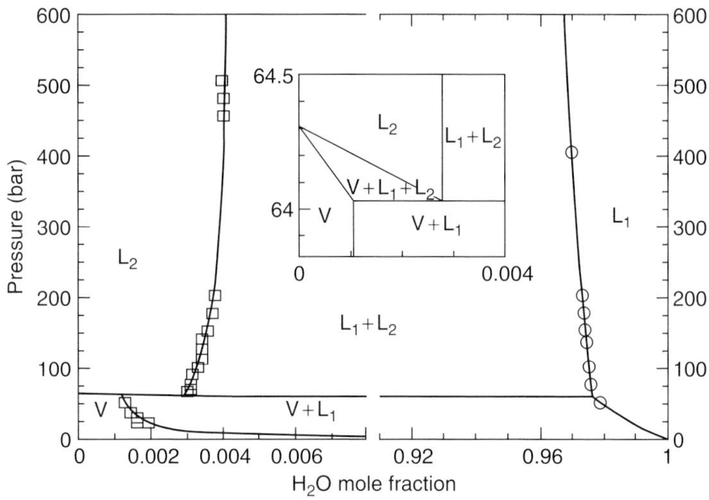

Figure 3.6 Pressure-mole fraction cross-section of the CO_2–H_2O phase diagram at 25°C. (Reproduced from Spycher *et al.*, copyright 2003, with permission of Elsevier). Solid curves are drawn to delimit the various phase coexistence regions: V is the vapor phase, L_1 is the H_2O-rich liquid phase, and L_2 is the CO_2-rich liquid phase. Literature data points are open circles for CO_2 solubility in H_2O and open squares for H_2O solubility in CO_2 (Wiebe and Gaddy 1940, 1941; Coan and King 1971; Gillepsie and Wilson 1982; King *et al.* 1992).

diagram, it can be noticed that isotherms crossing the L_1L_2V coexistence curve, below 31°C, exhibit a sharp discontinuity in the solubility of H_2O in the CO_2 phase. The sharp discontinuity in H_2O solubility at subcritical temperatures coincides with the phase change from a gaseous to a liquid CO_2-rich phase. The pressure interval over which the three phases coexist–H_2O-rich liquid, CO_2-rich gas, and CO_2-rich liquid – is very small. Above this saturation pressure, the H_2O solubility in the CO_2-rich phase increases with pressure and temperature. Above the critical temperature, the H_2O solubility trend with pressure becomes progressively smoother. The CO_2 solubility trend with pressure below the critical temperature reflects two solubility curves for two distinct phases: liquid CO_2 above saturation pressures, and gaseous CO_2 below these pressures. Above the critical temperature, the CO_2 solubility trend reflects the solubility curve for gaseous CO_2. At the critical point, the solubilities of liquid and gaseous CO_2 in water should be equal.

3.3 Experimental methods for carbon dioxide solubility measurement

Experimental methods for the investigation of high-pressure phase equilibria can be divided in two classes, depending on how the composition is determined: analytical methods (or direct-sampling methods) and synthetic methods (or indirect methods) (Dohrn and Brunner 1995).

3.3.1 Analytical methods

Analytical methods determine the compositions of the coexisting phases by taking samples from each phase and analyzing them outside the equilibrium cell at normal pressure, or by using physicochemical methods of analysis inside the equilibrium cell under pressure, (e.g. spectroscopic methods) (Kaiser *et al.* 1992). Withdrawing a large sample from an autoclave causes a considerable pressure drop, which disturbs the phase equilibrium significantly. This change of pressure associated with material withdrawal can be minimized by using variable volume cells (Staby and Mollerup 1991) or by blocking off a sampling cell from the equilibrium cell before reducing the pressure to analyze the sample (Dohrn *et al.* 1993). If only a small sample is withdrawn or if a relatively large equilibrium cell is used, the slight pressure drop may not affect the phase compositions significantly and the procedure may be valid. Small samples can be withdrawn using capillaries (Matos *et al.* 1989) or special sampling valves, using HPLC valves or fast-acting pneumatic valves (Lauret *et al.* 1994). Often sampling valves are directly coupled with analytical equipment (e.g. a gas chromatograph) (Danesh and Todd 1990). A simple method to minimize pressure variations due to temperature variations during sampling was proposed by Brunner *et al.* (1993). They designed a buffer autoclave connected to the equilibrium cell by means of a capillary, ending directly above the bottom of each autoclave. The equilibrium cell was totally filled with liquid, while 20% of the buffer autoclave's volume was filled by liquid and the rest by hydrogen as expansion gas. The technique implies the determination of the vapor–liquid equilibrium of the tested component with hydrogen to extrapolate subsequently the liquid–liquid equilibrium of the component of interest. With the presence of the buffer autoclave, pressure drop and variations were held to about 0.2 MPa at a total pressure of 20 MPa.

Depending on how the equilibrium is attained, analytical methods can be classified as constant-temperature methods, constant-pressure and -temperature methods, and constant-pressure methods.

3.3.1.1 Constant-temperature methods

An equilibrium cell is loaded with the system of interest at almost its final overall composition. The mixture is kept at the desired temperature. At the beginning of the experiment, the pressure is adjusted above or below the desired equilibrium value, depending on how equilibration changes the pressure. Mass transfer between phases is promoted by stirring the mixture, by rocking the autoclave, or by recirculating one or more phases to reduce the time necessary for equilibrium between the phases when the pressure reaches a plateau. The pressure can be readjusted by adding or withdrawing material or by changing the volume of the equilibrium cell. Usually equilibration is allowed for at least 30 min after the pressure plateau is sufficiently close to the desired value. Before analyzing the compositions of the coexisting phases, the mixture is given time without stirring, rocking, or recirculation for the separation of the phases (Waterling *et al.* 1991).

Constant-temperature methods that use stirring or rocking to ensure a rapid approach to equilibrium are also often called "analytical–static methods."

The recirculation of one or more phases has the advantage to allow a separate sampling volume to be filled isobarically with the recirculated phase. This procedure can avoid some separation problems that may occur when samples of light and heavy components are withdrawn through capillaries (Renon *et al.* 1989; Brunner and Teich 1994).

Knapp *et al.* (1990) reported a system where only the vapor phase is recirculated. It is withdrawn continually and passed back into the equilibrium cell through the equilibrium phase by the action of a magnetic pump. Using recirculation loops, samples can also be withdrawn by placing a sampling valve (Han *et al.* 1992) or a sampling autoclave (Dohrn *et al.* 1993) in the loop. The liquid phase is usually analyzed by taking samples through capillaries. Kim *et al.* (1989) designed a recirculation-type equilibrium cell for the determination of high-pressure vapor–liquid equilibria of several systems containing aromatic hydrocarbons as one component and supercritical CO_2 or ethane as the other component. In this system, both the vapor and liquid phases were recirculated with custom-built magnetic pumps with the advantage that sampling from both phases was possible without using capillaries but through gas chromatograph for direct analysis of the phases in equilibrium. Another possible design is to measure the density of the circulated phase by a vibrating tube densimeter installed in the recirculation loop. However, pump pulses could interfere with the measurement. Wendland *et al.* (1993) recommended turning off the pump during density measurements. Sometimes only the liquid phase is recirculated for the evaluation of the solubilities of gases in liquids (Chang 1992) or of liquid–liquid equilibria measurement (Hooper *et al.* 1988).

Constant-temperature methods involve relatively simple and inexpensive laboratory equipment and, if carried out carefully, they can produce very reliable results.

3.3.1.2 Constant-pressure and -temperature methods

These are often called "dynamic methods." One or more fluid streams are pumped continuously into a thermostated equilibrium cell. The pressure is kept constant during the experiment by controlling an effluent stream, usually of the vapor phase.

3.3.1.3 Continuous-flow methods

In a typical design, high-pressure metering pumps feed the preheated components into a mixer where the desired temperature is attained. The feed stream from the mixer is separated in an equilibrium cell into a vapor and liquid phase. Effluents from both phases are withdrawn continually, depressurized, accumulated, and analyzed, usually after the experiment. The pressure is adjusted by controlling the effluent stream of the top phase. The interface level between the fluid phases in the equilibrium cell can be adjusted with the bottom-phase expansion valve. Although the interface level is usually determined visually, nonvisual methods are successfully used utilizing an AC impedance bridge technique (Hochgeschurtz *et al.* 1993). Continuous-flow methods have the advantage that sampling does not disturb the equilibrium. If larger samples are needed for analyses, the run time of the experiment can be extended to accumulate more material. However, these methods can be used only for systems where the time needed to attain phase equilibrium is sufficiently short. The short residence time is an important factor to be considered when the equilibrium is evaluated for systems in which the high pressure and temperature conditions can induce chemical cracking or polymerization reactions, as demonstrated by Hutchenson *et al.* (1990).

3.3.1.4 Semi-continuous methods

In semi-continuous methods, only one phase flows while the other phase stays in an equilibrium cell. These methods are sometimes called "single-pass flow methods" or "pure-gas circulation methods." For the measurement of vapor–liquid equilibria, a gas stream from a high-pressure cylinder passes through two cells in series containing the liquid. The first cell serves as a pre-saturator while the second is the equilibrium cell. The outcoming effluent of the vapor phase is depressurized and directed to a trap where the condensed liquid is collected. The quantity of the gas coming out from the trap can be determined volumetrically with a wet test meter. Samples from the liquid phase are withdrawn through tubing, depressurized, and analyzed (Lee and Chao 1988). The major uncertainty is the effective attainment of thermodynamic equilibrium.

Semi-continuous methods can also be used to measure the solubility of a gas in a liquid. The experimental procedure is similar to the one for measuring vapor–liquid equilibria, but there is no need to determine the composition of the effluent from the vapor phase (Huang *et al.* 1988). Many investigators use a semi-continuous method to determine the solubility of a low-boiling (liquid or solid) substance in supercritical gas (Di Giacomo *et al.* 1989). Only the composition of the vapor phase effluent is analyzed by using a multiport sampling valve (Sako *et al.* 1991) or after expansion to atmospheric pressure. No samples from the condensed phase are taken.

3.3.1.5 Constant-pressure methods

An alternative to the direct measurements of pressure, temperature, liquid mole fraction, and vapor mole fraction data is measurement followed by a thermodynamic analysis. A fast and simple way is to use an ebulliometer, which is a one-stage total reflux boiler equipped with a vapor lift pump to spray slugs of equilibrated liquid and vapor onto a thermometer (Oslon 1989). A mixture of known composition is brought to boil at a controlled pressure. The compositions of the liquid and vapor phases change with time and reach a steady state which should differ insignificantly from the true equilibrium value. A thermodynamic analysis yields the phase compositions. Usually ebulliometry is used to measure low-pressure data. Ebulliometers have been proposed (Oslon 1989; Wisniewska *et al.* 1993) for pressures up to 3 MPa.

3.3.1.6 Example of experimental apparatus

An analytical method was used by Dohrn *et al.* (1993) to design an apparatus to measure CO_2 solubility in solutions containing glucose, water, and ethanol. The designed setup was provided with a vapor phase recirculation (Figure 3.7).

The core of the apparatus was a 500 cm^3 stainless steel autoclave serving as a mixing and separating vessel. It was equipped with capillary tubes to withdraw liquid samples at two different heights. At the bottom a porous filter (made of sintered 20 μm metal particles) provided a homogeneous gas distribution. At the top a stainless steel mesh prevented liquid to be entrained by the vapor phase. Temperature and pressure were measured by a calibrated NiCrNi thermocouple, covered with a 3.18 mm stainless

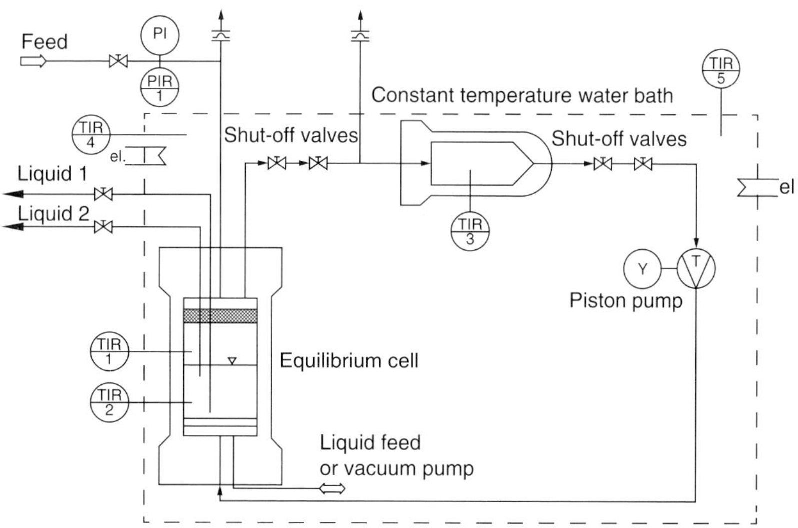

Figure 3.7 Diagram of the experimental apparatus designed by Dohrn *et al.* (1993) to determine the phase equilibria of ternary and quaternary systems of glucose, water, CO_2, and ethanol. Reproduced from Dohrn *et al.*, copyright 1993, with permission of Elsevier.

steel sheath, and a calibrated precision pressure transducer, respectively. Both were connected to a signal amplifier which, in turn, was connected to an analog-to-digital (A/D) converter linked to a personal computer for monitoring and recording. A 300 cm^3 autoclave was placed in the vapor phase loop to be used as a sample container for the gas phase. An air-driven double-acting piston pump was used to circulate the vapor phase which exited the mixing vessel at the top, flowed through the sample container. and then reentered the main autoclave at the bottom. The complete apparatus was enclosed in a water bath heated by two thermostats to ensure temperature uniformity. The sampling system used for the vapor and liquid phases consisted of two traps immersed in ice water and a dry ice–acetone bath, respectively; two glass flasks of different sizes to collect the gas; and a rotary vane pump for evacuating the section before the experiment. During the gas evacuation step, the pressure in the sampling line was measured by a Bourdon-tube and a Pirani-type precision vacuum gauge. The temperature was measured with a digital thermistor.

The apparatus was first evacuated, then the liquid feed was introduced into the mixing vessel and the CO_2 added to the system at the cylinder supply pressure. After waiting some time to equalize the system temperature and to evacuate the gas sample container, the system was carefully filled with gas by opening the interconnecting valves. The circulation then was started by switching on the circulating pump. To attain equilibrium, circulation was kept for at least a couple of hours after the test temperature and pressure were reached. Before sampling, the circulation was stopped for the separation of the phases in the mixing vessel to take place.

Three liquid samples were taken separately by means of a thermostated valve. Each sample was expanded in the evacuated sampling section where it separated into a liquid part, collected in traps placed in series, and a gaseous part which filled the available volume. The weight of the collected liquid from the sampling bomb was measured on an analytical balance. The amount of gas was estimated by applying the gas law considering that the gas occupied the entire volume of the sampling cell at an increased pressure whose volume was measured.

The sample of the vapor phase at equilibrium in the container was withdrawn by closing the interconnecting valves and taking the container out of the apparatus. The container was thermostated at low temperature to ensure that no liquid was entrained by the gas during the following expansion step. The volume of the CO_2 was measured by the pressure increase of a glass flask of 20 L capacity.

Tomasula and Boswell (1999) designed an apparatus based on the analytical method to measure CO_2 solubility in milk (Figure 3.8). CO_2, from a cylinder equipped with a dip tube and chilled to 6°C, was allowed to flow freely to the batch reactor or was pumped if the desired pressure exceeded approximately 5.9 MPa. The system pressure was monitored by a pressure transducer. Thermocouples were used to record system temperature, jacket temperature, and ambient temperature.

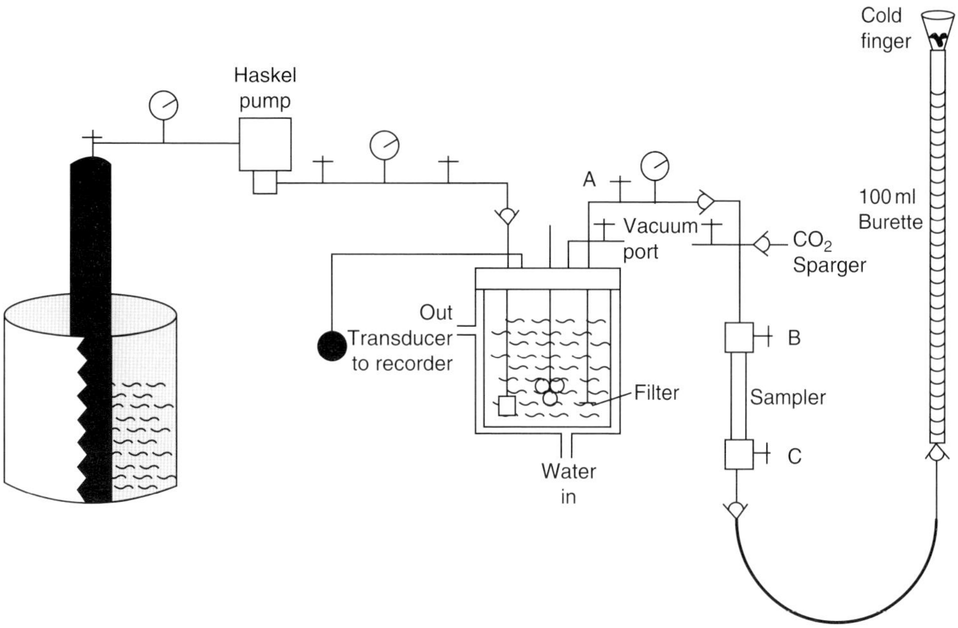

Figure 3.8 Diagram of the experimental apparatus designed by Tomasula and Boswell (1999) to measure CO_2 solubility in milk. Reproduced from Tomasula and Boswell, copyright 1999, with permission of Elsevier.

The vessel was filled with milk and the lid was secured. A vacuum pump was used to evacuate the piping loop of the reactor. After a run, the milk–CO_2 mixture was isolated between valves B and C (Figure 3.8). The sampler (volume = 11.0 cm^3) was built from a tube (nominal diameter = 1.3 cm, wall thickness = 0.12 cm). A 100 mL burette, with a cold finger to prevent the evaporation of water, was used to measure the amount of CO_2 dissolved in the milk. The solid content, as weight percent, of the liquid sample collected in the burette after each run was determined using a moisture analyzer. Barometric pressure was determined with a barometer, and the burette temperature was measured with a calibrated glass thermometer.

In a typical experiment, the reactor was filled with about 750 mL of milk. After removing all the air from the system, temperature was increased by hot water flowing into the jacket of the reactor, and CO_2 was pumped into the system until the desired pressure was reached. The equilibrium conditions were achieved in less than 2 hours. Then, a valve was opened and the liquid content of the reactor drawn through a filter tube. The expansion of the sample to ambient pressure was carried out in a burette connected to the valve. CO_2 solubility was evaluated by measuring the volume of CO_2 in the sampler at the temperature and pressure of the experiments. To calculate the volume of CO_2 left in the fluid in the burette, the Bunsen coefficient for CO_2 in water was multiplied by a factor taking into account the density of the milk to compensate for the barometric pressure and the vapor pressure of the milk. If the fluid in the burette was

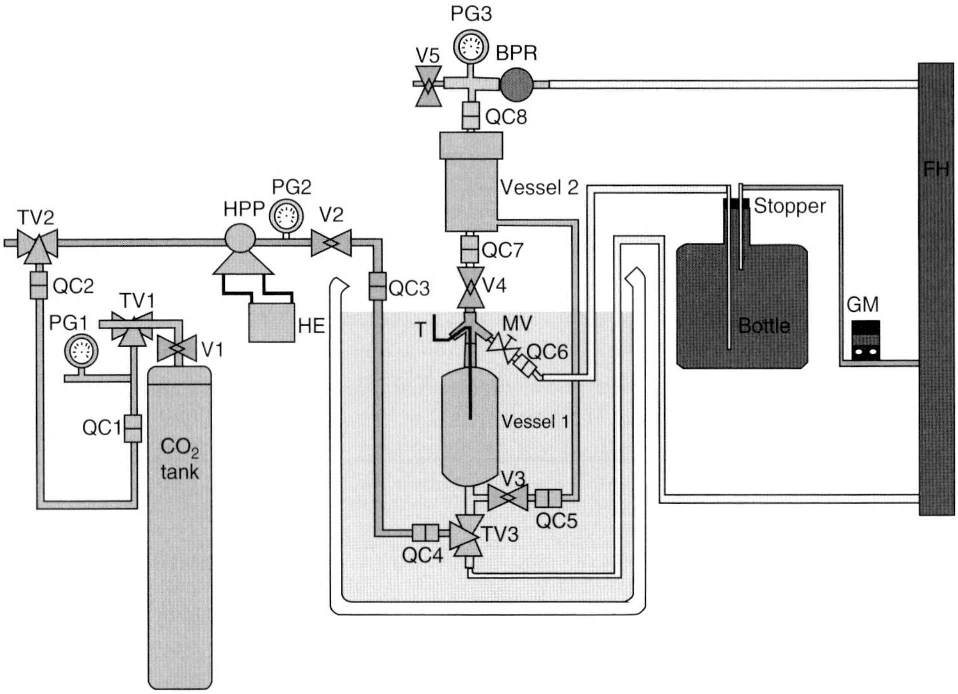

Figure 3.9 Diagram of the experimental apparatus designed to measure CO_2 solubility in juices and aqueous solutions addressing the composition of juices (Calix 2008). HPP: high-pressure pump; HE: heat exchanger; TV1, TV2, and TV3: manual valves, BPR: back pressure regulator; T: thermocouple; V1, V2, V3, V4, and V5: two-way valves; GM: gas meter; PG1, PG2, and PG3: manometers; QC1, QC2, QC3, QC4, QC5, and QC6: stainless steel connections; MV: metering valve; and FH: funnel hood.

white and the casein precipitation did not occur in the reactor after depressurization, the density of the milk was used to evaluate the correction factor. If casein precipitated and the fluid had a yellowish color, the density of the whey was used in the correction factor calculation. The volume of CO_2 was then subtracted from the volume indicated on the burette to give the volume of milk or whey in the sampler. The volume of dissolved CO_2 at the tested temperature and pressure was obtained by subtracting the volume of the milk or whey from the volume of the sampler.

An apparatus based on the analytical method was designed by Calix (2008) to measure CO_2 solubility in aqueous solutions and in commercial fruit juices such as apple juice and orange juice. The principle is based on the evaluation of the amount of CO_2 in solution in a liquid of known mass. The experimental apparatus was designed to accurately measure the amount of CO_2 in solution in water or any other liquid (Figure 3.9). The results were expressed in normal volume of CO_2 gas per unit volume of liquid.

The apparatus was made of two superimposed vessels. The lower vessel (Vessel 1) had a capacity of 150 mL and a maximum working pressure of

34.4 MPa at 121°C. The upper vessel (Vessel 2) had a maximum working pressure of 69 MPa at 22°C and a capacity of about 55 mL. The two vessels were also connected with a side arm. Gaseous CO_2 bubbled upward into the system from the bottom vessel while the liquid flowed down by gravity from the upper vessel to the lower vessel. The side arm connecting the upper to the bottom vessel allowed the recirculation of the liquid during the process. A high-pressure pump (Model X-10, Milton Roy, Ivyland, PA, US) and a heat exchanger were used to liquefy and pump CO_2 into the system. Vessel 1 was submerged in a water bath with a capacity of about 60 L kept at constant temperature with water flow recirculating from a high-precision water bath. A medium-flow high-pressure metering valve was used to control the CO_2 flow coming out from the lower vessel during the depressurization step which was measured using a gas flow meter. A plastic bottle was used as an expansion vessel to trap the liquid coming out with CO_2.

During the experimental run, about 500 mL of sample were pumped by the peristaltic pump through a plastic tube to the stem connection QC3 (Figure 3.9). After the desired temperature was reached, the system was purged with gaseous CO_2. V1 valve was opened and the high-pressure pump (HPP) turned on. V2 valve was opened very slowly and, if some gas with water was coming out of the system, the back pressure regulator (BPR) was used to increase the pressure, until a pressure level of 1.03–1.38 MPa above the desired pressure was reached. The HPP remained on for 1 hour and 10 min to reach saturation of CO_2 in the liquid. During this time, CO_2 continuously bubbled through the liquid (entering from the bottom of vessel 1, and exiting off the top of vessel 2 through the BPR) to saturate the liquid with CO_2. At the end of the experiment, the CO_2 flow was stopped. The connecting valves V4 and V3 were left open for a while, to let all the gaseous CO_2 to leave Vessel 1 and collect in the upper Vessel 2. The side arm allowed a back recirculation to Vessel 1 of saturated liquid and facilitated the lighter gaseous phase to rise from Vessel 1 to Vessel 2. This system was designed to work at pressures at which the density of the CO_2-rich phase was lower than that of the liquid.

To evaluate the liquid composition, the vessel 1 (assumed to contain only liquid and dissolved CO_2) was isolated under pressure from the system, and depressurized by opening the metering valve. The dissolved gas coming out from the system was measured by the gas flow meter. The liquid inside the vessel 1, the expansion bottle, and the plastic connections was collected and weighted. The mass of CO_2 was calculated from the conversion of the CO_2 volume by dividing it by 22.4 slt·mol^{-1} and multiplying by CO_2 molecular weight. CO_2 solubility was obtained as the ratio of grams of CO_2 in 100 grams of liquid.

An experimental equipment to determine CO_2 solubility in liquid foods was designed at the Department of Chemical and Food Engineering at the University of Salerno (Ferrentino *et al.* 2010b; Ferrentino 2009). The experimental setup, shown in Figure 3.10, was equipped with a vessel with

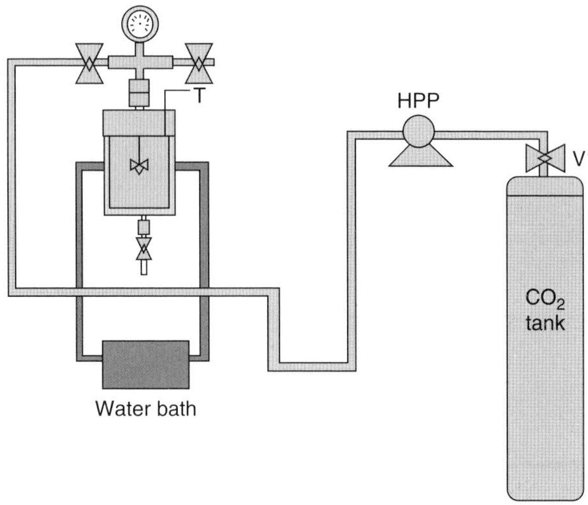

Figure 3.10 Diagram of the experimental apparatus designed to measure CO_2 solubility in aqueous solutions addressing the composition of juices and in liquid foods (Ferrentino 2009). HPP: high-pressure pump; V1, V2, V3, and V4: manual valves; PG2: pressure gauge; and T: thermocouple.

a maximum pressure of 20.0 MPa. The vessel (200 mL) was equipped with a four-bladed impeller magnetically coupled to a direct current (DC) motor allowing adjustable mixing speed. A type J thermocouple was used to measure the temperature inside the vessel. All the connections for the inlet and outlet lines, the stirrer, and the thermocouple were fitted on the removable vessel head. A pressure gauge, also on the reactor head, displayed the pressure inside the system. The CO_2 pump was a chromatography model with flow control based on volume displacement or on output pressure. Liquid CO_2 from a cylinder was pumped into the reactor through an on-off valve, kept closed after pumping for the time required for the experiments. The time needed for the reactor pressurization and heating was about 5 min. The vessel was equipped with a jacket in which warm water at constant temperature from an external thermostated water bath was recirculated to keep the system at the desired experimental temperature. A valve at the bottom of the vessel was used to withdraw the liquid.

The experiments were carried out in batch mode. The vessel was loaded with the liquid mixture. The external jacket was connected with the water bath and the system heated to the experimental temperature. Then, CO_2 was pumped to reach the desired experimental pressure. The temperature and the pressure inside the vessel were monitored by the thermocouple and the pressure gauge, respectively. After the pressure reached the desired level, the system was stirred for at least 1 hour to assure thermodynamic equilibrium. In all the tests, the same stirring speed (approx. 850 rpm) was used to facilitate the dissolution of CO_2 in the liquid phase.

At the end of the equalization period, a plastic syringe (60 mL syringe catheter) was connected to the bottom valve of the vessel through an on–off valve. Before connection, the piston of the syringe was pushed all the way to the front in order to remove the air from it. A small amount of liquid was drawn by opening the bottom valve. This operation was very quick and, due to the small amount of liquid withdrawn, the pressure and temperature of the pressurized reactor did not change significantly. The syringe expansion measured the sum of the volumes of CO_2 released to the gaseous state at ambient pressure and that of the final liquid sample. The syringe was then disconnected from the vessel and weighted on a high-precision balance to obtain the mass of the sampled liquid. From the density of the liquid mixture, the volume of the liquid was calculated and subtracted from the total volume measured from the syringe expansion to estimate the CO_2 volume. Multiplying this latter by the CO_2 density at the sampling conditions, the mass of CO_2 was obtained. The ratio between the mass of CO_2 and the mass of liquid represented the solubility, expressed in grams of CO_2 per 100 g of solution.

3.3.2 *Synthetic methods*

These methods are based on preparing a mixture of known composition and then seeking for the conditions at which a new phase forms. No sampling is necessary. The problem of analyzing fluid mixtures is replaced by the problem of "synthesizing" them (Deiters and Schneider 1986). After known amounts of the components have been placed into an equilibrium cell, values of temperature and pressure are adjusted to ensure mixture homogeneity. The temperature and pressure are, then, varied until the formation of a new phase is observed (Suppes and McHugh 1989). Each experiment yields one point on the pressure–temperature–liquid mole fraction phase diagram. Synthetic methods can be used where analytic methods fail, that is, when phase separation is difficult due to similar densities of the coexisting phases, as it happens near the critical state and in barotropic systems, in which, at certain conditions, the coexisting phases have the same density. Because no sampling is necessary, the experimental equipment can be inexpensive. Often, the experimental procedure is easy and quick (Schneider 1975). For multicomponent systems, experiments based on synthetic methods yield less information than those utilizing analytical methods, because the tie lines cannot be determined without additional experiments (e.g. refractive index measurements) (Bolz and Stephan 1991).

Synthetic methods can be classified based on the means of observation for the appearance of a new phase.

3.3.2.1 Visual synthetic methods

The appearance of a new phase is detected by visual observation of a cell provided with a transparent window. It results in a turbidity increase or in the appearance of a meniscus. It was proven that the visual synthetic method can be used not only for the determination of simple vapor–liquid equilibria, but also to study complicated phase behavior (e.g. multiphase equilibria) (Jangkamolkulchai and Luks 1989), to measure the solubility of gases in electrolyte solutions (Rumpf and Maurer 1993), or for solid–liquid and solid–fluid equilibria (Hamman and Sivik 1993). However, for isooptic systems, where the coexisting phases have approximately the same refractive index, visual observation is impossible.

3.3.2.2 Nonvisual synthetic methods

As an alternative to visual observation, other physical properties can be monitored to detect phase transitions. For example, if the total volume of a variable-volume cell can be measured accurately, the appearance of a new phase can be obtained from the abrupt change in slope on the pressure–volume plot more accurately than it can be done on the basis of visual observations (Japas *et al.* 1992). Fogh and Rasmussen (1989) used a microwave technique to detect phase transitions.

3.3.2.3 Synthetic methods using material balance

For systems with two degrees of freedom (e.g. binary two-phase equilibria or ternary three-phase equilibria), the compositions are fixed when pressure and temperature are given. Tanaka *et al.* (1993) measured the phase densities and the total volume for different cell loads to calculate the phase compositions using material balance.

3.3.2.4 Examples of experimental apparatus

In this section, examples of apparatus designed for measurements based on the synthetic method are reported. Most of these were used to evaluate the formation of equilibrium hydrates for CO_2 in aqueous solutions. Breland and Englezos (1996) performed experiments to evaluate the equilibrium of CO_2 in aqueous glycerol solutions. The apparatus is shown in Figure 3.11.

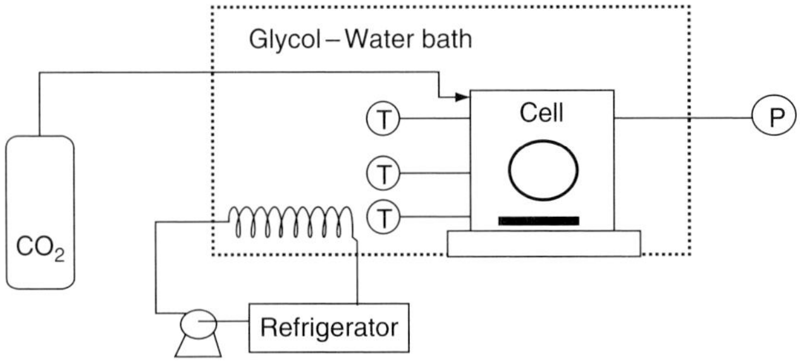

Figure 3.11 Diagram of the experimental apparatus designed by Breland and Englezos (1996) to determine the equilibrium conditions for carbon dioxide hydrate in pure-water and aqueous glycerol solutions. T: thermocouple; and P: pressure transducer. Reproduced with permission of the American Chemical Society.

A high-pressure cell, immersed in a temperature-controlled bath, was built from a 316 stainless steel rod. Two circular viewing windows were fitted onto the front and back. The top lid was sealed with a neoprene O-ring. A temperature-controlled bath contained 30 L of a solution consisting of approximately 50/50 (mass %) water and ethylene glycol. A motor-driven mechanism was used to stir the contents of the bath. The temperature of the bath was controlled by an external refrigerator-heater. Mixing of the sample in the cell was accomplished by a stirring bar magnetically coupled to rotating magnets placed directly underneath the cell driven by an electric motor. The temperature inside the cell was measured with three copper–constantan thermocouples placed in the cell: one near the top, one at the middle, and one near the bottom. The pressure in the cell was measured using a Bourdon tube Heisse pressure gauge.

The experimental procedure started by injecting 125 mL of the sample solution into the cell. The cell was sealed and the mixing of the solution initiated. The cell was pressurized and then purged with CO_2 three times to ensure that the atmosphere within the cell was pure CO_2. The system reached the set temperature controlled by the refrigeration unit and the ethylene glycol–water bath. Then, the cell was pressurized with CO_2 to a pressure that resulted in a large amount of hydrate forming. The pressure was decreased below the equilibrium point and the hydrate was allowed to decompose.

Chun *et al.* (1996) designed an apparatus based on the same principle to characterize the clathrate phase equilibria for the water+deuterium oxide+CO_2 and water+deuterium oxide+chlorodifluoromethane (R22) systems (Figure 3.12).

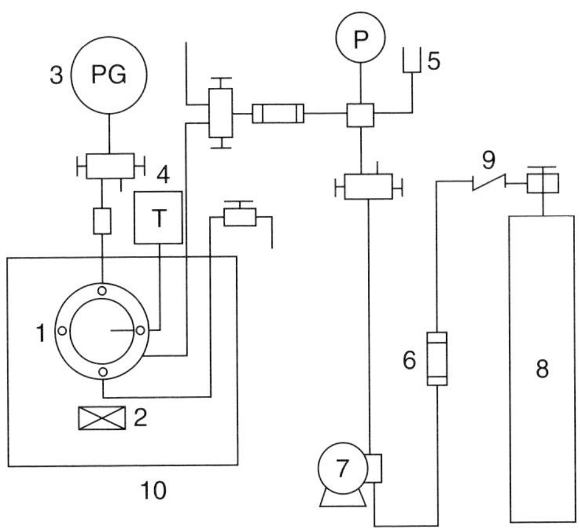

Figure 3.12 Diagram of the experimental apparatus designed by Chun *et al.* (1996) to determine the clathrate phase equilibria for the water + deuterium oxide + carbon dioxide and water + deuterium oxide + chlorodifluoromethane ternary systems. 1: equilibrium cell; 2: magnet; 3: pressure gauge; 4: thermometer; 5: rupture disk; 6: check valve; 7: high-pressure pump; 8: CO_2 cylinder; 9: line filter; and 10: water bath. Reproduced with permission of the American Chemical Society.

An equilibrium cell made of stainless steel with an internal volume of about 50 cm^3 was used. Two sight glasses at the front and back of the cell allowed visual observation of the phase transitions. The cell content was vigorously agitated by a magnetic spin bar. The water bath temperature was controlled by an externally circulating refrigerator-heater. The temperature in the cell was measured by a K-type thermocouple. A Heise gauge was used to measure the relatively high dissociation pressures of the water + deuterium oxide + CO_2 system. For the water + deuterium oxide + R22 system, a different pressure gauge was used since the expected dissociation pressures of this system are below 0.9 MPa.

The equilibrium cell was charged with about 20 cm^3 of the mixture of water and deuterium oxide. After the cell was pressurized to a desired pressure with either CO_2 or R22, the system was cooled to about 5 K below the anticipated clathrate-forming temperature. After the system temperature stabilized, clathrate nucleation was induced by agitating the magnetic spin bar. When the clathrates formed and the system pressure became constant, the cell temperature was increased at a rate of about 1 K/h until the clathrate phase was in coexistence with the liquid and vapor phases. The system temperature was then slowly increased at a rate of 0.2 K/h. The nucleation and dissociation steps were repeated at least twice to reduce hysteresis. When a small amount of crystal was visually observed, without significantly increasing or decreasing in size, and the system temperature was kept constant for at least 8 h after stabilizing the system pressure, the pressure was considered as the equilibrium clathrate dissociation pressure at the specified temperature.

3.4 Review of experimental results

The most extensive studies on CO_2 solubility have been performed on aqueous solutions. A number of papers have been published on the measurement of CO_2 solubility in water containing salts. In Table 3.3 a summary of experimental studies on CO_2 solubility in aqueous solutions is reported.

The phase equilibrium of aqueous solutions containing NaCl has been extensively studied due to increasing interest in CO_2 sequestration (US Department of Energy 1999) to fight global warming. This consists of injecting CO_2 into deep saline aquifers (Pruess and Garcia 2002). Predicting the sequestration potential and long-term behavior of man-made geologic reservoirs requires the calculation of pressure, temperature, and composition (P–T–X) properties of CO_2–H_2O–NaCl mixtures at depths where temperatures are below 100°C, but pressures may reach several hundred bars (Spycher *et al.* 2003). Early experimental gas solubility data for the binary systems CO_2–1% wt. NaCl water solution were reported by Nighswander *et al.* (1989). Solubility was slightly reduced by the inclusion of the salt. Also, this difference was smaller with increasing temperature. The solubility in aqueous solutions of NaCl was also measured by Bando *et al.* (2003). The experimental apparatus was designed to dissolve CO_2 in NaCl solution in a pressurized vessel based on the static–analytic method. CO_2 solubility was calculated from the evaluation of the collected mass of the sample and the pressure of the gas obtained by the depressurization of the sample in a known volume. Experiments were performed for NaCl solutions with mass fractions changing from 0.01 to 0.03 at a temperature in the range from 30°C to 60°C and pressure ranging from 10 to 20 MPa. With increasing mass fraction of NaCl in the aqueous solutions, CO_2 solubility decreased compared to that in pure water.

Studies on the solubility of CO in aqueous solutions containing strong electrolytes have also been performed. These systems have applications with supercritical CO_2. Kurz *et al.* (1996a) measured the simultaneous solubility of ammonia and CO_2 in aqueous solutions of ammonium sulfate (Na_2SO_4) and mixtures containing ammonium sodium sulfate ($(NH_4)_2SO_4$) at temperatures between 202 K and 282 K and total pressures up to 10 MPa. Results showed that the total pressure and the partial pressures of NH_3 and CO_2 above the salt solution were larger than those in the salt-free system (ammonia+CO_2+water). This was expected as NH_3 and CO_2 are both salted out by $(NH_4)_2SO_4$. Kurz *et al.* (1996b) also measured the simultaneous solubility of ammonia and CO_2 in aqueous solutions of sodium chloride in the temperature range of 313–393 K at total pressures up to 3 MPa, showing that the addition of NaCl to the NH_3–CO_2–H_2O system caused CO_2 solubility to decrease at fixed temperature and pressure.

Rumpf *et al.* (1997) studied CO_2 solubility in aqueous solutions containing sodium nitrate ($NaNO_3$) or ammonium nitrate (NH_4NO_3) at temperatures from 313 K to 433 K and pressures up to 10 MPa. They compared CO_2 solubility in solutions containing about 6 mol·kg^{-1} and 10 mol·kg^{-1} $NaNO_3$ or NH_4NO_3 with results of the salt-free system (water + CO_2) showing that CO_2 is salted out by these compounds. Rumpf *et al.* (1998) measured CO_2 solubility in aqueous solutions containing acetic acid and sodium hydroxide in the temperature range of 313–433 K and total pressures up to 10 MPa with molalities of acetic acid or sodium hydroxide in the liquid phase of about 4 and 1 mol/kg, respectively. CO_2 solubility decreased in aqueous solutions of acetic acid compared to that of pure water due to the presence of another acid in solution, and the same trend was observed for sodium hydroxide aqueous solutions.

Xia *et al.* (1999) evaluated the solubility of CO in aqueous solutions containing single salts (sodium acetate and ammonium acetate) with molarities in the liquid phase of about 4 and 6 mol/kg in the temperature range 313–433 K at total pressures up to 10 MPa. The presence of sodium acetate caused a large decrease in the solubility of CO_2 which increased with increasing salt molarity in the aqueous solution. The same effect was observed with ammonium acetate.

Table 3.3 Summary of experimental study on CO_2 solubility in complex solutions.

System	Pressure (MPa)	Temperature (°C)	Reference	Year
Water–carbon dioxide–glucose–fructose	8–17	45–80	D'Souza and Teja	1988
Water–carbon dioxide–glucose	0–30	50–70	Dohrn *et al.*	1993
Water–carbon dioxide–glucose–ethanol	0–30	50–70	Dohrn *et al.*	1993
Water–carbon–dioxide–acetone	1.8–9.5	20, 40, 60	Wendland *et al.*	1994
Water–carbon dioxide–glucose–acetone	4, 6, 8	40, 50, 60	Pfohl *et al.*	1996
Water–carbon dioxide–ammonium sulfate	0–10	40–120	Kurz *et al.*	1996
Water–carbon dioxide–ammonium sulfate–sodium sulfate	0–10	40–120	Kurz *et al.*	1996
Water–carbon dioxide–potassium carbonate	0.005–2	25, 50	Park *et al.*	1997
Water–carbon dioxide–potassium carbonate–poly(ethylene glycol)	0.005–2	25, 50	Park *et al.*	1997
Water–carbon dioxide–sodium bicarbonate	0–58	50, 100, 130	Gao *et al.*	1997
Water–carbon dioxide–acetic acid	0–10	40–160	Rumpf *et al.*	1998
Water–carbon dioxide–sodium hydroxide	0–10	40–160	Rumpf *et al.*	1998
Water–carbon dioxide–sodium acetate	0–10	40–160	Xia *et al.*	1999
Water–carbon dioxide–ammonium acetate	0–10	40–160	Xia *et al.*	1999
Carbon dioxide–milk	0–6.9	25, 38, 50	Tomasula and Boswell	1999
Water–carbon dioxide–sodium chlorite	10–20	30–60	Bando *et al.*	2003
Water–carbon dioxide–alcohol–sugar–proteins–free aminoacids	0.1–0.6	7–27	Descoins *et al.*	2006
Water–carbon dioxide–fructose	7.6, 10.3	40	Calix	2008
Water–carbon dioxide–malic acid	7.6, 10.3	40	Calix	2008

Water–carbon dioxide–sucrose–fructose–glucose–malic acid	7.6–15.9	40	Calix *et al.*	2008
Water–carbon dioxide–sucrose–fructose–glucose–citric acid	7.6–15.9	40	Calix *et al.*	2008
Carbon dioxide–orange juice	7.6–15.9	40	Calix *et al.*	2008
Carbon dioxide–apple juice	7.6–15.9	40	Calix *et al.*	2008
Carbon dioxide–grapefruit juice	7.6–15.9	40	Ferrentino *et al.*	2009
Carbon dioxide–guava puree	7.6–15.9	40	Plaza *et al.*	2009
Carbon dioxide–*Hibiscus sabdariffa* beverage	7.6–15.9	40	Ramirez-Rodrigues *et al.*	2009
Water–carbon dioxide–glucose	7.5–15	40, 50	Ferrentino *et al.*	2010
Water–carbon dioxide–malic acid	7.5–15	40, 50	Ferrentino *et al.*	2010
Water–carbon dioxide–glucose–malic acid	7.5–15	40, 50	Ferrentino *et al.*	2010
Water–carbon dioxide–glucose–malic acid–sucrose–citric acid–pectin	7.5–15	40, 50	Ferrentino *et al.*	2010
Carbon dioxide–apple juice	7.5–15	40, 50	Ferrentino *et al.*	2010
Water–carbon dioxide–sodium phosphate monobasic	7.5–15	35, 40, 50	Ferrentino *et al.*	2010

Park *et al.* (1997) measured the solubility of CO in aqueous potassium carbonate (K_2CO_3) solutions and potassium carbonate–poly(ethylene glycol) (PEG) solutions at 298.2 and 323.2 K with a CO_2 partial pressure between 2 and 5 MPa. The presence and the increase of K_2CO_3 in the aqueous solution decreased the solubility of CO. The addition of PEG to 5% K_2CO_3 solutions reduced the solubility of CO_2 at a constant temperature and pressure.

Gao *et al.* (1997) measured CO_2 solubility in aqueous sodium bicarbonate ($NaHCO_3$) solutions showing that, with increasing the $NaHCO_3$ concentration from 2% to 7% the CO_2 solubility decreased at a constant pressure and temperature.

Phase equilibrium data were obtained by Dohrn *et al.* (1993) for aqueous solutions with a composition close to that of liquid foods. The high-pressure apparatus was designed to obtain reliable phase equilibrium data for glucose–water–CO_2 and glucose–water–ethanol–CO_2 systems at temperature up to 343 K and pressure up to 30 MPa. CO_2 solubility decreased due to the presence of the glucose, while adding ethanol as a polar cosolvent resulted in a substantial increase of the solubility of glucose in the vapor phase at high pressure.

More complex solutions have also been tested. Phase equilibria in a five-component system (glucose + fructose + water + ethanol + CO_2) were measured at temperatures between 318.15 and 353.15 K and pressures between 8 and 17 MPa using a high-pressure vapor–liquid recirculation apparatus (D'Souza and Teja 1988). The purpose was to understand phase equilibria in order to increase the selectivity and the solvent power of supercritical CO with ethanol as a cosolvent used to separate the two sugars. The authors demonstrated that the total weight of solute (glucose + fructose + water + ethanol) extracted in the vapor phase increased with increasing pressure at constant temperature and decreased with increasing temperature at constant pressure, while the concentrations of fructose and glucose in the liquid phase were nearly equal at all temperatures and pressures tested.

Phase equilibria for glucose + acetone + water + CO_2 system were investigated experimentally in the three-phase region at temperatures of 313, 323, and 333 K and pressures of 4, 6, and 8 MPa (Pfohl *et al.* 1996). The results showed the dependence of all component mole fractions on temperature and pressure. As in the ternary system without glucose (Wendland *et al.* 1994), increasing the pressure at constant temperature or decreasing the temperature at constant pressure led to a middle phase similar to the gas phase. Glucose fractions showed the expected behavior. Glucose was never detected in the gas phase, and its content in the water-rich phase was nearly constant.

Descoins *et al.* (2006) reported the construction of a small laboratory-scale carbonation unit that allowed the study of CO_2 solubility in various hydro-alcoholic media with different compositions. The apparatus was a real industrial unit and measured the instantaneous concentration of CO_2 in the carbonation process. The effervescence properties of carbonated drinks depend on the concentration of the carbon dioxide dissolved and on its ability to be released by the liquid phase to form bubbles. Descoins *et al.* (2006) showed that while sugar reduced CO_2 solubility and, consequently, carbonation free amino acids and proteins increased the concentration of CO_2 in the solution.

The experimental equipment designed by Tomasula and Boswell (1999) allowed measuring CO_2 solubility in milk at temperatures of 25°C, 38°C, and 50°C and at pressures up to 6.9 MPa. These data are useful for the design and control of processes which use CO_2 for precipitation of casein from milk. They found that the solubility of CO in milk increased with pressure and decreased with increasing temperature, a trend similar to the one observed for the solubility of CO in pure water. At 25°C the solubility of CO_2 in milk was lower than the solubility in water over the entire pressure range studied and no precipitation of casein was observed. At 38°C and 50°C and pressures higher than 5.5 MPa, complete precipitation of casein from milk was observed and the solubility of CO in milk was approximately the same as that in water.

Solubility of CO_2 in fructose solutions was measured at 40°C and two levels of pressure (7.58 and 10.34 MPa) and two fructose concentrations (5% and 10%) with the experimental apparatus designed by Calix (2008). With increasing fructose concentration, the solubility of CO_2 decreased. Similar results were obtained by Descoins *et al.* (2006) who stated that at 0.6 MPa and pH of 3.1, an increase of sugar concentration (0, 10, and 40 g/L of sugar) led to a decrease of the average level of dissolved CO_2 due to the different orientation of the water molecules around the solute particles.

Solubility of CO_2 in malic acid solutions was also measured at 40°C and pressures of 7.58 and 10.34 MPa (Calix 2008). The experiments were performed to test the effects of this substance, one of the main components of apple juice, on CO_2 solubility by changing its concentration up to 5% aqueous solution. There was a significant difference between CO_2 solubility in water (0% malic acid concentration) and in 5% malic acid solutions at 7.58 MPa. No significant differences in CO_2 solubility between solutions at 0.5% and 5% malic acid were detected at the same pressure and temperature conditions. Increasing the pressure up to 10.34 MPa gave similar results.

More complex solutions were also studied. The purpose was to evaluate CO_2 solubility in solutions approximating the composition of real juices. The comparison of CO_2 solubilities between a real and a model system could give the possibility to develop thermodynamic numerical predictions for the real systems. A first attempt was carried out by Calix *et al.* (2008) to evaluate CO_2 solubility in: (1) commercial pasteurized orange and apple juice, (2) in aqueous solutions approximating the composition of apple juice (3 grams sucrose, 6 grams fructose, 4 grams glucose, and 0.01 grams malic acid), and in (3) aqueous solutions close to the composition of orange juice (4.5 grams sucrose, 3.5 grams fructose, 3.5 grams glucose, and 0.01 grams citric acid). A difference was detected between CO_2 solubility values measured in water and orange juice, apple juice, and solutions approximating the composition of both juices. The differences were noticeable with increasing pressure and were estimated to be about 1.3 g of CO_2/100g of solution due to the presence of sugars and acids. The differences in CO_2 solubility between the orange juice and the aqueous solution approximating its composition were not significant, and were attributable to other minor components present in the orange juice.

Calix *et al.* (2008) simulated the thermodynamic equilibrium of the solutions with the process simulation software Aspen Plus. The chosen equation was the electrolyte non-random two-liquid (ELECNRTL) model with the Redlich–Kwong equation of state for aqueous and mixed-solvent applications. This thermodynamic model was not able to simulate CO_2 solubility in both water and more complex solutions.

Ferrentino *et al.* (2009b) measured CO_2 solubility in grapefruit juice. Knowing the value of CO_2 solubility in the juice allowed one to avoid the current practice of using excess CO_2 in continuous DPCD processes treatments. The exact amount of CO_2 to be used in DPCD processing is generally unknown since CO_2 solubility data are unavailable at different pressures, temperatures, and juice compositions. To evaluate the effectiveness of the process in the pasteurization of grapefruit juice, experiments were carried out in a continuous DPCD system by changing pressure at constant temperature (40°C) and CO_2 level (5.7%) obtained from the solubility tests. The implemented process parameters resulted in 5 log reductions in terms of yeasts and molds and total aerobic microorganisms at 34.5 MPa and 7 min of treatment time.

CO_2 solubility was also measured in guava puree (Plaza *et al.* 2009) and in a *Hibiscus sabdariffa* beverage (Ramirez-Rodrigues *et al.* 2009) to optimize the CO_2 % level to be used in the continuous DPCD system.

Experiments were carried out by Ferrentino *et al.* (2010c) to determine CO_2 solubility in ternary systems (e.g. water–CO_2–glucose and water–CO_2–malic acid), in quaternary systems (e.g. water–CO_2–glucose–malic acid), in more complex solutions approximating the composition of a given juice (e.g. water–CO_2–malic acid–citric acid–sucrose – fructose–pectin), and in real juices (e.g. apple juice). The experimental conditions were 40°C and 50°C and pressures between 7.5 and 15.0 MPa. Also, the effect of changing the concentration of some compounds (glucose from 4 to 12 g in 100 g solution and malic acid from 0.01 to 2.68 g in 100 g solution) on CO_2 solubility was analyzed. The CO_2 solubility values decreased significantly for the water–glucose system compared to the pure water system at the same conditions of pressure and temperature. Moreover, an increase in the glucose concentration from 4 to 12 g in 100 g of solution caused a marked drop in the CO_2 solubility value.

For the malic acid–water system, the results did not show a significant dependence of CO_2 solubility value on malic acid concentration as reported by Calix (2008). Also, comparing CO_2 solubility values with those in pure water, insignificant differences were detected in the range of pressures tested. The results obtained in more complex solutions approximating the composition of real apple juice (water, citric acid, malic acid, sucrose, glucose, and pectin) were compared to CO_2 solubility measured in a commercial apple juice. For both systems, CO_2 solubility was lower than that in pure water. CO_2 solubilities of the model solution and commercial juice were very close, and the small differences were attributed to other minor components present in the real juice. The results confirmed those reported by Calix *et al.* (2008), performed in the semicontinuous apparatus.

The process simulation software Aspen Plus was used to assess the ability of three different thermodynamic models to describe the vapor–liquid equilibrium of the solutions discussed above. The ELECNRTL model and the completely predictive Soave–Redlich–Kwong (PSRK) model were not able to describe the equilibrium in the conditions of temperature and pressure tested. On the other hand, the Peng–Robinson–Wong–Sandler (PRWS) model based on the Peng–Robinson equation of state (PR-EOS) with the Wong and Sandler mixing rule provided good predictions for all the aqueous solutions investigated. The combined use of simple solubility experiments and model equilibrium calculations could be a promising approach in the experimental interpretation and design of applications involving the DPCD technology.

Solubility experiments were also performed on sodium phosphate monobasic aqueous solutions (Ferrentino *et al.* 2010a). The objective was to characterize the phase equilibrium of these solutions and also to explain the microbial inactivation mechanisms of the DPCD process. CO_2 solubility was demonstrated to be a strong function of sodium phosphate monobasic concentrations. Experimental results showed that increasing sodium phosphate concentration CO_2 solubility and, in turn, the amount of CO_2 dissolved in the solution and in contact with the microbial cells decreased. Since the amount of CO_2 dissolved in liquid foods is the forcing factor to cause microbial death (Ferrentino *et al.* 2010b), the possibility to express the inactivation kinetics of microorganisms suspended in sodium phosphate monobasic solutions at different concentrations with a unified model based on CO_2 solubility and temperature, instead of pressure, temperature, and composition of the suspending medium, was demonstrated.

The proposed kinetic model was much simpler than the one obtained correlating the microbial inactivation to pressure and temperature and, more important, was independent of the suspending medium composition. The results indicated that the coupling CO_2 solubility, which can be measured experimentally or predicted with simulation software, with inactivation models depending on solubility and temperature could be a powerful instrument for the interpretation of microbial inactivation data and for the design of DPCD processes and equipment.

3.5 Conclusions

The knowledge of CO_2 solubility is an important factor to increase the applicability of several processes using CO_2 at high pressures. A large amount of experimental and theoretical solubility values have been reported in the context of addressing problems related to the simulation of petroleum reservoirs, enhanced oil recovery, transportation and storage of natural gas, study of geological processes, and supercritical fluid extraction processes. Only recently the increased importance of the DPCD treatment, proposed as nonthermal pasteurization for liquid foods, has focussed the attention on the study of phase equilibria between CO_2 and liquid foods under conditions of temperature and pressure applied during this process. Considering that in a continuous treatment the process parameters are pressure, temperature, residence time, and the amount of CO_2 dissolved in the juice, the knowledge of CO_2 solubility will be a great importance for the development of the technology at industrial scale.

4 Effects of Dense Phase Carbon Dioxide on Vegetative Cells

Osman Erkmen

Abstract: In recent years several papers have been published on the effects of dense phase CO_2 on different insect pests, bacteria, viruses, and fungi The use of pressurized CO_2 for inactivating vegetative cells has significantly increased over the years mainly due to its treatment efficacy in significantly reducing the numbers of vegetative cells in liquid foods, liquid media, and solutions typically in the range of 2–8 log/mL of liquid food. In the present chapter the experimental results describing the effect of dense phase carbon dioxide on vegetative cells will be reported, describing the potential of different gases used for inactivating microorganisms, the effect of dense phase CO_2 on microorganisms, the factors affecting the sensitivity of microorganisms to the treatment, the mechanisms of microbial inactivation, the shape of the survival curves of the vegetative cells, and the quantification of inactivation.

Keywords: vegetative cells; inactivation mechanisms; microbial inactivation.

4.1 Introduction

Consumer demand for high-quality natural foods has awakened a growing interest in nonthermal preservation techniques to inactivate microorganisms and enzymes. Compared with conventional processes, these techniques avoid drawbacks such as loss of flavor, denaturation of nutrients, toxic side reactions, and changes in the physical and chemical properties of foods. Nonthermal techniques must preserve food quality and enhance the safety of foods.

The use of dense phase carbon dioxide (DPCD) seems to be a most promising technique since it operates at mild conditions (such as lower temperature and pressure), uses small vessels, requires less pumping power (thus reducing process costs), is easier to control and manage pressure, is more adaptable to continuous processing and consequently encourages new applications for the technology (Murakami *et al.* 2004; Garcia-Gonzalez *et al.* 2007).

Dense Phase Carbon Dioxide: Food and Pharmaceutical Applications, First Edition. Edited by Murat O. Balaban and Giovanna Ferrentino.

Table 4.1 Properties of gases used in high-pressure processing.

Gas	T_c (°C)	P_c (MPa)	D[1]	Solubility (g/L)[2]	Critical density (g/cm^3)
Carbon dioxide	31.1	7.38	0	1.80	0.47
Argon	−122.3	4.91	0	0.043	0.54
Nitrogen	−147.0	3.39	0	0.018	0.31
Nitrous oxide	36.42	7.25	0.17	1.20	0.45
TFE	54.85	4.06	2.1	1.10	—

[1] Dipole moment.
[2] In H_2O at 25°C and 0.1 MPa.

The phase diagram of carbon dioxide (CO_2) indicates that CO_2 can exist in several states with different characteristics depending on the temperature and pressure: solid, liquid, gas and supercritical fluid. The term 'dense phase' is used to indicate that supercritical CO_2 remains in a fluid state, yet is denser with respect to gaseous CO_2. DPCD is the application of CO_2 at the supercritical state. DPCD was first addressed by Fraser (1951) and Foster *et al.* (1962). They reported the disturbance of bacterial cells by the rapid release of pressurized CO_2. Several articles were published in the early 1980s, pointing out the effect of DPCD on different insect pests, bacteria, viruses and fungi (Spilimbergo *et al.* 2002). The number of publications on the use of pressurized CO_2 for inactivating microorganisms has significantly increased over the years (Parton *et al.* 2003; Spilimbergo and Bertucco 2003; Damar and Balaban 2006; Zhang *et al.* 2006b; Calvo *et al.* 2007; Garcia-Gonzalez *et al.* 2007).

This chapter is structured as follows: gases used for inactivating microorganisms, the effect of DPCD on microorganisms, factors affecting the sensitivity of microorganisms to DPCD, the mechanisms of microbial inactivation, the characterization of CO_2 states and survival curves, and quantifying inactivation.

4.2 Gases used for inactivating microorganisms

Different gases have been used in the inactivation of microorganisms under pressure. The majority of gases used in the HP treatments include CO_2, nitrogen (N_2), nitrous oxide (N_2O), ethylene, argon (Ar) and tetrafluoroethane (TFE) (Table 4.1). N_2, Ar and TFE have less microbicidal efficacy at higher pressures; only N_2O and CO_2 have microbicidal effects. Ar, N_2, N_2O and CO_2 were used for the inactivation of *E. coli* with pressure from 1.7 to 6.2 MPa at 37°C for 5 min (Fraser 1951). Ar, N_2 and N_2O were less effective than CO_2 on this bacterium. Enomoto *et al.* (1997a) reported the microbicidal effects of N_2O at 4 MPa and 40°C after 4 h treatment. The high density and low polarity of supercritical N_2O can favor solubilility in water and easy diffusion in the cell walls and membranes; its critical parameters are close to those of CO_2. Wei *et al.* (1991) reported that CO_2 treatment at 35°C and

6.2 MPa for 2 h completely inactivated *Listeria monocytogenes*, while N_2 treatment at the same conditions had no effect. Nakamura *et al.* (1994) indicated inactivation of *Saccharomyces cerevisiae* by 7 log after 3 h exposure to CO_2 or N_2O at 4 MPa and 40°C, while both N_2 and Ar, which have poor solubility in water, had no effect on the yeast at the same conditions. Dillow *et al.* (1999) indicated that TFE treatments at the same low temperature and pressure did not reduce viable cells as successfully as CO_2.

In the supercritical state the physical properties of CO_2 are altered: its density increases, its viscosity decreases and its diffusivity increases. These give it excellent dissolving and penetrating power, enhancing entry through the cell walls and membranes. For use as a sterilant, CO_2 remains the most preferred gas due to the fact that it is nontoxic, has no special handling and ventilation requirements, is chemically inert (does not react with polymers), has extremely low or zero surface tension (quickly penetrates through the complex structure and porous materials), is nonflammable, is physiologically safe, is readily available, can be easily removed by simple depressurization and out-gassing, is easily controlled, is an ingredient of foods and beverages (carbonated products), does not leave residues in the treated product, is accessible and is cheap. It is considered 'generally regarded as safe' (GRAS), which means that it can be used in food products. Its critical point (7.38 MPa and 31.1°C) is moderately low, so the organic thermolabile products may be processed with no risk of thermal denaturation or decomposition. N_2O's critical parameters are very close to those of CO_2. N_2O may also be effective in killing microbial cells due to its small dipole moment and high solubility in water. N_2 and Ar don't have the unique properties of gas-like diffusivity and liquid-like density due to their critical points. TFE has a similar critical point with CO_2, but it has different chemical properties (dipole moment).

4.3 Effect of DPCD on vegetative microorganisms

The degree of microbial inactivation in the application of DPCD varies widely among treatments. Much of this variation is due to the alteration of process parameters such as pressure, temperature, treatment time and others.

4.3.1 Effect of DPCD on bacterial cells

Bacteria are generally categorized as Gram-positive and Gram-negative, according to their responses to the Gram staining reaction. The different responses come from different peptidoglycan content of the cell walls. Gram-positive cell walls have very thick peptidoglycan layers (90% of the cell wall), which make the cell walls strong and robust. Gram-negative cells have complex cell wall structures and much thinner peptidoglycan layers (10% of the cell wall). Therefore, the Gram-positive cells are

Table 4.2 Effects of DPCD treatments on Gram-positive bacterial cells.

Bacterial species	Process parameters	Suspending medium	Reduction (cfu)	References
Bacillus subtilis	7.4 MPa, 38°C, 2.5 min	Phosphate-buffered saline (PBS)	>7 log	Spilimbergo *et al.* (2002)
	20.5 MPa, 34°C, 2 h	Growth medium (6-cycle)	5 log	Dillow *et al.* (1999)
	20.5 MPa, 60°C, 4 h	Growth medium (6-cycle)	8 log	
	20.5 MPa, 60°C, 2 h	Growth medium (3-cycle)	1 log	
	8.9 MPa, 40°C, 40 min	Juice	3	Parton *et al.* (2003)
	7.8 MPa, 35°C, 60 min	0.85 NaCl + 0.1% peptone	0.7 log	Wu *et al.* (2007)
Enterococcus faecium	10 MPa, 40°C, 80 min	PBS buffer	5 log	Dillow *et al.* (1999)
Lactobacillus casei	10 MPa, 40°C, 80 min	PBS buffer	5 log	
Lactobacillus brevis	7.8 MPa, 35°C, 60 min	0.85 NaCl + 0.1 % peptone	7.3 log	Wu *et al.* (2007)
Lactobacillus innocua	20.5 MPa, 34°C, 30 min	Growth medium (3-cycle)	3 log	Dillow *et al.* (1999)
	20.5 MPa, 34°C, 30 min	Growth medium (6-cycle)	9 log	
Lactobacillus dextranicum	20.7 MPa, 35°C, 15 min	Nutrient broth	9 log	Lin *et al.* (1993)
	31.05 MPa, 42°C, 15 min	PBS	7 log	Sirisee *et al.* (1998)
Lactobacillus plantarum	15 MPa, 35°C, 60 min	Buffer	7 log	Smelt and Rijke (1992)
	13.8 MPa, 30°C, 30 min	MRS broth	6 log	Hong *et al.* (1999)
Staphylococcus aureus	20 MPa, 35°C, 2 h	Peptone saline (PS) buffer	4.8 log	Kamihira *et al.* (1989)
	20.5 MPa, 34°C, 30 min	Growth medium (3-cycle)	3 log	Dillow *et al.* (1999)
	20.5 MPa, 34°C, 30 min	Growth medium (6-cycle)	7 log	
	20 MPa, 34°C, 10 min	Sterile water	3.5 log	Spilimbergo *et al.* (2003)

Table 4.3 Effects of DPCD treatments on Gram-negative bacterial cells.

Bacterial species	Process parameters	Suspending medium	Reduction (cfu)	References
Escherichia coli	20 MPa, 35°C, 2 h	PS	7.5 log	Kamihira *et al.* (1987)
	10 MPa, 35°C, 14 min	Sterile water	9 log	Kobayashi *et al.* (2007)
	15 MPa, 35°C, 60 min	Buffer	8 log	Smelt and Rijke (1992)
	10 MPa, 40°C, 80 min	PBS buffer	5–6 log	Dellaglio *et al.* (1999)
	10 MPa, 35°C, 50 min	Nutrient broth	6 log	Erkmen (2001)
	31 MPa, 35°C, 40 min	PBS	3.5 log	Sirisce *et al.* (1998)
	20 MPa, 34°C, 10 min	Sterile water	2.5 log	Spilimbergo *et al.* (2003)
	7.8 MPa, 35°C, 60 min	0.85 NaCl + 0.1% peptone	8 log	Wu *et al.* (2007)
Listeria monocytogenes	21.7 MPa, 45°C, 60 min	Milk	8.6 log	Lin *et al.* (1994)
Listeria dunnifii	20.5 MPa, 34°C, 36 min	Growth medium (6-cycle)	4 log	Dillow *et al.* (1999)
Proteus vulgaris	20.5 MPa, 34°C, 36 min	Growth medium (3-cycle)	8 log	
Pseudomonas aeruginosa	7.4 MPa, 38°C, 2.5 min	PS	7 log	Spilimbergo *et al.* (2002)
	20.5 MPa, 34°C, 36 min	Growth medium (3-cycle)	6 log	Dillow *et al.* (1999)
Salmonella salford	20.5 MPa, 34°C, 36 min	Growth medium (3-cycle)	3 log	
	20.5 MPa, 40°C, 2 h	Growth medium (3-cycle)	6 log	
Salmonella typhimurium	13.7 MPa, 35°C, 2 h	Chicken meat	1.5 log	Wei *et al.* (1991)
Serratia mercescens	7.4 MPa, 38°C, 10 min	PBS	7.3 log	Spilimbergo *et al.* (2003)
Coliforms	10 MPa, 35°C, 13.3 min	Sterile water	9 log	Kobayashi *et al.* (2007)

generally stronger, less likely to be broken mechanically and less permeable than the Gram-negative cells. A summary of the effect of DPCD on Gram-positive and Gram-negative bacteria from literature is given in Tables 4.2 and 4.3, respectively. There are differences in sensitivity to DPCD between Gram-positive and Gram-negative bacteria, while both are sensitive to DPCD treatment. Generally, inactivation of Gram-positive bacteria is more difficult than inactivating Gram-negative bacteria.

Kamihira *et al.* (1987) reported a 4–6 log reduction in the number of *E. coli* and *Staphylococcus aureus* treated with DPCD for 2 h at 20 MPa and 35°C in a suspension. Lin *et al.* (1994) demonstrated that DPCD inactivated Gram-positive *L. monocytogenes* cells by 7 log in nutrient broth (NB). *L. monocytogenes* was completely inactivated after 60 min exposure to 6 MPa CO_2 pressure at 45°C (Erkmen 2000c). DPCD (10 MPa and 40°C) inactivated Gram-negative *E. coli* by 6 log after 50 min in phoshate buffer saline (PBS), while both *Enterococcus faecium* and *Lactobacillus casei* were inactivated by 6 log at the same conditions (Dellaglio *et al.* 1999). Dillow *et al.* (1999) studied the effect of DPCD (20.5 MPa and 34°C for 30 min) on two Gram-positive bacteria (*Listeria innocua* and *S. aureus*) and five Gram-negative bacteria (*Salmonella salford, Pseudomonas aeruginosa, E. coli, Proteus vulgaris* and *Legionella dunnifii*). Generally, the Gram-positive bacteria showed higher or at least equal resistance to the Gram-negative species. But Gram-negative *S. salford* showed only a 3 log reduction, while two Gram-positive bacteria, *L. innocua* and *S. aureus*, were reduced by 9 and 7 log, respectively. Scanning electron microscope (SEM) images of Gram-negative *E. coli* cells showed partial damage to the cell wall and a rupturing effect on the cell membrane, while the image of Gram-positive *S. aureus* cells showed no cell wall damage. In a semicontinuous system, Gram-positive *Bacillus subtilis* cells in PBS were completely inactivated (7 log) after 2.5 min with exposure to DPCD at 7.4 MPa and 38°C (Spilimbergo *et al.* 2002), while the same number of Gram-negative *Serratia marcescens* was inactivated after 30 min (Garcia-Gonzalez *et al.* 2007). Bacterial cells differ in their resistance to inhibitory effects of DPCD depending on the cell envelope and its permeability to CO_2.

Lin *et al.* (1993) reported that the effect of DPCD on *Leuconostoc dextranicum* was due to a specific interaction between CO_2 and the cells, not to high pressure itself, since the use of N_2 at the same conditions did not lead to any inactivation. Dillow *et al.* (1999) achieved complete inactivation of a wide variety of bacterial species at moderate temperature in a batch system containing suspension with and without biodegradable polymers.

Haas *et al.* (1989) reported that pressurized N_2 treatment at the same conditions used for DPCD (13.7 MPa and 35°C for 2 h) neither reduced the *Listeria* and *Salmonella* species nor affected the pH, therefore the bacterial inactivation was also due to the reduction of pH with the use of CO_2. Debs-Louka *et al.* (1999) observed that the pH end-points of the DPCD-treated suspension were lower by 1.5 pH units with respect to the control sample (initial pH 6.8) but pH did not vary when different experimental conditions were applied. Hong and Pyun (1999) applied DPCD (8 MPa and 40°C) on *Lactobacillus plantarum*, an acid-tolerant bacterium typically found in dairy products. They observed that microbial inactivation was mainly dependent on the transfer rate of CO_2 into cells and improved by increasing the temperature and pressure. Erkmen (2000d) indicated the strong synergism between temperature and pressure in the inactivation of *E. coli* with DPCD treatment.

Erkmen (1997a, 2001a, 2000a, 2000b, 2000c, 2000e) reported a series of studies regarding the effect of DPCD on a range of foodborne microorganisms in a batch system in different suspending materials, some of them pathogenic and psychrotrophic, which are well known for their ability to grow on foods and to cause spoilage and foodborne diseases. In these studies, two stages were observed on the survival curves. The early stage was characterized by a slow rate of reduction in microbial numbers,

Table 4.4 Effects of DPCD treatments on fungal species.

Fungal species	Process parameters	Suspending medium	Reduction (cfu)	References
Saccharomyces cerevisiae	10 MPa, 35°C, 2 h	PS	3.9 log	Kamihira *et al.* (1987)
	13.8 MPa, 35°C, 10 min	Growth medium	7 log	Lin *et al.* (1992a)
	25 MPa, 35°C, 15 min	PS	6 log	Ishikawa *et al.* (1995)
	7.5 MPa, 40°C, 70 min	Potato dextrose broth	6 log	Erkmen (2002)
	48.3 MPa, 35°C, 5 min	Grape juice	6.5 log	Gunes *et al.* (2005)
	9.0 MPa, 38°C, 18 min	Sabouraud dextrose agar	7 log	Parton *et al.* (2007)
Candida utilis	10 MPa, 33°C, 5 min	PS	6 log	Isenschmid *et al.* (1995)
Kluyveromyces fragilis	10 MPa, 33°C, 5 min	PS	6 log	
Kloeckera apiculata	27.6 MPa, 35°C, 5 min	Grape juice	5.2 log	Gunes *et al.* (2005)
Pichia awry	9.0 MPa, 38°C, 15 min	Sabouraud dextrose agar	6 log	Parton *et al.* (2007)
Aspergillus niger	20 MPa, 35°C, 2 h	PS	5 log	Kamihira *et al.* (1987)

which then sharply decreased at a later stage. The inactivation rate increased with increasing temperature, pressure and exposure time, and was also dependent on the initial number of cells, the type of bacterial species and the type of suspending materials. Carbohydrates, fats and other organic compounds in foods increased the resistance of bacteria to DPCD treatment.

4.3.2 Effect of DPCD on vegetative forms of fungi, pests and viruses

A summary of the effects of DPCD on vegetative forms of moulds and yeasts from literature is given in Table 4.4. The antimicrobial effect of DPCD represents a promising alternative for the pasteurization of foods by significant inactivation of fungi at moderate temperature and pressure. The fungicidal effects of DPCD on *S. cerevisiae* was dramatically enhanced with increasing pressure, temperature and length of treatment time; the temperature and pressure tended to synergistically act on each other (Nakamura *et al.* 1994; Enomoto *et al.* 1997b; Erkmen 2002). Viabilities of *Kluyveromyces fragilis*, *S. cerevisiae* and *Candida utilis* under DPCD treatment were dependent on the temperature and dissolved CO_2 concentration, and the inactivation of yeast cells was due to the removal of vital components from the cell membrane rather than cell rupture (Isenschmid *et al.* 1995). They also indicated that subcritical CO_2 was less harmful to yeasts than supercritical CO_2, and the resistances of the yeasts to DPCD treatment were in the following order: *K. fragilis* > *S. cerevisiae* > *C. utilis. S. cerevisiae* was inactivated by 6 log after 60 and 70 min exposure to DPCD at 10 and 7.5 MPa, respectively, at 40°C (Erkmen 2002). Shimoda *et al.* (2002) inactivated *Aspergillus niger* spores by 3 log with DPCD at 19 MPa and 46°C after 1.7 min in physiological saline (PS), while Kamihira *et al.* (1987) inactivated this mould by 5 log at 20 MPa and 35°C after 2 h in PS.

CO_2 pressures of 1.0 and 5.0 MPa are enough to kill insect eggs, larvae or beetles after explosure from 10 to 20 min; this strong effect may be connected with gas action as a respiratory analeptic (Stahl *et al.* 1986). The most common insects and their eggs present in rice can be completely eradicated with CO_2 at low pressure (2.5 MPa) (Capilla *et al.* 2003).

DPCD sterilization was used for increasing the safety of human allograft tissue and biodegradable polymers (Fages *et al.* 1998; White *et al.* 2006). Fages *et al.* (1998) used DPCD (25 MPa and 50°C for 10 min) to inactivate four types of viruses: human immunodeficiency virus type-1 (HIV-1), Sindbis virus, Polio Sabin type-I virus and Pseudorabtes virus (PRV). The reduction factors (log_{10}) for four viruses were 14.2 for HIV-1, 18.2 for Sindbis virus, 24.4 for Polio Sabin virus and 17.6 for PRV. The DPCD process was effective in inactivating viruses on human tissues and provided a level of virus inactivation similar to that obtained by conventional cleaning methods. The DPCD process can become a new practice for the inactivation of viruses, as an alternative or in addition to the processes presently used in medicine and food processing.

The components of viruses are proteins, nucleic acids and sometimes lipids. Self-assembly of components inside cells produces infectious viral particles. Protein aggregation with noncovalent forces is primarily responsible for holding virus particles together. Protein–protein and protein–lipid interactions of viruses can be disrupted with DPCD (e.g. 138 MPa or higher). Thus, viral particles undergo partial or total disassembly under the DPCD process. Under atmospheric conditions, most viruses are chemically quite stable. At DPCD treatment, chemical reactions might occur on viruses at a significant rate that do not normally occur under atmospheric conditions (Fages *et al.* 1998). The fluid interactions with the envelope cause inactivation, and the decay of enveloped viruses is much higher than that of non-enveloped ones (Bouzidi *et al.* 1998). Therefore, partial or total disassembly of viral particles, interactions and chemical reactions might be responsible for the inactivation of viruses by DPCD treatment.

4.4 Factors affecting the sensitivity of microorganisms to DPCD

Microbial cells have complex chemical systems composed of different organic and inorganic components. The rates of microbial inactivation due to the application of DPCD vary among treatments. Much of these variations are due to the many parameters that can be altered during the DPCD precess. Typical parameters include the physical state of the CO_2, pressure, temperature, treatment time, microbial type, initial microbial number, type of system, pressure cycling, agitation, physical–chemical properties of the suspending material and others.

4.4.1 Effect of CO_2 physical states

The physico-chemical properties of CO_2 depend on temperature and pressure. At or over its critical point, CO_2 has a single phase and exhibits a liquid-like state with increasing density and solubility, which becomes a more effective solvent with high extraction power on the cells. It also exhibits a gas-like state with decreasing viscosity and increasing diffusivity, which facilitate quick diffusion and penetration through complex cell wall and membrane structures, and cytoplasm. In addition, the very low surface tension of CO_2 incrases its penetration ability through microbial cells (Garcia-Gonzalez *et al.* 2007). Therefore, pressurized CO_2 at the supercritical state has more lethality on microbial cells than subcritical states such as liquid or gas. Gunes *et al.* (2005) studied the inactivation of *E. coli* in diluted apple cider using three states of CO_2 under pressure. They noted that the CO_2–product concentration was the most important factor affecting the inactivation rate of *E. coli*, which was more sensitive to DPCD than the increase of temperature and pressure. Total psychrotrophic microbial reductions ranged from 1.14 to 5.36 log with carbonation of suspension by 66 and 132 g/kg, respectively, at 35°C with CO_2 pressure from 10.3 to 20.7 MPa (Werner and Hotchkiss 2006). The effects of DPCD on microorganisms can be enhanced by increasing the temperature, pressure and carbonation of the suspension with CO_2.

4.4.2 Effect of temperature and pressure

Each microorganism has specific optimum, maximum and minimum growth temperatures. As the temperature increases above the optimum, proteins start to denature, the cytoplasmic membrane collapses (i.e. disintegrates, crumples and breaks down) and above the maximum these changes occur at higher rates and are sufficient to kill or lyse microbial cells (Erkmen and Bozoglu 2008a). The temperature and pressure, by changing the state of the CO_2, strongly affect the microbicidal efficiency of CO_2.

In general, increasing CO_2 pressure accelerates microbial inactivation. At a higher pressure, a shorter exposure time is needed to inactivate the same level of microbial cells. A 14.3 min D-value was obtained for *L. monocytogenes* under 1.5 MPa CO_2 pressure at 45°C, while a 7.3 min D-value was found at 6.0 MPa under the same conditions (Erkmen 2000c). The pressure may not act directly on the cells. Instead, it may contribute to change in the physical state of CO_2, to increase its solubility in the water phase of suspension and cell cytoplasm, reinforce the contact between cells and CO_2 and accelerate its penetration into the microbial cells. These effects of pressure do not increase indefinitely and are limited by the saturation solubility of CO_2 in the suspension (Sims and Estigarribia 2003). Spilimbergo (2002) indicated that an increase of pressure over 10 MPa cannot influence much the solubility of CO_2 in water.

Temperature plays a major role in the inactivation of microorganisms by directly disturbing biological systems and indirectly affecting the microbicidal action of CO_2 as well as pressure. The influence of temperature on the microbicidal activity of CO_2 appears at two levels. Firstly, temperature acts on the structure of water and CO_2 hydrates, which are increasingly destroyed, and the physical state of CO_2, such as a decrease in viscosity, which facilitates its diffusion in the cells. Secondly, temperature may stimulate the diffusion of CO_2 into cells by affecting the integrity of the cellular membrane, such as by increasing its fluidity.

As the temperature and pressure increase to or exceed a critical point, CO_2 becomes supercritical with a high diffusion coefficient, and low viscosity and density (Oulé *et al.* 2006). CO_2 may diffuse immediately in the cells at higher temperature and pressure. The survival curve may appear as a single phase of cell mortality, comparable to the effect of thermal treatment at higher temperature. The increased temperature and pressure can reduce the duration of treatments. But microorganisms generally are more resistant to pressure than to the temperature of the DPCD process.

Hong and Pyun (1999) reported that the inactivation of *L. plantarum* at 30°C at 7 MPa CO_2 pressure was better than at 40°C. This probably results from higher density of CO_2 at 30°C (0.27 g/mL) than at 40°C (0.2 g/mL). Erkmen (2000d) indicated that inactivation of *E. coli* at 7.5 MPa CO_2 pressure was higher with the temperature increase from 20°C to 30°C than the temperature increase from 30°C to 40°C. Therefore, CO_2 treatment under pressure should not be processed at temperatures far above its critical point, because the density of the solvent and its solubilization capacity decrease quite rapidly, which reduces the ability of CO_2 to extract low-volatility materials from cells (Lin *et al.* 1993; Hong and Pyun 1999; Lucien and Foster 1999). An increase in temperature and pressure from an economical point of view increases the efficiency of the process significantly with respect to operation, but also increases investment and operating costs of the DPCD process.

4.4.3 Effect of CO_2 concentration

The penetration of CO_2 into microbial cells depends on CO_2 concentration in the liquid medium. The highest microbial killing rate is achieved when the concentration of dissolved CO_2 saturates in suspension. Solubility of CO_2 in water increases with increasing pressure and decreasing temperature. Solubilization of CO_2 increases rapidly with increasing pressure up to 5.0–7.5 MPa, but an increase of pressure beyond 7.5 MPa gives little additional increase in the solubility of CO_2 or the microbial killing effect (Garcia-Gonzalez *et al.* 2007).

S. cerevisiae and *Torulopsis versatilis* were decreased by 6 log with the treatment at 1.0 and 2 kg/h flow rate of CO_2, respectively, under 6 MPa CO_2 pressure at 35°C, while to achieve the same killing rate on *Zygosaccharomyces rouxii*, a treatment of 2.0 MPa at 3 kg/h of CO_2 flow rate was required at the same temperature (Shimoda *et al.* 1998). The antimicrobial effectiveness of pressurized CO_2 varied with strains of microorganisms besides its concentration in suspending solution. *T. versatilis* and *Z. rouxii* are osmophiles. Their cell walls might provide a barrier against the penetration of CO_2 or might be strong enough to withstand more turgor pressure. In a batch system, the survival of microorganisms (*E. coli, Lactobacillus brevis* and *S. cerevisiae*) was decreased slowly with 0.8 g/cm^3 of CO_2 density, and at the higher density (>0.8 g/cm^3) survivors decreased sharply (Shimoda *et al.* 1998). On the other hand, in a continuous system, the microorganisms were inactivated effectively with CO_2 density from 0.16 to 0.9 g/cm^3. The reduction rate of microorganisms is releated not only to temperature and pressure but also to dissolved CO_2 concentration.

4.4.4 Effect of agitation

The inactivation rate of microorganisms with the DPCD process can be increased by agitation. The suspension in a DPCD vessel can be mixed, stirred, recirculated or agitated using any techniques prior to and during CO_2 treatment. Agitation can enhance the transfer and solubilization rates of pressurized CO_2 in the liquid phase of suspension, and increase the contact of CO_2 with microbial cells. This can enhance and provide easier penetration of CO_2 through cell membranes which can disorder the membrane. An increase in agitation speed accelerates the cell inactivation by the DPCD process (Oulé *et al.* 2006). Tsuji *et al.* (2005) observed higher reduction rates of the number of microbial cells with increasing agitation speed in the vessel filled with CO_2 under pressure. On the other hand, Lin *et al.* (1992) reported a significant decrease in the inactivation rate of microbial cells with agitation compared to without agitation.

4.4.5 Effect of water content

Water can enhance the microbicidal effect of CO_2, in combination with other factors such as temperature, pressure, phase and concentration of CO_2. Water can swell the cells, which increases the permeability of cell walls and hydration of CO_2. Microbial inactivation with DPCD strongly depends on the water content (or water activity) of the suspension. Vegetative cells with low water content can be poorly inactivated, and their resistance to inactivation can increase with decrease of water in the suspension. Dillow *et al.* (1999) showed that the inactivation rate of *E. coli* was increased by the addition of water into suspension. They dried an *E. coli* culture and submitted it to the DPCD sterilization at 14 MPa and 34°C for 2.5 h. One ml of water was added to a second dried sample (8.8 log/mL) and exposed to DPCD pressure under the same conditions. One hundred percent inactivation was achieved for the wetted cells, while most of the cells survived in dried sample. Kumagai *et al.* (1997) reported that the inactivation rate of *S. cerevisiae* under pressurized CO_2 was zero with water content below 0.2 g/g dry matter and the inactivation rate increased with increasing water content in suspension. They also noted the high correlation between unfreezable water and inactivation of microbial cells. The water absorbed by the cell is important in the inactivation of microbial cells with DPCD, but it is not the water in environment. The cell walls and membranes can be expanded with water to become more penetrable by CO_2, and this enhances the modification of biological barriers. The water content within the medium and cell can play a major role in the inactivation of microorganisms with the DPCD process. Another reason for wet microbial cells' susceptibility to CO_2 pressure is probably due to the solubility of CO_2 and the formation and dissociation of carbonic acid (H_2CO_3) in water. The dissociation liberates H^+ ions that subsequently reduce the pH of the suspension and cytoplasm, which inactivates enzymes, and precipitates cellular compounds and ions in the cytoplasm. These changes enhance the microbicidal action of pressurized CO_2.

4.4.6 Effect of pressurization and depressurization rates

CO_2 penetration through the cell wall is the rate-controlling step in the inactivation of microorganisms under pressure. If the mass transfer rate of CO_2 under pressure is increased, this increases the inactivation rate of microorganisms. Fast pressurization and depressurization may enhance the transfer of CO_2 into the cells, the extraction of materials from the cells, and the bursting and mechanical rupturing of cell structures. Pressurized CO_2 rapidly expands through cells. When pressurized CO_2 was rapidly depressurized, this would mechanically break the cell like a popped ballon (Ballestra *et al.* 1996). Nakamura *et al.* (1994) analysed the explosively depressurized yeast cells under SEM and observed mechanical rupture of the cell structures. They concluded that explosive depressurization always gives higher inactivation than a lower rate of depressurization. They also proposed that the rupture of cells happens during pressurization due to the swelling of cells by CO_2.

Enomoto *et al.* (1997a) indicated that the slower depressurization would not always lead to a lower reduction of the cells since inactivated *S. cerevisia* cells were reduced by 7 and 7.2 log after explosive (4.8 MPa/min) and slow (0.33 MPa/min) depressurization of CO_2, respectively, from 4 MPa at 40°C after 4 h exposure. They also detected nearly the same amount of protein from treated suspension with either fast or slow depressurization CO_2. They postulated that the depressurization rate is not essential in cell inactivation. Kumagai *et al.* (1997) reported that other causes of cell rupture can be present, since a slow depressurization rate (e.g. 8 MPa/h) can still achieve as high an inactivation as a fast rate. Lin *et al.* (1992) indicated that the explosive depressurization of CO_2 from 6.8 and 20.6 MPa influenced the survival of yeast cells to a lesser extent compared with slow depressurization. The slower depressurization results in a longer contact time of the cells with the pressurized CO_2, and the sterilization effect of explosive depressurization is much lower than expected.

4.4.7 Effect of pressure cycling

Pressure cycling can enhance the inactivation of microorganisms. This involves the repeated depressurization and pressurization of CO_2 in treatment systems. Pressure cycling can enhance the inactivation of microbial cells by enhancing the cell rupture and mass transfer through cell membrane. Fraser (1951) achieved more than 90% inactivation with cell rupture after two pressure cycles, while 15% cells inactivated after a single pressurization. Dillow *et al.* (1999) reported that, for DPCD treatment of *S. aureus* at 34°C and 20.5 MPa for total treatment time 30 min, the degree of *S. aureus* inactivation was increased from 3 log with three pressure cycles to 9 log with six pressure cycles. Enomoto *et al.* (1997b) reported that when the total time of the pressurization process (at 4 MPa and 40°C) was the same, an operation with a longer treatment time and smaller number of cycling was usually more effective for killing the yeast cells than that with a shorter treatment time and more frequent cycling. They reported that the inactivation of *S. cerevisiae* was 4 log after CO_2 pressure treatment for 2 h without pressure cycling, whereas four sets of the pressure cycling for 30 min resulted in 5 log reduction. Lin *et al.* (1993, 1994) observed a higher degree of inactivation with a pressure-cycling process of CO_2. On the other hand, Hong *et al.* (1997) cycled CO_2 pressure on *Lactobacillus* sp. in MRS broth, but they observed same inactivation rate between pressure cycling and a single pressurization under the same conditions.

4.4.8 Effect of microbial type

Microbial sensitivity to DPCD treatment varies greatly among species. However, comparisons of microbial inactivations with DPCD are difficult because processes differ by systems, treatment media, pressure, temperature and so on. In general, Gram-positive bacteria are expected to be more resistant to DPCD than Gram-negative bacteria due to the differences in the composition of their cell walls. The cell walls of osmophilic microorganisms may provide a barrier layer against CO_2 penetration or may be strong enough to withstand more turgor pressure. Sims and Estigarribia (2002) observed that Gram-positive *Bacillus cereus* cells were more resistant to DPCD treatment than Gram-negative *E. coli* and *P. vulgaris*.

Microbial species are very relevant in the degree of inactivation of microorganisms. Balaban *et al.* (2001) reported that DPCD treatment at 35°C reduced *E. coli* by 4 log, while *Salmonella* and *Listeria* cells were reduced by 6 log. *S. cerevisiae* was totally inactivated by 6.7 log in 6 min with DPCD treatment at 38°C and 9 MPa in liquid medium, whereas at the same operation conditions *Pichia awry* was inactivated by 6.7 log in 9 min (Parton *et al.* 2003). Erkmen (2000b, 2000d, 2001a, 2002) reported the inactivation of *Brochothrix thermosphacta, Yersinia enterocolitica, E. coli* and *S. cerevisiae* by DPCD depending on the pressure, temperature, pH and suspending media. Among the microorganisms, *S. cerevisiae* had higher resistance than other bacteria.

One of the major concerns about a new preservation technology for the food industry is the occurrence of resistant vegetative bacteria to inactivation. There is limited information on the ability of microorganisms to becoming resistant to DPCD. Hong *et al.* (1999) cultured isolates derived from *L. plantarum* cells treated at 13.8 MPa and 30°C after an extended exposure, and subjected these cells again to pressurized CO_2 treatment at the same conditions. They observed little difference in sensitivity between previously treated and first-treated microbial cells.

4.4.9 Effect of initial microbial number

The effect of DPCD treatment on microorganisms can be influenced by their initial number. Under the same conditions of DPCD treatment, the highest degree of inactivation can be obtained with the lowest initial microbial number than the highest initial number (Erkmen 2000e). So at higher initial microbial numbers, longer exposure times to DPCD treatment are needed to achieve the same log reduction that can be achieved at low initial number. Parton *et al.* (2003) indicated that 4 log of *S. cerevisiae* were completely inactivated after 40 min exposure to DPCD (at 9 MPa and 38°C), while more than 60 min was needed to achieve complete inactivation of 7 log yeast cells. Erkmen (2000c) reported that pressurized CO_2 treatment at 6 MPa and 35°C completely inactivated 2.21 log of *L. monocytogenes* cells in 10 min, while it completely inactivated 6.98 log of cells in 75 min. Cellular components (proteins, lipids, carbohydrates etc.) released from lysed cells can protect other microbial cells. Microbial cells are more exposed to a pressurized CO_2 effect with a low initial microbial number and hence are more sensitive to inactivation.

4.4.10 Effect of physical and chemical properties of suspension

The inactivation of microorganisms during DPCD treatment is strongly affected by the physical and chemical properties of the suspending media and foods. Components and properties of the suspension, such as acidity, fat and oil, water, proteins and others, can influence the DPCD inactivation process. Hong and Pyun (1999) reported that pressurized CO_2 treatment of 25 min in acetate buffer (pH 4.5), 35 min in sterile distilled water (pH 6.0) and 60 min in phosphate buffer (pH 7.0) were required to achieve 5 log reduction of *L. plantarum* cells. Erkmen (2000d) indicated that about 95, 80 and 50 min were required for complete inactivation of 7.5 log of *E. coli* cells with pressurized CO_2 at 10 MPa and 30°C in NB with pH 6.75, 5.50 and 4.50, respectively. The initial environmental pH greatly affects the inactivation rate of HPCD treatments since acidity can act synergistically with pressure and temperature. The low initial pH can increase cell permeability to CO_2; therefore, more inactivation is achieved (Damar and Balaban 2006).

Wei *et al.* (1991) indicated that *L. monocytogenes* in broth was completely inactivated with CO_2 at 13.7 MPa and 35°C after 2 h, while same number of *L. monocytogenes* was inactivated after twofold times in whole egg at the same conditions. Sterilization effects of DPCD on *E. faecalis* were observed after 24 h in whole milk, 16 h in skimmed milk, 3 h in orange juice and 4 h in carrot juice, while it completely inactivated after 80 min in PS (Erkmen 2000a). Fat or oil content of suspension increases microbial resistance to DPCD treatment by changing the structure of the cell walls and membranes (Lin *et al.* 1994; Hong and Pyun 1999) and by depressing CO_2 penetration into cells (Erkmen 2000b).

Water content of suspension can enhance the microbicidal effect of pressurized CO_2. Microorganisms in the medium (or food) with low water content can show low or no inactivation behavior with DPCD treatment. Yeast cells (8 log) in the suspension with more than 70% water content were completely inactivated with CO_2 pressure (4 MPa and 40°C) after 5 h, while in growth medium with less than 40% water content they were reduced by 0.5 log (Nakamura *et al.* 1994). Pressurized CO_2 at 6 MPa and 45°C completely inactivated *L. monocytogenes* after 17.5 min in PS, while 60 min was needed in PS containing 1% brain heart infusion broth (Erkmen 2000c). The microbicidal effect of pressurized CO_2 on microorganisms was strengthened by adding ethanol (2 %) or acetic acid (0.5%) into suspension (Kamihira *et al.* 1987).

4.4.11 Effect of culture conditions and growth phases

Microbial inactivation efficiency of DPCD treatment can be influenced by culture conditions and phase of growth. Hong and Pyun (1999) observed a higher resistance of *L. plantarum* cells to CO_2 pressure when they prepared samples for pressure process by incubation at 37°C than at 20°C. On the other hand, Lin *et al.* (1993) observed small differences in the numbers of inactivated *L. dextranicum* cells exposed to pressurized CO_2 treatment between cells prepared by incubation at optimum-growth temperature and lower than optimum-growth T. They also indicated that *L. monocytogenes* cells incubated at 10°C were more sensitive to pressurized CO_2

than those at 27°C. More microorganisms can be inactivated when they are grown at lower temperatures than the optimum growth temperature. Microbial inactivation at DPCD treatment can also increase by lowering or increasing initial pH of the suspension from optimum.

Generally, young cells are more sensitive to DPCD treatment than mature cells. Stationary-phase cells are more pressure resistant than log-phase cells. Hong and Pyun (1999) reported a higher inactivation rate for log phase *L. plantarum* cells exposed to HPCD treatments at 30°C and 6.9 MPa compared to the stationary-phase cells. Microbial cells that enter the stationary phase of growth synthesize proteins and other biomolecules that protect them against various adverse environmental conditions including high temperature, oxidative stress, high pressure and others.

4.4.12 Injured microorganisms

When microorganisms are subjected to environmental stress, they can undergo cellular injury. Injured cells cannot grow on selective media, but uninjured cells can grow on selective media. The number of sublethally injured microbial cells can be determined by difference between the counts obtained on the nonselective and selective media (Erkmen 2007). Erkmen (2000a, 2000c, 2001a) and Hong and Pyun (2001) determined various degrees of sublethal injury on different microbial species depending on CO_2 pressure and suspending medium. They indicated that the counts of microorganisms on the nonselective media were higher than those obtained on the selective media. Sirisee *et al.* (1998) indicated that injured cells of *S. aureus* from DPCD treatment (31 MPa and 42.5°C for 10 min) started to recover after 2 h of incubation in the nonselective medium, whereas *E. coli* did not recover after 30 h of incubation. Kincal *et al.* (2005) observed an increase in the bacterial number during refrigerated storage of the DPCD-treated orange juice, while at the beginning of storage no culturable organisms were present on selective medium. Two possible explanations can be given for the growth of microorganisms in DPCD-treated foods: (1) microbial cells can repair injured cell structure during storage, or (2) postcontamination can occur during or after processing. Formation of cellular injury on microbial cells from DPCD treatment has three main consequences for food processing and research. Firstly, cell injury must always be taken in account with a DPCD inactivation system. Secondly, injured cells might not be detected when selective media are used for the enumeration of survivors after DPCD treatment and this leads to an over-estimation for the lethality of microorganisms from treatment. Thirdly, the occurrence of injured cells after DPCD treatment has great importance for pathogenic and spoilage microorganisms in foods because under appropriate conditions, these cells can repair themselves and grow in food, and this can result in foodborne diseases, food spoilage and economical loss.

4.4.13 *Effect of combination processes*

A combination of a DPCD treatment with conventional or emerging food preservation techniques, such as pulsed electric fields (PEF), high hydrostatic pressure (HHP), antimicrobials, additives, temperature and others, is possible. This approach, generally known as 'hurdle technology', may enhance the lethal effects of the DPCD process and reduce the severity of treatment time needed to obtain a given level of microbial inactivation. A synergistic effect on the inactivation of microorganisms can be observed with the combination treatments. A combined treatment of 4.9 MPa HPCD and 600 MPa HHP for 10 min at 25°C completely inactivated aerobic bacteria (AB) in carrot juice, while the individual treatment with both methods under the same conditions resulted in higher survival of AB (Park *et al.* 2002). Lin *et al.* (1992) reported that trace amounts of sulphur dioxide (30 ppm) were effective in the inactivation of *S. cerevisiae* cells together with pressurized CO_2. Spilimbergo *et al.* (2003a) showed that the survival of microorganisms (*E. coli, S. aureus* and *B. cereus*) was decreased by PEF treatment, and a further treatment of microorganisms exposed to PEF with DPCD caused complete inactivation.

Food preservation using combined methods involves successive or simultaneous applications of various treatments. Combined treatments are advantageous, because many individual treatments alone may not be adequate to ensure food safety or stability. Combining the DPCD process with other food preservation techniques can (1) enhance the lethal effects of process on microorganisms, (2) reduce the treatment time needed to inactivate a given level of microorganisms, (3) inactivate injured cells, (4) allow the use of milder treatment conditions and (5) improve food quality and safety at milder process conditions.

4.4.14 *Effect of type of system*

The type of system used for the DPCD process can affect the microbial inactivation rate. Pressurized systems allowing better contact of CO_2 with the suspending medium (or food) are more effective in microbial inactivation since this provides more rapid saturation of the solution with CO_2. Usually, pressurized CO_2 in batch systems requires a longer treatment time for microbial inactivation than in continuous systems. Pressurized CO_2 in a semicontinuous process is more efficient for microbial inactivation than a batch process. Spilimbergo *et al.* (2003b) indicated that treatments of 60 min were necessary for complete inactivation of different bacterial species by using a batch system, while 10 min treatment was enough by a semicontinuous system at the same process parameters. Ishikawa *et al.* (1995c) obtained more than four times higher inactivation in the numbers of *L. brevis* and *S. cerevisiae* by using a micro-bubbling filter system at the HPCD process rather than a batch one. They improved mass transport in the semicontinuous micro-bubble reactor, resulting in higher efficiency in the inactivation of microorganisms.

Microbial inactivation can be related to CO_2 concentration introduced into the water phase of suspension depending on the type of system. Shimoda *et al.* (1998) applied pressurized CO_2 on microorganisms (*E. coli, S. cerevisiae* and *T. versatilis*) in a continuous-flow system and a batch one. They found that the continuous-flow system effectively inactivated microorganisms due to more saturation of CO_2 in suspension compared with the batch system. Ishikawa *et al.* (1995c) achieved a greater efficiency in the inactivation of *L. brevis* under DPCD treatment in a semicontinuous system than in a batch one. Debs-Louka *et al.* (1999) inactivated *E. coli, S. cerevisiae* and *E. faecalis* cells deposited onto solid materials with HPCD treatment. The type of DPCD pressurized system is important in the inactivation of microorganisms with respect to exposure to CO_2, temperature and pressure controls and others.

4.4.15 Treatment time and inactivation kinetics

Treatment time for the inactivation of microorganisms under DPCD can vary from a few seconds to several days. Inactivation increases with an increase of exposure time to treatment, but the time period required to completely inactivate microorganisms depends on other factors, such as the type of microorganism (e.g., bacteria, yeasts or moulds), the form of microorganism (e.g., vegetative or spore) and treatment conditions. Debs-Louka *et al.* (1999) treated *E. coli* and *E. faecalis* cells on a solid hydrophilic medium with pressurized CO_2. They indicated a linear correlation between microbial inactivation and exposure time. In microbial inactivation with DPCD, several types of survival curves have been observed (Figure 4.1). The inactivation curve can show a two-stage inactivation kinetic. Erkmen (2000c) reported that about 45 min was required for the first stage in the inactivation of *L. monocytogenes* at 45°C and 1.5 MPa CO_2 pressure, this was reduced to 10 min under 6.0 MPa CO_2 at the same conditions. The inactivation rate of *L. monocytogenes* increased at the second stage. The results suggest that *L. monocytogenes* cells could be initially stressed or sensitized by CO_2 under pressure, and are rapidly reduced in the second stage. Lin *et al.* (1992, 1994), Hong *et al.* (1997), Ballestra and Cuq (1998) and Erkmen (1997, 2000c, 2000d, 2000e, 2002) also observed similar slow-to-fast two-stage inactivation kinetics for other microorganisms with DPCD treatments. Lin *et al.* (1991) and Enomoto *et al.* (1997b) reported two distinct stages in the inactivation of yeast cells with the DPCD treatment. The rate of cell inactivation is slow at the first stage and increases drastically at the later stage.

Even though the slow-to-fast two-stage kinetic behavior is dominant at the DPCD process, several studies (Kamihira *et al.* 1987; Ishikawa *et al.* 1995c; Hong and Pyun 1999; Hong *et al.* 1999; Debs-Louka *et al.* 1999)

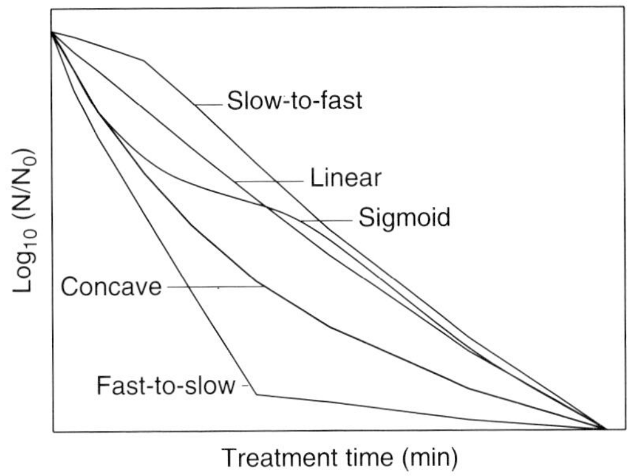

Figure 4.1 Schematic representation of commonly observed survival curves during HPCD inactivation of vegetative microbial cells.

showed fast-to-slow two-stage inactivation kinetics. Debs-Louka *et al.* (1999) observed fast-to-slow kinetic curves in the inactivation of *E. coli* under pressurized CO_2. They reported 60 min for the first stage for the inactivation of 4 log of cells, and 360 min for the second stage for the inactivation of remaining cells (1.5 log). Both kinds of kinetic curves were observed by Hong and Pyun (2001). They indicated that when the temperature and pressure were increased to higher levels, the kinetic behavior of the system can change from slow-to-fast to fast-to-slow. Some studies showed only one fast linear curve in the inactivation of microorganisms at higher temperature and pressure of CO_2 treatment (Ballestra *et al.* 1996; Kumagai *et al.* 1997; Erkmen 2000b, 2000e; Shimoda *et al.* 2002). Kamihira *et al.* (1987) performed experiments on the inactivation of microorganisms with DPCD (20 MPa) and showed fast-to-slow kinetics, while the experiments under lower pressure (from 1.2 to 10 MPa) showed slow-to-fast kinetics. Ishikawa *et al.* (1995c) indicated that bubbling increased contact between CO_2 and liquid media, and between CO_2 and the microbial cells. They observed fast-to-slow inactivation kinetics. Variations in kinetic behavior of microorganisms at DPCD treatment may be due to the differences in the efficiency of contact between CO_2 and suspension, of contact between CO_2 and the microbial cells, of variation in temperature and pressure level, and others.

4.5 Mechanisms of microbial inactivation by DPCD

Several mechanisms have been proposed to explain the antimicrobial action of CO_2 on microorganisms. Some can be considered as primary, while others are complementary. Figure 4.2 schematically illustrates how pressurized CO_2 exerts its lethal action on microbial cells. The following are the general mechanisms of inactivation of microorganisms with the

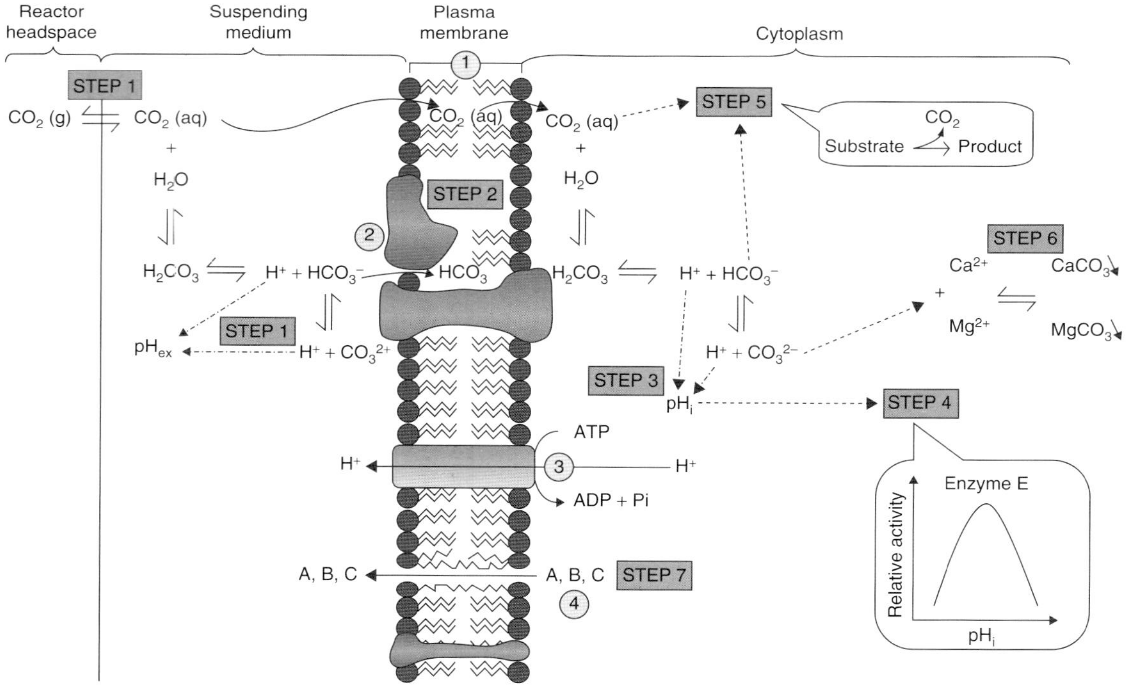

Figure 4.2 A schematic diagram for the different steps of the inactivation mechanism of pressurized CO_2 exerting lethal action on bacteria. 1: phospholipid bilayer, 2: integral membrane proteins, 3: a plasma membrane H^+-ATPase and 4: intracellular substances. (Reproduced from Garcia-Gonzalez *et al.*, copyright 2007, with permission of Elsevier.)

DPCD process: (1) extracellular and intracellular pH decrease (acidification), (2) inhibition of biological activities, (3) cell rupture, (4) modifications of cell walls and membranes, (5) precipitation of proteins and ions, (6) extraction of cellular substances and (7) disordering the intracellular electrolyte balance.

Most of these inactivation steps may not occur consecutively, but rather take place simultaneously in a very complex and interrelated manner.

4.5.1 Solubilization of CO_2 under pressure into suspension

CO_2 can dissolve in the liquid phase of foods (or media) to form H_2CO_3, which further dissociates to give bicarbonate (HCO_3^-) and hydrogen (H^+) ions. Finally bicarbonate dissociates to H^+ and carbonate (CO_3^{2-}) ions according to the following equilibrium reactions at 25°C:

$$CO_2\ (g) \leftrightarrow CO_2\ (aq)$$

$$[CO_2]_{aq} = H^+ \times p_{CO_2} \text{ with } [H^+] = 3.3 \times 10^{-2}\ \text{mol/ (L atm)} \tag{4.1}$$

$$CO_2\ (aq) + H_2O\ (l) \leftrightarrow H_2CO_3\ (aq)\ pK_1 = 2.80 \tag{4.2}$$

$$H_2CO_3\ (aq) \leftrightarrow H^+\ (aq) + HCO_3^-\ (aq)\ pK_2 = 3.60 \tag{4.3}$$

$$HCO_3^-\ (aq) \leftrightarrow H^+\ (aq) + CO_3^{2-}\ (aq)\ pK_3 = 10.25 \tag{4.4}$$

The notations (g), (l) and (aq) refer to the state of the gas, liquid and aqueous solution respectively. HP allows better contact of CO_2 with water, and facilitates saturation (hydration) of the solution with CO_2 (CO_2–H_2O) and solubilization of CO_2 in water. Solubilization of CO_2 in water can acidify suspension by the liberation of H^+ ions. The low extracellular pH (pH_{ex}) may have little function in the inhibition of microorganisms since, according to the chemiosmotic theory, the cytoplasmic membrane is impermeable to protons and other charged molecules, and cells tend to transport charged molecules to the exterior and synthesize basic components by expending cellular energy in the form of ATP. Cell membranes are impermeable to ions, and permeable to CO_2 and weak acids (Damar and Balaban 2006; Erkmen and Bozoglu 2008b). The effects of low pH_{ex} on microbial cells due to hydration of CO_2 can be summarized as follows: (1) at lower pH_{ex}, cells use too much energy that they may not able to generate, and (2) besides high pressure, the lower pH_{ex} can also contribute to an increase in the permeability of cell walls and membranes to facilitate the penetration of CO_2 into microbial cells (Lin *et al.* 1993, 1994).

Acidification of the suspension causes chemical modifications on the phospholipid bilayer of membranes to increase their permeability to CO_2 which can rapidly diffuse through the cell. Enhanced concentration of CO_2

in cells can easily interact with biomolecules and water in the cytoplasm, and compromise its biological balance. All of these factors provide additional contributions to the antimicrobial effects of pressurized CO_2 on microorganisms.

4.5.2 Cell membrane modification

The cell membrane consists of a double layer of phospholipids with hydrophobic outer and hydrophilic inner layers. Unhydrated CO_2 can easily diffuse into the cellular membrane and accumulates in the lipophilic inner layer of a phospholipid bilayer structure (Isenschmid *et al.* 1995). It can extract an amount of constituents from the cell membrane that can be sufficient to modify the structure of the membrane. CO_2 has lipophilic solvent characteristics. The high affinity between CO_2 and the membrane phospholipids facilitates accumulation of an amount of CO_2 (unhydrated) in the lipophilic phase. The accumulated CO_2 (up to a critical level) in the lipid phase may structurally and functionally modify the membrane by disordering the hydrocarbon lipid chains and rupturing lipid–protein interactions and an order loss of the lipid chain (a process known as the 'anesthesia effect'). This may increase the fluidity and enhance the permeability of the membrane (Hong and Pyun 2001; Damar and Balaban 2006; Garcia-Gonzales *et al.* 2007). Besides anesthetic action, an increase in the membrane permeability to water can also occur. An additional effect on the membrane function may be due to the presence of HCO_{3-} ions, which may act on the charged phospholipid head groups and the proteins at the surface of the membrane to alter the surface charges of the cells, thereby altering the membrane function. Therefore, pressurized CO_2 first penerates into the cells to build up the density to a critical level, then it removes cellular constituents to disturb or alter the structure of the biomembrane or imbalance the biological system, thus promoting inactivation.

4.5.3 Cytoplasmic leakage

Hong and Pyun (1999) reported that SEM and TEM analysis of DPCD-treated cells of *L. plantarum* had intact cell wall and membrane modifications with possible cytoplasmic leakage. They observed enlargement in the periplasmic space between the cell wall and membrane, and empty spaces in the cytoplasm. Hong and Pyun (2001) showed that DPCD-treated (7 MPa and 30°C for 10 min) microbial cells had irreversible cellular damage including loss of salt tolerance, leakage of UV-absorbing substances, release of intracellular ions and impaired proton permeability. Shimoda *et al.* (1998) indicated that cell bursting was induced by explosive pressurization of CO_2 in the cells (*E. coli*, *L. brevis* and *S. cerevisiae*). The SEM observation of *S. cerevisiae*, treated at 6 MPa and 35°C, indicated cell bursting with possible leakage of cytoplasm. Another evidence for cell bursting was a great amount of UV-absorbing substances in the treated solution.

4.5.4 Intracellular pH decrease

Due to the increased membrane permeability, pressurized CO_2 can easily penetrate through the cell membrane and be accumulated in the cytoplasm. CO_2 dissolves in the liquid phase of cytoplasm to form carbonic acid which further dissociates up to H^+ and CO_3^- ions. DPCD can penetrate through the cell walls and membranes of microorganisms at a higher rate than in its gaseous or liquid state. DPCD is a better solvent and has lower surface tension than its liquid state, and diffuses more quickly through cell membranes and cytoplasm. The relative concentrations of both the undissociated (aqueous) form and dissociated products of CO_2 are controlled by homeostatic systems of cells in order to maintain cytoplasmic pH (pH_i). This is essential for normal cell viability and cellular activity. Most microbial cells control and regulate pH_i by cellular homeostatic systems: (1) cytoplasmic buffering characteristics, (2) proton pumping and (3) producing acids or bases. The cytoplasm of most microorganisms has a relatively high buffering capacity, which can control variations of pH_i.

Microorganisms can survive in an acidic environment with pH_i remaining near neutrality. To compensate for acidic surroundings, cells try to maintain a pH gradient between the internal and external environment with a proton pump system. Membrane-bound H^+-ATPase as the homeostatic system pumps protons (H^+ ions) from the cytoplasm to the environment to retain pH gradient ($\Delta pH = pH_i - pH_{ex}$) and electrochemical (membrane potential) gradient across the membrane (Garcia-Gonzalez *et al.* 2007). The sum of both gradients is known as the 'proton motive force' (PMF). If enough CO_2 enters into the cytoplasm up to a critical level and its components dissociate to corresponding ions in the aqueous phase of cytoplasm, pH_i will be reduced, exceeding the cell's buffering capacity. Therefore, PMF across the cell membrane can collapse (Hong and Pyun 1999). In addition, as a result of the lower pH_{ex}, the cells also may be unable to maintain the resulting large ΔpH. Therefore, the impairment of cellular activity may result from both a low pH_i and a large ΔpH.

Acidification of cytoplasm results in the denaturation of cellular enzymes (Erkmen and Bozoglu 2008b). Dillow *et al.* (1999) indicated that the reduction of pH is the main source of microbial inactivation, and pressure alone (up to 120 MPa) is generally insufficient to cause the same microbicidal effect. Spilimbergo *et al.* (2005) used flow cytometry in combination with a fluorescent probe to determine the pH_i of *B. subtilis* cells. The pH_i of cells decreased to a value of 3.3 upon exposure to DPCD (8 MPa and 30°C for 5 min), and a 5 log reduction of cells was detected. They reported that cells exposed to DPCD were unable to maintain pH_i with a favorable cytoplasmic pH homeostasis. The inactivation of microorganisms with DPCD can be the result of decreasing pH_i and pH_{ex}, and enzyme denaturation. Moisture is essential to achieve an effective inactivation level on microorganisms since CO_2 must dissolve in an aqueous solution at a sufficient concentration, and dissociate into corresponding ions to lower pH_{ex} and pH_i of microbial cells.

4.5.5 Key enzyme inactivation

The metabolic activities of enzymes and enzyme structures mainly depend on the pH_i. Enzymes, which make up most of the proteins in the cytoplasm, have maximum activity at their optimum pH and their activity declines sharply on either side of the optimum. Low pH_i can cause inhibition and/or inactivation of key enzymes essential for metabolic and regulatory processes of cells, such as glycolysis, amino acid and peptide transport, active transport of ions and proton translocation (Garcia-Gonzalez *et al.* 2007). Thus, a loss over biological control of the pH_i of cells may be detrimental to metabolisms and microbicidal cellular functions of microorganisms.

Low pH_i from microbicidal DPCD process inactivates microorganisms by inhibiting essential metabolic processes with denaturating and precipitating enzymes at their acidic isoelectric point. The activity of some enzymes from *E. coli* at their acidic isoelectric points, such as alkaline phosphatase and galactosidase, disappeared after treatment with DPCD, whereas those with basic isoelectric points, such as acid phosphatase, were slightly affected (Daniels *et al.* 1985; Zhang *et al.* 2006a). Hong and Pyun (2001) showed that among 13 different enzymes from *L. plantarum*, some such as cystine arylamidase, galactosidase and glucosidase, lost their activities after DPCD treatment whereas others, such as lipase, leucine arylamidase and acid and alkaline phosphatases, were slightly affected. At this treatment, more than 90% of *L. plantarum* was inactivated. Spilimbergo (2002) demonstrated a selective inactivation of several enzymes from *B. subtilis* cells after pressurized CO_2 treatment. DPCD inactivation of enzymes is selective, since DPCD may activate some enzymes and some enzymes retain their activity (Hong and Pyun 2001).

4.5.6 Inhibitory effect of molecular CO_2 and HCO_3^- on metabolism

Effects of DPCD treatment on the metabolic pathways of microorganisms can occur at several levels. The primary elements for the activity of enzymes are intracellular concentrations of substrates, products and cofactors besides optimum pH and temperature. The concentrations of HCO_3^- during DPCD treatment are important in the regulation of enzymatic activity. This anion regulates enzymatic activity by affecting a general anion-sensitive site on the enzymes (Damar and Balaban 2006).

HCO_3^- and dissolved CO_2 can effect carboxylation and decarboxylation reactions in microbial cells. Carboxylation reactions are important for the glucogenesis and the synthesis of amino acids and nucleic acids. Glucogenesis is a metabolic pathway that results in the generation of glucose from noncarbohydrate carbon substrates such as pyruvate, lactate, glycerol and gluconic amino acids. The ratio of dissolved CO_2–HCO_3^- will determine the relative rates of these reactions. Dissolved CO_2 under pressure can inhibit decarboxylation reactions by inactivating the decarboxylase enzymes and breaking the metabolic chain (Spilimbergo and Bertucco 2003). At low pH_i, protein-bound arginine may interact with CO_2 to form a carbonate complex and this inactivates the enzymes containing arginine (Damar and Balaban 2006).

4.5.7 Intracellular precipitation and electrolyte imbalance

Lethal damage on the biological systems can occur when the pressurized CO_2 is accumulated at a critical level in the cytoplasm of the microbial cells. CO_3^{-2} can precipitate together with inorganic electrolytes (such as Ca^{2+}, Mg^{2+} and similar ions), and this can be aided by pressure release (Lin *et al.* 1993; Spilimbergo and Bertucco 2003). These inorganic electrolytes are important in the regulation of many cellular activities and aid in maintaining the osmotic balance between cells and their surroundings. Changes in osmotic regulation can have deleterious effects on cell volume. The cytoplasmic free Ca^{2+} is a buffering ion together with other intracellular constituents (Gangola and Rosen 1987), particularly proteins. Ca^{2+} (and Mg^{2+}) binding proteins are important molecules in the intracellular regulation of biological activities (Aitken 1990). Certain types of Ca^{2+} and Mg^{2+} associated proteins can be precipitated by CO_3^-, depending on the binding site of the ions on the proteins (Damar and Balaban 2006). Collapsing PMF across the membrane due to lower pH_{ex} can disorder cytoplasmic Ca^{2+} levels, because in procaryotes Ca^{2+} extrusion is catalyzed by a Ca^{2+}–H^+ antiporter system (Gangola and Rosen 1987). A lethal effect of DPCD on the cell can be enhanced by precipitations of ions and proteins, loss of cellular regulation and disordering the osmotic balance.

4.5.8 Extraction of vital cellular constituents

Increasing permeability due to increasing fluidity of membrane and relatively high solvating ability of DPCD allows extraction of cellular constituents (such as phospholipids and hydrophobic compounds, proteins, fatty acids and UV-absorbing substances) to an extent that will be sufficient to disturb the biological system (Lin *et al.* 1992, 1993; Garcia-Gonzalez *et al.* 2007). The removal of biomolecules can also be stimulated by a depressurization of the applied pressure, since depressurization can lead to a rapid transfer of membrane and intracellular materials out of the biological system (Lin *et al.* 1992, 1993; Oulé *et al.* 2006). Transfer of intracellular materials out of the biological system may also be improved by cycling the CO_2 pressure. Isenschmid *et al.* (1995) indicated that yeast cells were mainly inactivated due to an 'anaesthesis effect' of pressurized CO_2. Ulmer *et al.* (2002) detected fatty acids in liquid CO_2 with gas chromatography and estimated that approximately 10% fatty acids in the cell membranes were extracted by liquid CO_2.

Ballestra *et al.* (1996) observed deformation on the cell wall and leakage of the cytoplasmic membrane and cytoplasm of *E. coli* under SEM. Hong and Pyun (2001) observed that *L. plantarum* cells treated with pressurized CO_2 at 7 MPa and 30°C for 10 min showed irreversible cellular membrane damage including leakage of UV-absorbing substances, release of intracellular ions (such as Mg^{2+} and K^+) and uptake of Phloxine-B dye which indicates no actively growing viable cells.

4.5.9 Physical cell rupture

Morphological changes on the microbial cells caused by DPCD may differ based on the treatment conditions, CO_2 release rate, type of microorganism and others. Nakamura *et al.* (1994) examined CO_2 pressure-treated (4 MPa and 40°C) *S. cerevisia* cells under SEM and observed that some cells were completely burst, whereas some had wrinkles or holes on their cell surfaces. They also detected that the total amount of proteins released from 8 log of *S. cerevisiae* cells subjected to HPCD treatment was 6.86 and 7.05 mg after fast (33 MPa/min) and slow depressurization (4.8 MPa/min), respectively. Enomoto *et al.* (1997b) proposed the rupture of the yeast cells exposed to CO_2 pressure (4 MPa and 40°C) based on SEM examination. Bertoloni *et al.* (2006) found that different intracellular enzymes of *E. coli* were moved into the extracellular environment after DPCD treatment at 8.5 MPa and 40°C. Pressurized CO_2 expands and saturates rapidly through the cells, and a part of the cell membrane can be mechanically broken (explosive rupture) like a popped balloon (Nakamura *et al.* 1994; Ballestra *et al.* 1996). Castor and Hong (1995) claimed that supercritical fluid extraction of intracellular components led to cellular materials' recovery in suspension (i.e. nucleic acids) with lysis of cells. Hong and Pyun (1999) indicated modifications on the cell membranes treated with CO_2 pressure: fractures in the membranes, enlargement of periplasmic space between the walls and the cytoplasmatic membranes, leakage from cytoplasm and empty spaces in the cytoplasm. Deformation and loss of integrity of the membranes increase permeability and associate with the extraction of biomolecules. The leakage from membranes and cytoplasm can rupture cell walls and membranes, and that results in lysis of microorganisms.

Dillow *et al.* (1999) observed that the cell walls of some Gram-negative *E. coli* cells treated with DPCD were ruptured, yet some *E. coli* cells did not suffer cell wall damage. They also found by SEM analysis of *S. aureus*

(Gram-positive) and *P. aeruginosa* (Gram-negative) that cell walls remain unchanged after the treatment with CO_2 pressure at 20.5 MPa and 25°C for 60 min. Hong and Pyun (1999) showed that *L. plantarum* treated with pressurized CO_2 (6.8 MPa and 30°C for 60 min) were completely inactivated, but any cell rupture and degradation of proteins were not observed from SEM and transmission electron microscope (TEM) analysis of inactivated cells. White *et al.* (2006) exposed *S. typhimurium* cells to DPCD treatment at 9.65 MPa and 35°C for 60 min in distilled water. The physical structure of inactivated *S. typhimurium* remained intact when viewed by SEM. On the other hand, structural analysis of inactivated and nontreated *S. typhimurium* cells in TEM revealed little difference except that lipid bilayers in the inactivated cells appeared to be 'roughened and the internal structures of the inactivated cells appeared less distinct compared to the untreated cells. Other researchers (Kamihira *et al.* 1987; Wei *et al.* 1991; Lin *et al.* 1992, 1994) reported that the highly pressurized CO_2 can extract vital constituents including phospholipids and hydrophilic compounds from the cell walls and membranes, which lethally disturb or alter the balance of biological systems to prompt microbial inactivation without rupturing the cell walls.

4.6 Characterization of CO_2 states and survival curves

There might be relationships between different states of CO_2 and survival curves (Figure 4.1) in the inactivation of microorganisms with DPCD. One of the survival curves was reported as a slow-to-fast curve by Ballestra *et al.* (1996) and Erkmen (2000d) for inactivation of *E. coli* and by Erkmen (1997, 2000a, 2000b, 2001a, 2002) for inactivation of other microbial species. A fast-to-slow survival curve was observed by Kamihira *et al.* (1987) and Debs-Louka *et al.* (1999) for the inactivation of *E. coli*. A log-linear survival curve was reported for the inactivation of *E. faecalis* (Debs-Louka *et al.* 1999), *S. cerevisiae* (Kumagai *et al.* 1997; Shimoda *et al.* 2001) and *B. subtilis* (Ballestra and Cuq 1998). Two different curves (concave and log-linear curves) were reported by Hong and Pyun (1999, 2001) for *L. plantarum*. A sigmoid (S-shape) curve was indicated by Isenchmid *et al.* (1995) for the inactivation of yeast species.

For a temperature up to 8°C and pressure below 4.5 MPa, the CO_2–H_2O system appears as CO_2 hydrates and the CO_2–vapor phase. At temperatures over 8°C, water-rich and CO_2 vapor–rich phases coexist. At these states, CO_2 may have very little effect on microbial cells. Increasing pressure, while the temperature is below 9°C, favors the formation of CO_2 hydrates and the changes of the physical state of CO_2 from vapor to the CO_2 liquid–rich phase (North *et al.* 1998; Oulé *et al.* 2006), which can cause a microbicidal effect. When the temperature and pressure increase to 9.9°C and 4.5 MPa (triple point), all species (hydrates, water-rich, CO_2 liquid–rich and vapor-rich phases) are present in the system. Above this point, the CO_2

liquid–rich phase and water-rich phase coexist (North *et al.* 1998). At these states, a higher amount of liquid CO_2 can enhance a microbicidal effect. As the temperature and pressure rise to 31.1°C and 7.38 MPa (critical point), CO_2 passes from the liquid to the supercritical phase (Oulé *et al.* 2006). At or above the supercritical point, the microbicidal effect of CO_2 is increasingly pronounced. Therefore, the temperature and pressure determine CO_2 states and the rate of inhibitory effects of CO_2 on microorganisms.

The hydration of $CO_{2(aq)}$ to H_2CO_3 is a very slow process ($[H_2CO_3] / [CO_{2(aq)}] = 1.6 \times 10^{-3}$). H_2CO_3 represents a small portion (0.16%) of the $CO_{2(aq)}$ (Oulé *et al.* 2006). H_2CO_3 is a relatively weak acid, which would be unable to cause a significant decrease in extracellular pH. Therefore, the acidification of the suspension may not be the principal factor in the microbial inactivation. During the treatment of microbial cells with vapor CO_2 pressure at 4.5 MPa below 8°C, CO_2 can diffuse through the membranes due to its hydrophilicity and liposolubility characteristics. Its accumulation in membrane can cause an 'anaesthetic effect' on the membrane. CO_2 can also decrease membrane viscosity due to rendering solubility of the phospholipids (Isenschmid *et al.* 1995; Oulé *et al.* 2006). These alterations in the cell membrane can cause cell surface modifications that result in a slight collapsing of cell walls and membranes, and cause stress on the cells (injury). This effect can be reversible. As the contact between cells and CO_2 is broken, the cells probably synthesize new biomolecules to repair damage on the cell walls and membranes, to continue cellular division and growth (Erkmen and Bozoglu 2008a). These effects of vapor CO_2 on cells are microbistatic, and lower numbers of cells may be inactivated with prolonged exposure of cells to CO_2 pressure; only the slow stage (lag stage) in the inactivation curve may be characterized. In slow-to-fast two-stage curves, the first (lag) stage can be characterized by a slower rate of microbial inactivation. The first stage can be related to the stress and sensitivity of microbial cells to pressurized CO_2.

As the temperature and pressure increase, CO_2 changes from vapor to the liquid state and then to the supercritical state. In these states, the effects of CO_2 would not be limited to the cellular membrane since it can penetrate the cell to cause irreversible damage, and the effect is microbicidal. In the liquid state, the microbial inactivation with pressurized CO_2 involves two distinct stages (slow to fast). The first (lag) stage may consist of stressing and sensitivity of the cells initially as explained for vapor CO_2. 'Stress' and 'sensitivity' can be defined as the biological or physical disorders provoked by some attack on a cell. Penetration of liquid CO_2 into the cells is slow and does not significantly modify their morphology at a faster rate to cause lethality but acidifies the cytoplasm. After stressing and sensitivity of cells, logarithmic (faster) inactivation of cells starts due to the inhibition of biological functions (Oulé *et al.* 2006): the (1) alterations in the membrane structure, (2) accumulation of toxic compounds in cells, (3) changes in

DNA, which prevent DNA-dependent processes (transcription, translation and recombination) and cause modifications in gene expression, and (4) inhibition of essential metabolic systems by denaturating enzymes and proteins.

DPCD is characterized as the cell death stage, and can be represented on the log survival curve as a linearly descending (logarithmic) inactivation. The inactivation of microorganisms with DPCD involves the diffusion of CO_2 in the cells, the supercritical extraction of membrane substances, the perforation of the cellular envelope and the loss of cellular vital constitutes to the extracellular environment in addition to the effects of liquid CO_2. The cells may become flaccid, completely collapsed, emptied of their cytoplasmic content, shrunken and wrinkled, which result in holes on the surface of the cellular envelope. The acidification of cytoplasm with CO_2 (low pH_i) can affect the enzymatic system by inhibition or inactivation of the activities of essential enzymes involved in metabolic or regulatory processes (Hutkins and Nannen 1993). They can disturb or alter balance of the biological system and cause the death of cells with or without mechanical rupture of the cell wall. The effects of DPCD can be characterized by a linear inactivation curve. In a fast-to-slow two-stage survival curve, a fast initial rate of inactivation can appear at higher temperatures and pressures. In the first stage, inactivated and lysed cells (which contact with CO_2 first) may release biomolecules (proteins, lipids, carbohydrates, etc.) and these biomolecules can protect cells that contact with CO_2 later. Therefore, that protection might reduce the inactivation effect, and the slow stage can appear in the survival curve of DPCD treatment. But the inactivation of microorganisms with DPCD cannot be fully characterized by the fast-to-slow curve. Since, depending on processing factors, the stressing of microorganisms at the beginning of DPCD treatment might be slow, a slow-to-fast inactivation curve can also appear.

In highly supercritical conditions, the effect of CO_2 on microbial cells is strong and spontaneous. The cells do not pass through a stressed period (no lag phase). CO_2 would diffuse immediately through the membranes and inside the cells, which causes acidification of cytoplasm, the perforation of the cytoplasmic membrane, the supercritical extraction of intracellular substances, the loss of vital cellular constituents and the fragmentation of the cell envelope. The effect of DPCD treatment at this condition is strong and the cells are completely destroyed. It breaks the cell envelope, and the cell wall and membrane fragments can be released into the suspension. Strong agitation or a fast depressurization may enhance the breaking of the cell envelope. In the highly supercritical conditions, the survival curve can be characterized by only a single phase: cellular death (Oulé *et al.* 2006). Therefore, when the temperature and pressure are increased over 31.1°C and 7.38 MPa, the microbial survival curve in the DPCD treatment can be modified to a linear inactivation curve. This can be characterized by shortening the duration of the slow stage in the inactivation curve or contacting CO_2 with all cells at the same time and shortening or eliminating slow stage in both slow-to-fast and fast-to-slow inactivation curves.

4.7 Quantifying inactivation

After exposure of microbial cells to DPCD treatment in suspension, several techniques can be used to quantify the number of surviving microorganisms and to characterize changes in the structure of microbial cells. One technique is a standard agar-plate counting method to indicate the surviving number of microbial cells in suspension. The treated suspension can be serially diluted and inoculated to agar plates (Erkmen 2007). The number of colonies on each Petri plate is counted after incubation, and then the colony count is converted to the survival cell number in the treated suspension.

SEM and TEM can be used to observe changes on the surface and in the internal structure of the microbial cells after DPCD treatment. Burst cells, wrinkles and holes on the cell surface, as well as intact cell surfaces, can be observed from microscopic analysis. Internal components, such as lipids, amino acids and peptides released from DPCD-treated cells, can be detectable from suspension by UV absorbance and protein analysis methods. UV studies reveal an increase in an absorbance with the release of the cell components during DPCD treatment. Proteins can be analysed using a bicinchoninic acid assay kit. Other techniques, such as microscopic counting to detect the number and size of microbial cells, and an enzyme activity assay, can also be used to determine the effects of DPCD treatment on microorganisms, and dye uptake of pressurized cells can be used to indicate membrane damage.

4.8 Conclusions

A critical objective of any sterilization or pasteurization process would be the development of a process that can destroy high levels of microorganisms. DPCD treatment is able to significantly inactivate vegetative cells in liquid foods, liquid media and solutions. DPCD can reduce microbial numbers from 2 to 8 log/mL of suspension. The standard criterion for pasteurization requires at least 6 log reduction per ml of suspension. DPCD-treated cells showed irriversible cellular damage including loss of salt tolerance, leakage of UV-absorbance substances, release of intracellular ions and collapse of proton permeability. Simultaneously, CO_2 can also pass through the membrane and dissolve inside the cell enough to exceed the buffering capacity of the cytoplasmic pool and lower pH_i. Low pH_i can inhibit key enzymes involved in essential metabolic and regulatory processes such as glycolysis, amino acids and peptide transport, proton translocation and active transport of ions. CO_2 and its dissociated product (CO_3^{2-}) can precipitate cellular inorganic electrolytes and proteins in the cytoplasm. These effects of DPCD are microbicidal on microbial cells. DPCD treatment can also cause a reversible inhibitory effect on microbial cells (injury). Injury must always be taken into account with proton translocation DPCD process. The microbicidal activity of DPCD mainly requires the presence of water, which facilitates its entry into microbial cells. Process parameters, such as temperature, pressure cycles or a combination of methods, can shorten the treatment time, provoke a more severe microbicidal effect of DPCD and allow the application of DPCD at milder process conditions.

5 Effects of Dense Phase Carbon Dioxide on Bacterial and Fungal Spores

Patricia Ballestra

Abstract: Bacterial and fungal spores are different from vegetative cells, having a complex and resistant structure which includes a coat, a cortex, and a core wall. Spores are less sensitive to chemical or physical antimicrobial treatments compared to vegetative forms. Few data are available for the application of DPCD technology to bacterial and fungal spores. Some studies indicate that dense phase CO_2 alone at moderate temperatures is insufficient to stimulate germination and therefore the inactivation of spores' latent forms to the extent required for food pasteurization and sterilization purposes. The aim of the present chapter is to focus on the application and the subsequent lethal effects induced by pressurized CO_2 on spores. The literature results regarding spore inactivation levels will be presented as a function of the process parameters (pressure and temperature) and characteristics of the microbial suspension. In addition, a proposed inactivation mechanism will be presented, followed by new experimental findings suggesting that a combination of treatments may increase the effectiveness of this technology in inactivating spores.

Keywords: bacterial and fungal spores; resistance to dense phase CO_2; combination treatments.

5.1 Introduction

Dense phase CO_2 processing (DPCD) is a promising alternative to heat treatments traditionally applied to pasteurize or sterilize liquid and solid foodstuffs. This technology utilizes the antimicrobial properties of CO_2 under pressure (<50 MPa) at ambiant or mild temperatures to stabilize sensitive foods. Under these conditions, depending on processing parameters, pressurized CO_2 can be gaseous, liquid (<31.1°C) or supercritical (above 31.1°C and 7.38 MPa), and each of these states is characterized by its own physical properties. Supercritical CO_2 (SC–CO_2) exhibits a high

Dense Phase Carbon Dioxide: Food and Pharmaceutical Applications, First Edition. Edited by Murat O. Balaban and Giovanna Ferrentino.

density (close to liquids), a high diffusivity (close to gases), a low viscosity (as gases) and a very low surface tension, which (1) gives it excellent solvating power and (2) facilitates its penetration into complex structures. Gaseous and liquid CO_2 are considered as subcritical states. The gaseous CO_2 density increases with increasing pressure or decreasing temperature but remains lower than that of both other states.

In DPCD technique, food is contacted with either sub- or supercritical CO_2 for a certain amount of time in a batch, semibatch, or continuous system. In the first case, the sample is placed into and heated in the pressure vessel. When the experimental temperature is reached, CO_2 is injected at the selected pressure. Solubilization of CO_2 can be enhanced by sample agitation. At the end of the treatment, a CO_2 outlet valve is opened to release the gas. Ishikawa *et al.* (1997) have added to their batch system a stainless steel filter (pore size, 10 µm) in the CO_2 inlet flux to disperse microbubbles of pressurized CO_2 in the aqueous medium.

In the semicontinuous system, CO_2 flows continuously through the pressure vessel. The concentration of dissolved CO_2 in the sample is increased by microbubbling CO_2 entering into the chamber (Spilimbergo *et al.* 2002).

More recently, continuous systems have been developed, through which both CO_2 and the liquid food flow. In the continuous microbubble DPCD, solutions are saturated with pressurized and microbubbled CO_2 before flowing through a heater to reach the desired temperature (Shimoda *et al.* 2001). In the flow DPCD, CO_2 and the liquid food are mixed before passing through a high-pressure pump and flowing through heating coils (Kincal *et al.* 2005; Damar and Balaban 2006). In another system, developed by Sims and Estigarribia (2002), pressurized CO_2 flows through a membrane contactor and diffuses into the liquid food pumped continuously with a high-performance liquid chromatography pump. In such a system, CO_2 saturation levels are reached instantaneously by means of the large contact area between gas and liquid.

The ability of DPCD to inactivate microorganisms has been known since the 1950s (Fraser 1951; Witter *et al.* 1958). However, it is with the work published in 1987 by Kamihira (Kamihira *et al.* 1987) that the antimicrobial effect of dense CO_2 started to be investigated systematically. These authors were the first to compare the inhibitory effect of pressurized CO_2 at the gaseous, liquid or supercritical state on *Escherichia coli* and *Staphylococus aureus* cells. Since then, many studies have shown that DPCD is effective in killing vegetative forms of pathogenic and spoilage bacteria, yeasts and moulds. Its lethal effect generally increases with treatment time, pressure and temperature. Microbial inactivation achieved by DPCD ranges from 2 to 12 logs, using pressures below 50 MPa and temperatures between 5°C and 60°C, mostly in the 25–35°C range, treatment times being significantly different depending on the system used (Damar and Balaban 2006). Antimicrobial effectiveness of CO_2 is also affected by (1) the nature of food or suspending medium, (2) the type and the physiological state of treated microorganisms, (3) the state of pressurized CO_2 and (4) the type of system used for DPCD treatments (Spilimbergo and Bertucco 2003; Damar and Balaban 2006; Garcia-Gonzalez *et al.* 2007). Many reports related to DPCD research on vegetative cells' inactivation have been published and are reviewed in chapter 4 of this book. Survival curves of vegetative microbial cells exposed to CO_2 pressure are now well characterized, and a mechanism of CO_2 bactericidal action is proposed.

The aim of this chapter is to focus in the lethal effect of pressurized CO_2 on bacterial and fungal spores. Generally, spore forms are less sensitive to chemical or physical antimicrobial treatments when compared to vegetative forms. DPCD technology is not an exception to this rule. This may explain the limited attention given to CO_2 effect on spores. Only few data are available in this area, especially for fungal spores.

5.2 Inactivation of bacterial spores by DPCD

Dormant endospores are produced by some bacteria, such as *Bacillus* or *Clostridium* species, to ensure their survival in case of environmental stress. Bacterial spores are quite different from vegetative cells. Their envelope consists of a complex and resistant structure including a coat, a cortex and a core wall. Unlike the vegetative cells, bacterial spores exhibit both inner and outer membranes (Madigan *et al.* 2002; Driks 1999). Their fluidity is much less than that of corresponding vegetative forms (Ishihara *et al.* 1999). The core has normal cell structures but is very dehydrated and metabolically inactive (Madigan *et al.* 2002). These specific characteristics render bacterial spores highly resistant to physical treatments such as heat, high pressure, radiation, drying and chemical agents used to kill microorganisms in food (Gould 1983).

5.2.1 *Effect of temperature*

Many studies investigating inactivation of spores by DPCD showed that gaseous or supercritical CO_2 (<30 MPa), when applied at ambient temperature (20–40°C), failed to significantly inactivate bacterial spores. Generally, DPCD had to be performed at higher temperatures, above a threshold value depending on the studied bacterial spores, to achieve a killing effect. A summary of the literature including different species of bacterial spores, treatment conditions and the inactivation achieved is given in Tables 5.1 and 5.2.

Enomoto *et al.* (1997a) showed that static gaseous pressurized CO_2 (5.9 MPa) did not induce inactivation of *Bacillus megaterium* spores when it was applied at temperatures below 50°C, whereas it produced a significant bacterial spore destruction at 60°C. Similarly, Ballestra and Cuq (1998) observed that temperatures above 80°C were needed to inactivate *Bacillus subtilis* spores with 5 MPa of static gaseous CO_2. Furukawa *et al.* (2003)

Table 5.1 Effect of gaseous pressurized CO_2 (DPCD) on bacterial spore inactivation.

Solution	Microorganism*	Process conditions	Reduction	System	References
Nutrient broth	*C. perfringens*	5.5 MPa 70°C 2 h		Batch	Haas *et al.* 1989
pH 4.0			0.8 log		
pH 3.0			7.5 log		
pH 2.5			7.8 log**		
Distilled water	*B. megaterium*	5.9 MPa 60°C 24 h 5.9 MPa 60°C 30 h	5.8 log 7 log	Batch	Enomoto *et al.* 1997a
Ringer solution	*B. subtilis*	5.0 MPa 80°C 60 min	3.5 log	Batch	Ballestra and Cuq 1998
Phosphate buffer solution pH 7.4	*B. subtilis*	7 MPa 75°C 24 h	7 log**	Batch with micro bubbles	Spilimbergo *et al.* 2002
Distilled water	*B. coagulans* *B. licheniformis*	6.5 MPa 35°C 2 h	0.05 log 0.1 log	Batch	Furukawa *et al.* 2003

* different *Bacillus* or *Clostridium* species.
** total inactivation.

showed that CO_2 pressurized at 6.5 MPa and 35°C in a batch system induced germination of *Bacillus coagulans* or *Bacillus licheniformis* spores but was not sufficient to kill them.

Other authors studied the effect of higher CO_2 pressures on spore viability. Kamihira *et al.* (1987) did not succeed in inactivating spores of *Bacillus subtilis* and *Geobacillus stearothermophilus* after 2 h of static supercritical CO_2 treatment at 35°C at 20.3 MPa. Watanabe *et al.* (2003a) and Furukawa *et al.* (2006) obtained similar results with many species of *Bacillus* spores with CO_2 batch systems using treatment parameters close to those used by Kamihira *et al.* (1987). Moreover, Watanabe *et al.* (2003a) suggested that temperatures above 85°C were needed to significantly inactivate spores of *G. stearothermophilus* with supercritical CO_2 at 30 MPa. Similarly, Werner and Hotchkiss (2006) reported that the use of supercritical CO_2 in a continuous system at 35°C or 40°C did not result in any inactivation of *Bacillus cereus* spores suspended in milk, even if very high CO_2 pressures were applied (up to 48.3 MPa).

Thus, in most cases, DPCD technology required mild temperatures (>60°C) to promote spore inactivation, except for a system using relatively high pressures of microbubbled CO_2 (Ishikawa *et al.* 1997) and systems using pressurization–depressurization cycles (Dillow *et al.* 1999; Spilimbergo *et al.* 2002).

Ishikawa *et al.* (1997) achieved 4-log reduction in spore populations of different *Bacillus* species with microbubbled CO_2 applied in a batch system at 30 MPa and 40°C for 30 min. They showed that the use of a filter, acting as a gas sparger, to microbubble CO_2 improved the inactivation effect by 3 logs, probably because the CO_2 concentration in spore suspensions was increased to a near saturation level. However, Spilimbergo *et al.* (2002),

Table 5.2 Effect of supercritical CO_2 (DPCD) on bacterial spore inactivation.

Solution	Microorganism*	Process conditions	Reduction	System	References
Physiological saline or distilled water	*B. subtilis*	20.3 MPa 35°C 2 h	0.3 log	Batch	Kamihira *et al*. 1987
Physiological saline	*Bacillus spp.*	30 MPa 40°C 30 min	1 log	Batch	Ishikawa *et al*. 1997
Physiological saline	*Bacillus spp.* *B. polymyxa* *B. cereus* *B. subtilis*	30 MPa 40°C 30 min 30 MPa 45°C 60 min 30 MPa 50°C 60 min 30 MPa 55°C 60 min	4 log 6 log** 6 log** 6 log**	Batch with micro bubbles	Ishikawa *et al*. 1997
Nutrient broth	*B. cereus*	20.5 MPa 34°C 2 h 20.5 MPa 60°C 2 h 20.5 MPa 60°C 4 h	1 log 5 log 8 log**	Batch with 6 cycles ($-P \geq 10$ MPa)	Dillow *et al*. 1999
Phosphate buffer solution pH 7.4	*B. subtilis*	7.4 MPa 40°C 30min 20 MPa 40°C 60 min	0.9 log 0.9 log	Semi-continuous with micro bubbles	Spilimbergo *et al*. 2002
Phosphate buffer solution pH 7.4	*B. subtilis*	15 MPa 36°C 30 min	3.5 log	Semi-continuous with micro bubbles with 15 cycles $-P = 8$ MPa	Spilimbergo *et al*. 2002
Distilled water	*B. subtilis* *B. coagulans* *B. cereus* *B. licheniformis* *Geobacillus stearothermophilus*	30 MPa 35°C 27.8 h 30 MPa 35°C 2.73 h 30 MPa 35°C 2.22 h 30 MPa 35°C 3.03 h 30 MPa 35°C 6.42 h 30 MPa 95°C 2 h	1 log 1 log 1 log 1 log 1 log 5 log	Batch	Watanabe *et al*. 2003a
Distilled water	*B. coagulans* *B. licheniformis*	20 MPa 35°C 60 min 20 MPa 35°C 60 min	0.2 log 0.3 log	Batch	Furukawa *et al*. 2006
Wet spores strips	*B. pumilus*	27.5 MPa 60°C 4 h	3 log	Batch	Zhang *et al*. 2006a
Milk	*B. cereus*	48.3 MPa 40°C 10 min $[CO_2] = 132$ g/kg	0 log	Continuous	Werner and Hotchkiss 2006

* different *Bacillus* species except for *Geobacillus stearothermophilus*.
** total inactivation.

who used a semicontinuous system with microbubbled supercritical CO_2, did not observe significant decrease in microbial count of *B. subtilis* spores (<1-log) when treatments were applied at 20 MPa at 40°C. These conflicting results could be explained by differences in spores' physiological state in addition of differences in treatment performances (such as pressure, carbonation level, pressurization and depressurization rate and treatment suspension). Ishikawa *et al.* (1997) isolated spores from a mixed vegetative and spore cells culture by heating at 80°C for 20 min. Such a treatment would most likely have initiated spore germination and thus increased the sensitivity of cells to microbubbled supercritical CO_2.

Interestingly, Spilimbergo *et al.* (2002) showed that the application of pressure cycles was beneficial to DPCD treatments performed at ambient temperature since the generation of cycles permitted to obtain significant inactivation of *B. subtilis* spores (3.5-log reduction with 15 cycles of pressurization–depressurization between 8 MPa and 15 MPa for 30 min). However, higher temperatures were needed to achieve complete inactivation of bacterial spores. For instance, a temperature of 60°C was required to obtain 8-log reduction of *B. cereus* spores with a 4-hr treatment at 20.5 MPa CO_2 with 6 cycles (Dillow *et al.* 1999).

Similarly, other studies reported that increasing temperature in appropriate conditions enhanced the antimicrobial effect of pressurized CO_2 on bacterial spores (Enomoto *et al.* 1997a; Ishikawa *et al.* 1997; Ballestra and Cuq 1998; Spilimbergo *et al.* 2002; Watanabe *et al.* 2003a).

Ballestra and Cuq (1998) showed that the inactivation rate of *B. subtilis* spores exposed to 5 MPa of CO_2 pressure increased with increasing temperature from 80°C (sublethal temperature) to 90°C (low lethal temperature). A same reduction level (3.5-log) was obtained in shorter time (50 min at 90°C compared to 60 min at 80°C). In this range of temperatures, the antimicrobial effect of pressurized CO_2 was particularly extensive. For instance, the decimal reduction time D at 90°C was 14 min in presence of pressurized CO_2 compared to 66 min without CO_2. Above 90°C, the antimicrobial effect of CO_2 was less significant compared to the lethal effect of heat (unpublished data).

Watanabe *et al.* (2003a) reported that the inactivation rate of *Geobacillus stearothermophilus* spores treated with DPCD at 30 MPa at sublethal temperatures increased significantly with increasing temperature. D-values, in presence of DPCD, were reduced from 175 min at 75°C to 29.9 min at 95°C. The application of DPCD at 95°C was very effective for inactivating *G. stearothermophilus* spores, whereas a thermal treatment at 95°C did not induce any significant bacterial spore reduction.

5.2.2 Effect of pressure

Several studies investigating inactivation of bacterial spores by DPCD have been published in which different pressures and different states of CO_2 have been used. However, no investigator has systematically studied the influence of pressure on the effectiveness of DPCD in killing bacterial spores. Nevertheless, on the basis of the results obtained with vegetative forms, it could be supposed that the antimicrobial effect of pressurized CO_2 on bacterial spores (1) increases with the pressure and (2) is higher when CO_2 is under supercritical state (other conditions being equal). The results of Zhang *et al.* (2006a) are directed to this end. These authors observed a significant increase in log reduction of *Bacillus pumilus* spores with increasing pressure of supercritical CO_2. They achieved a reduction of 3.06-log at 27.5 MPa compared to 1.91-log reduction at 10.3 MPa, both after 4 h-treatment at 60°C of wetted spore strips. In these conditions, CO_2 density increased with pressure from 0.65 g/mL to 0.90 g/mL which enhanced the solvating power of supercritical CO_2 and would explain the enhanced killing of *Bacillus pumilus* spores. Similarly, Ballestra and Cuq (1998) observed that gaseous pressurized CO_2 was more efficient in killing *B. subtilis* spores when applied at 5 MPa compared to 2.5 MPa (unpublished data). Surprisingly, Enomoto *et al.* (1997b) showed that increasing CO_2 pressure above 5.9 MPa resulted in a decrease of bactericidal effect of pressurized CO_2 (at 60°C) on *Bacillus megaterium* spores. These authors suggested that the inhibition of lethal effect of CO_2 observed at higher pressures could be explained by the aggregation of bacterial spores. But, even if the clumping of *Bacillus* spores was shown to effectively occur during CO_2 treatments, it did not increase significantly with an increase of pressure from 6.5 to 30 MPa (Furukawa *et al.* 2006). The particular pressure-dependant inactivation pointed by Enomoto *et al.* (1997b) was only reported once, and the explanation still remains unclear.

5.2.3 Effect of pH and a_w of the treatment medium

The effectiveness of pressurized CO_2 for bacterial spores inactivation was shown to vary with pH and a_w of the treatment medium. Haas *et al.* (1989) studied the effect of CO_2 applied at 5.6 MPa at 70°C for 30 min on *Clostridium sporogenes* spores suspended in nutrient broth acidified with HCl at different pH values ranging from 4 to 2.5 and they showed that the sporicidal effect of pressurized CO_2 was increasing with decreasing pH value. Kamihira *et al.* (1987) and Zhang *et al.* (2006a) reported that the antimicrobial effect of pressurized CO_2 on bacterial spores was antagonized by lowered a_w. These authors showed that DPCD technology was ineffective in killing dry spores.

5.2.4 Susceptibility of different bacterial spores

The resistance of bacterial spores towards DPCD technology varies greatly among species. Watanabe *et al.* (2003a) compared the resistance of different species of bacterial spores to SC–CO_2 by calculating CO_2 treatment *D*-values and showed that *B. subtilis* spores were by far the most resistant (ahead of *G. stearothermophilus*), while *B. cereus* spores were the most

Table 5.3 Effect of pressurized CO_2 (DPCD) combined with non-thermal preservation methods on bacterial spore inactivation.

Treatment combination	Microorganism*	Process conditions	Reduction	References
pulse electric field (PEF) + DPCD	*B. cereus*	PEF pre-treatment 20 pulses at 25 KV/cm then DPCD treatment 20 MPa 40°C 15 h or 20 MPa 40°C 24 h (spores in distilled water)	1.5 log 3.0 log	Spilimbergo *et al.* 2003b
DPCD with H_2O_2 (200 ppm)	*B. pumilus*	27,5 MPa 40°C 4 h 27,5 MPa 60°C 2 h 27,5 MPa 60°C 4 h (spore strips)	4.7 log 4.5 log 6.3 log**	Zhang *et al.* 2006a
DPCD with ethanol (70%)	*B. pumilus*	27,5 MPa 40°C 4 h (spore strips)	0.27 log	Zhang *et al.* 2006a
DPCD with isopropanol (70%)	*B. pumilus*	27.5 MPa 40°C 4 h (spore strips)	0.17 log	Zhang *et al.* 2006a
DPCD with H_2O_2 (200 ppm)	*B. anthracis*	27,6 MPa 40°C 4 h (spore strips)	5.7 log	Zhang *et al.* 2007

* different *Bacillus* species.
** total inactivation.

sensitive among the five species tested. These results agree with those of Ishikawa *et al.* (1997). Further comparisons are difficult to make because treatment systems and operating conditions differ in many studies. Nevertheless, it appears that the resistance of the spores to CO_2 treatment cannot be correlated with resistance to heat and pressure. Such results suggest that there are different mechanisms for inactivation of bacterial spores by CO_2, heat and pressure (Watanabe *et al.* 2003a).

5.2.5 *Effects of combination treatments*

Combination of DPCD treatment with nonthermal preservation methods has been suggested to reduce the temperature required to inactivate bacterial spores. This approach is generally known as 'hurdle technology'. An overview of the investigated combinations to kill bacterial spores by DPCD is given in Table 5.3.

Spilimbergo *et al.* (2003b) studied the sporicidal effect of a SC–CO_2 treatment applied at very mild temperature immediately after a pulsed electric field (PEF) pretreatment. A synergistic effect between SC–CO_2 (20 MPa/40°C) and PEF (20 pulses at 25 kv/cm), was observed but a relatively long residence time (24 h) was necessary to decrease significantly the count of *B. cereus* spores. The authors reported that PEF partially stressed the structure of bacterial spores which could render them more vulnerable to the subsequent SC–CO_2 treatment.

Zhang *et al.* (2006a) investigated the effects of SC–CO_2 containing various additives on killing *B. pumilus* spores at temperatures between 40°C and 80°C, and pressures between 10.3 MPa and 27.5 MPa. These authors showed that addition of 200 ppm of H_2O_2 to SC–CO_2 was highly effective in killing dry spores of *B. pumilus*. Complete destruction (6.28-log) was obtained after 4 h-CO_2 treatment at 60°C 27.5 MPa with H_2O_2, whereas no significant reduction was observed in the absence of H_2O_2. Also, these authors showed that the addition of ethanol or isopropanol (70%) was ineffective in enhancing the killing of bacterial spores. These results are in accordance with those of Kamihira *et al.* (1987) showing that the addition of ethanol (at a weight ratio of 2%) during 2 h-CO_2 treatment at 20.3 MPa 35°C induced only a very slight increase of destruction rate of wet or dry spores of *G. stearothermophilus*.

In further studies, Zhang *et al.* (2007) showed that the application of SC–CO_2 combined with H_2O_2 induced an increase in permeability of *B. anthracis* spores which could explain in part the increased inactivation of bacterial spores. The authors considered that SC–CO_2/H_2O_2 approach was promising as a method to achieve dry sterilization for medical devices. However, its applications in food industry might be limited.

5.2.6 Mechanisms of bacterial spore inactivation

In general, survival curves of vegetative microorganisms exposed to pressurized CO_2 show two apparent distinct stages (see chapter 4 of this book). The first one is characterized by a slow rate of inactivation, which increases sharply at the later stage. An increase of pressure and/or temperature reduce(s) the duration of the earlier stage and lead(s) to an increase of inactivation rate of the later stage.

Ballestra and Cuq (1998) also observed two distinct phases in survival curves of *B. subtilis* spores subjected to CO_2 under a pressure of 5 MPa at 80°C (Figure 5.1). Their results showed that an increase of temperature from 80°C to 90°C reduced the duration of the earlier stage until it disappeared but did not lead to an increased inactivation rate in the later stage.

In the mechanism of DPCD inactivation of vegetative cells that has been proposed by Garcia-Gonzalez *et al.* (2007), the first stage in the destruction of bacterial spores subjected to pressurized CO_2 could correspond to (1) the solubilization of CO_2 in the external liquid phase leading to the formation of carbonic acid which further dissociates to give bicarbonate, carbonate and H^+ ions lowering extracellular pH, and (2) the diffusion and the accumulation of CO_2 into the cells.

Although solubility of CO_2 decreases with increasing temperature, higher temperature can increase the diffusivity of CO_2 and the fluidity of cell membranes that facilitate penetration of CO_2 into the spores. Thus, the disappearance of this first stage at, or above, 90°C could be explained in

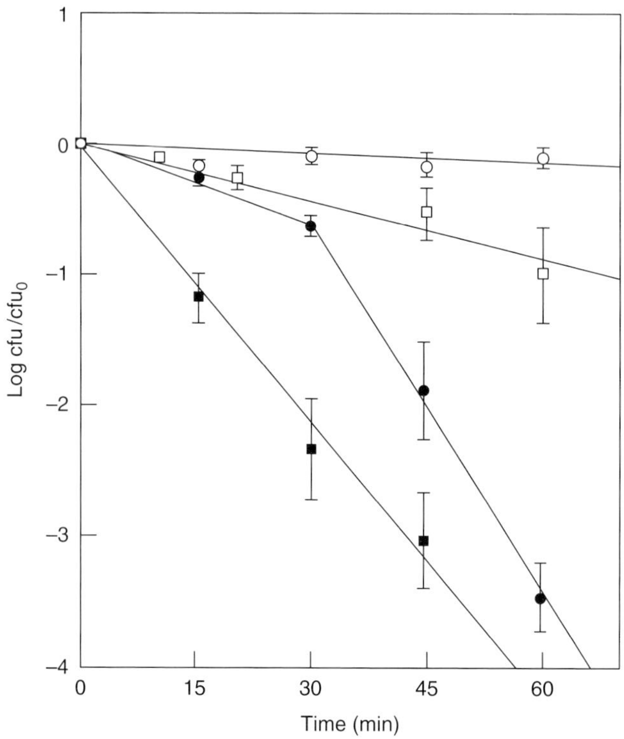

Figure 5.1 Influence of CO_2 under 5 MPa pressure on the thermal inactivation of *B. subtilis* spores at 80°C (●, ○) and 90°C (■, □). Filled symbols = 5 MPa; open symbols = control. (Data reproduced from Ballestra and Cuq, copyright 1998, with permission of Elsevier.)

part by easier penetration of CO_2 into *B. subtilis* spores. The following steps resulting in spore inactivation should differ from those leading to the death of vegetative cells, both forms of bacteria being characterized by different physiological states.

In the case of *Bacillus* spp. spores, two phases in the kinetics of thermal inactivation are often observed. Mathematical models have been developed to account for this. Even if the models are different, authors agree that the occurrence of two stages in thermal destruction would be due to the activation of dormant spores. Activation consists of breaking the spore dormancy and triggering the initiation of germination. Activation can be induced by heating the spores (e.g., 70°C for 30 min or 80°C for 20 min) and does not result in any detectable changes in either the spore structure or composition. Germination follows activation. This step is quite fast (minutes) and is determined by a rupture of spore coat, the rehydration of core, a loss of cortex components, the increase in metabolic activity and the loss of resistance to environmental stress (Madigan *et al.* 2002). The model of Sapru *et al.* (1992) considered concurrent activation of dormant spores and distinct inactivations of dormant and activated spores; it also indicated that the inactivation reaction would be the limiting phenomenon. Another mathematical model, that of Shull (Abraham *et al.* 1990), considers that the transitions from dormant to activated to dead spores are sequential. Validation of such a model implies that the limiting phenomenon is activation of dormant spores.

Considering similarities in the shape of survival curves of bacterial spores exposed to heat alone or heat combined with pressurized CO_2, Ballestra and Cuq (1998) suggested that inactivation of bacterial spores with pressurized CO_2 occurred in two steps, the first one including penetration of CO_2 into the cells and activation of dormant spores, the second one corresponding to inactivation of activated bacterial spores (destruction of dormant spores being not excluded). Watanabe *et al.* (2003a) also proposed a biphasic mechanism to explain destruction of *Bacillus* spores with pressurized CO_2. Moreover, on the basis of CO_2 treatment Z-values and activation energies, these authors suggested that the mechanism of inactivation should be different depending on whether the treatment temperature was mild or sublethal.

Further studies suggest that pressurized CO_2 would play a role in the activation of bacterial spores. Thus, Watanabe *et al.* (2003b) and Furukawa *et al.* (2003) reported that pressurized CO_2 applied at 6.5 MPa at 35°C was sufficient to activate and thus initiate germination of bacterial spores making the resulting cells more sensitive to heat inactivation. Considering the ability of pressurized CO_2 to activate bacterial spores and referring to the bacterial spore destruction model proposed by Shull (Abraham *et al.* 1990), it can be supposed that at higher temperatures, the application of pressurized CO_2 could increase the rate of heat activation of bacterial spores. Then, activated spores could start to germinate and become more sensitive to the metabolic disturbance induced by the combined effect of the large amount of antimicrobial CO_2 present in both cells and external medium, and the heat applied during treatment (Ballestra and Cuq 1998). In other respects, Watanabe *et al.* (2003a) suggested that spores that germinated during CO_2 treatment at high temperature (95°C) were finally inactivated by heat, the lethal effect of heat being enhanced with the reduction of pH caused by CO_2.

5.3 Inactivation of fungal spores by DPCD

Fungal spores are reproductive units produced asexually or sexually by lower and higher fungi. They are characterized by a slow rate of metabolism and can endure environmental stress (such as cold, heat or dryness). When conditions are favorable, fungal spores germinate and grow into hyphae. Some of them have a need for activation before germination can be initiated. As for the bacterial spores, activation can be triggered by heat.

Table 5.4 Effect of pressurized CO_2 (DPCD) on fungal spore inactivation.

Medium	Microorganism*	Process conditions	Reduction	System	References
Physiological saline solution or distilled water	*A. niger* (conidia)	20.3 MPa 35°C 2 h (SC–CO_2)	5.8 log	Batch	Kamihira *et al*. 1987
Nutrient broth pH 3.5	*P. roquefortii* (conidia)	5.5 MPa 45°C 2 h 5.5 MPa 45°C 4 h (gaseous CO_2)	5 log 6 log**	Batch	Haas *et al*. 1989
Ringer solution a_w = 0.99 a_w = 0.90 (+ saccharose)	*A. niger* (conidia)	5 MPa 50°C 11 min 5 MPa 50°C 46 min (gaseous CO_2)	1 log 1 log	Batch	Ballestra and Cuq 1998
Ringer solution	*B. fulva* (ascospores)	5 MPa 80°C 85.5 min 5 MPa 85°C 14.1 min (gaseous CO_2)	1 log 1 log	Batch	Ballestra and Cuq 1998
Physiological saline solution	*A. niger* (conidia)	10 MPa 44°C 1.74 min 10 MPa 52°C 0.16 min (SC–CO_2)	1 log 1 log	Continuous	Shimoda *et al*. 2002
Cocoa powder + 5 % water	*A. niger* *A. ochraceus* (conidia)	30 MPa 80°C 30 min 30 MPa 80°C 30 min (SC–CO_2)	5.3 log** 3 log**	Semi-continuous	Calvo *et al*. 2007

* different *Penicillium* or *Aspergillus* species.
** total inactivation.

Fungal spores are particularly resistant to heat. The ascospores of *Byssochlamys fulva* or *Byssochlamys nivea* can survive for nearly 1 h at 90°C (Bayne and Michener 1979), and are the cause of serious spoilage of canned fruit products. The conidia of *Aspergillus niger* are also thermo-resistant but less than the ascospores (Baggerman 1984). They cause spoilage of fruits, vegetables and dried fruits. Ascospores are sexually produced by ascomycetes, whereas conidia are asexually produced by higher fungi.

A limited number of studies investigated the antimicrobial effect of pressurized CO_2 on fungal spores. Most of them focused on moulds conidia, particularly those of *Aspergillus niger* (Kamihira *et al.* 1987; Haas *et al.* 1989; Ballestra and Cuq 1998; Shimoda *et al.* 2002; Calvo *et al.* 2007). Only one report deals with the effect of pressurized CO_2 on ascospores, those of *Byssochlamys fulva* (Ballestra and Cuq 1998). A summary of these studies is given in Table 5.4. The results clearly show that ascospores are more resistant to CO_2 treatment than conidia.

The application of gaseous CO_2 at pressures close to 5 MPa at very mild temperatures such as 45–50°C was sufficient to significantly or totally inactivate condidia of *Penicillium roquefortii* or *Aspergillus niger* in a limited time (<2 h) (Haas *et al.* 1989; Ballestra and Cuq 1998). Ballestra and Cuq (1998) showed that the antimicrobial effect of pressurized CO_2 (5 MPa) on *A. niger* conidia was particularly extensive at 50°C ($D_{50°C/CO_2}$ = 11 min compared to $D_{50°C}$ > 200 min) but nonexistent at 60°C ($D_{60°C/CO_2}$ = 1.3 min and $D_{60°C}$ = 1.2 min). The lethal effect of heat was found to mask that of pressurized CO_2. These results showed a significant interaction between temperature and pressure in CO_2 processing.

The use of SC–CO_2 at ambient or very mild temperatures also led to a significant inactivation of *A. niger* conidia (Kamihira *et al.* 1987; Shimoda *et al.* 2002). Kamihira *et al.* (1987) achieved a 5.8-log reduction with CO_2 treatment applied at 20.3 MPa 35°C for 2 h. More recently, Shimoda *et al.* (2002) used a DPCD continuous system, in which CO_2 is quickly dissolved in an aqueous medium to the required level, to investigate separately the effects of dissolved CO_2 (dCO_2) concentration, treatment pressure and temperature on the death kinetics of *A. niger* spores. They showed that the use of such a system enhanced the lethal effect of SC–CO_2 on moulds conidia. *D*-values, in physiological solution carbonated to a nearly saturated level and pressurized at 10 MPa, were 1.74 min at 44°C and 0.16 min at 52°C. In this case, the concentration of dCO_2 was 24.4 γ for 44°C and 22.9 γ for 52°C (γ is the Kuenen gas absorption coefficient; e.g. a γ value of 24 means that there were 24 volumes of CO_2 (in a normal state) in 1 volume of liquid). Shimoda *et al.* (2002) reported that the rate of inactivation of *A. niger* conidia increased with increasing dCO_2 concentration independently of the treatment pressure. They also showed that the log D values linearly increased or decreased with the increase in the treatment temperature (between 46°C and 52°C) depending on whether dCO_2 concentration was low ($\gamma \leq 10$), mild ($\gamma > 10$) or high ($\gamma > 20$). The authors suggested that this significant interaction between temperature and dCO_2 concentration might be caused by the difference in the levels of temperature dependence on the distribution of CO_2 between the medium and spores and the thermal sensitivity of the spores under high-pressure carbonation.

In addition, it was shown that the antimicrobial effect of pressurized CO_2 on *A. niger* conidia was antagonized by lowered a_w (Kamihira *et al.* 1987; Ballestra and Cuq 1998). This could be explained by a lower amount of CO_2 dissolved in the aqueous phase of treatment medium at lower a_w.

Ballestra and Cuq (1998) investigated the antimicrobial effect of gaseous pressurized CO_2 (5 MPa) on ascospores of *Byssochlamys fulva*. They showed that DPCD required a temperature above 80°C to significantly inactivate the ascospores (similar to the inactivation of *B. subtilis* spores). All survival curves they obtained in the presence or absence of CO_2 were linear. At 80°C, the lethal effect of CO_2 was particularly pronounced ($D_{80°C/CO_2}$ = 85.5 min compared to $D_{80°C}$ > 350 min). At 85°C, the antimicrobial effect of pressurized CO_2 was masked by the lethal effect of heat (similar to the inactivation of *A. niger* conidia).

The mechanism of inactivation of ascospores and conidia by CO_2 treatment is not clearly explained. It should exhibit common steps with the mechanism of bacterial spore destruction. The possible activation of fungal spores by pressurized CO_2 needs to be investigated.

5.4 Conclusion

Bacterial and fungal spores are resistant to DPCD technology. The level of inactivation is strongly dependent on (1) the temperature applied during treatment and (2) the concentration of CO_2 dissolved in aqueous phase of treatment medium. In most cases, DPCD technology requires mild or higher temperatures to promote spore inactivation. The synergistic effect of pressurized CO_2 with heat is extensive at sub- or low-lethal temperatures but disappears at higher temperatures. In addition, spore inactivation by DPCD is enhanced by any effect that increases the level and rate of CO_2 solubility and therefore penetration of CO_2 into the spores.

Some hypotheses have been proposed to explain the lethal action of pressurized CO_2 on bacterial spores. However, more studies are needed to determine the exact mechanism of bacterial and fungal spore inactivation.

6 Effects of DPCD on Enzymes

Murat O. Balaban

Abstract: The number of studies on enzyme inactivation by pressurized CO_2 is increasing, especially in fruit and vegetable juice processing, where enzymes cause quality deterioration if not inactivated. Among these enzymes, pectinesterase causes cloud loss in some fruit juices; polyphenol oxidase causes undesirable browning in fruits, vegetables, juices, and some seafoods; lypoxygenase causes chlorophyll destruction and off-flavor development in frozen vegetables; and peroxidase has an important role in the discoloration of foods, and is used as an index of heat treatment efficacy in fruit and vegetable processing. Carbon dioxide, at atmospheric pressure or under higher pressure, can affect the activity of some enzymes. This effect can be an increase or a decrease depending on the enzyme and on the treatment conditions. Enzyme inactivation by high-pressure CO_2 can be related to many causes such as pH lowering, conformational changes of the enzyme, and inhibitory effects of molecular CO_2 on enzyme activity. The aim of the present chapter is to discuss the effect of the DPCD treatment on different enzyme classes such as lipases, phosphatases, dehydrogenases, oxidases, amylases, and others in a way to define the factors (e.g., enzyme species, water content in the solution, and pressure and temperature of the treatment system) influencing the stability and activity of enzymes exposed to pressurized CO_2.

Keywords: enzyme inactivation; quality deterioration; alteration of protein structure; pressurized CO_2.

6.1 Introduction

Subcritical and supercritical carbon dioxide has gained considerable attention in the green chemistry industry for its advantage as a nontoxic, nonflammable and inexpensive gas. There is considerable interest in both using the compressed CO_2 environment as a reaction medium, in which case the maintenance of the activity of the enzyme is desired; and to use dense phase carbon dioxide (DPCD) to inactivate undesirable enzymes (e.g. in foods to preserve quality). Enzymes are complex, folded polypeptide molecules. Their folded structure, or their tertiary structure, is held together by hydrogen bonding, ionic and hydrophobic

Dense Phase Carbon Dioxide: Food and Pharmaceutical Applications, First Edition. Edited by Murat O. Balaban and Giovanna Ferrentino.

interactions and disulfide (S–S) bridges (Osajima 2000). Enzymes with disulfide bridges are more stable compared with those without such bonding. The effect of pressure on the catalytic activity of an enzyme depends on the magnitude and the sign of the differences in volume between the activated and ground states for the reaction steps involved (i.e. the activation volumes ΔV for each step). A reaction step which proceeds through a reduction in volume is accelerated by an increase in pressure, and vice versa. Enzymatic reactions in aqueous media have activation volumes generally in the range from -50 to $+50\,cm^3/mol$ (Fontes *et al.* 1998).

Carbon dioxide, at atmospheric pressure or under higher pressure, can affect the activity of some enzymes. This effect can be an increase or a decrease depending on the enzyme, and on the treatment conditions. Rezaei *et al.* (2007) discussed the effects of pressure and temperature on enzymatic reactions in supercritical fluids as related to enzyme activity, enantioselectivity, reaction rates and activation volumes.

Different enzyme classes such as lipases, phosphatases, dehydrogenases, oxidases, amylases and others are well suited for the reactions in DPCD. The stability and the activity of enzymes exposed to DPCD depend on the enzyme species, the water content in the solution and the pressure and temperature of the reaction system. The three-dimensional structure of enzymes may be significantly altered under extreme conditions, causing their denaturation and consequent loss of activity. If the conditions are less adverse, the protein structure may be largely retained. Minor structural changes may induce an alternative active protein state with altered enzyme activity, specificity and stability (Wimmer and Zarevucka 2010).

Inactivation of many enzymes that affect food quality by DPCD has been shown by several researchers (Balaban *et al.* 1991a, b; Chen *et al.* 1992a, 1993a; Tedjo *et al.* 2000; Park *et al.* 2002; Zhi *et al.* 2008). A summary of the literature including the enzymes, the activity loss achieved and the treatment conditions is given in Table 6.1. DPCD can inactivate certain enzymes at temperatures where thermal inactivation is not effective (Balaban *et al.* 1991a, b). Among these enzymes, pectinesterase (PE) causes cloud loss in some fruit juices; polyphenol oxidase (PPO) causes undesirable browning in fruits, vegetables, juices and some seafood; lypoxygenase (LOX) causes chlorophyll destruction and off-flavour development in frozen vegetables; and peroxidase (POD) has an important role in discolouration of foods and is used as an index of heat treatment efficacy in fruit and vegetable processing. The number of studies on enzyme inactivation by DPCD is increasing, especially in fruit and vegetable juice processing where these enzymes cause quality deterioration if not inactivated.

Enzyme inactivation by DPCD could be due to many causes such as pH lowering, conformational changes of the enzyme and inhibitory effect of molecular CO_2 on enzyme activity.

Table 6.1 Summary of studies on inactivation of enzymes by DPCD.

Enzyme	Source of enzyme	Pressure (MPa)	Time	Temp. (°C)	System	Loss of activity (%)	References
Acid protease	*Aspergillus niger*	25	30 min	35	Semibatch	75	Ishikawa *et al.* (1996)
	Aspergillus niger	10 MPa 15.6 g/min CO_2	13.9 min	50	Continuous	86.1	Yoshimura *et al.* (2002)
Alkaline protease	*Bacillus subtilis*	25	30 min	35	Semibatch	100	Ishikawa *et al.* (1996)
Alkaline phoshatase	Bovine intestinal mucosa	1 bar 200 mL/min CO_2	30 min	50	Semicontinuous	81	Fadiloglu *et al.* (2004)
Myrosinase	Canola flakes	62.1	3 h	75	Batch	90	Dunford and Temelli (1996)
	Commercial	22	5 min	65	Batch	99	Yang *et al.* (2011)
Alcohol dehydrogenase	Commercial (62–68% water)	20	2 h	35	Batch		Taniguchi *et al.* (1987)
Acid phosphatase	*E. coli* (ATCC 11775)	10	13 min	35	Continuous	47	Kobayashi *et al.* (2009)
Alkaline phosphatase	*Pseudomonas aeruginosa*	10	13 min	37			Mun *et al.* (2011)
	E. coli (ATCC 11775)	10		35	Continuous	69	Kobayashi *et al.* (2009)
Leucine arylamidase	*E. coli* (ATCC 11775)	10	13 min	35	Continuous	87	Kobayashi *et al.* (2009)
Thermolysin	Commercial (62–68% water)	20	2 h	35	Batch	−1	Taniguchi *et al.* (1987)
Glocoamylase	*Rhizopus delemar*	25	30 min	35	Semibatch	87	Ishikawa *et al.* (1996)
β-glucuronidase	*E. coli* (ATCC 11775)	10	13 min	35	Continuous	100	Kobayashi *et al.* (2009)
Naphthol-AS-BI-phosphhydrolase	*E. coli* (ATCC 11775)	10	13 min	35	Continuous	14	Kobayashi *et al.* (2009)
N-acetyl-b-glucosaminidase	*Lactobacillus plantarum*	7	10 min	30	Batch	90.3	Hong and Pyun (2001)
N-acetyl-b-glucosaminidase	*Lactobacillus plantarum*	7	10 min	30	Batch	92.8	Hong and Pyun (2001)
Lipase	*Rhizopus japonicus*	25	30 min	35	Semibatch	100	Ishikawa *et al.* (1996)
	Crude porcine pancreatic powder	15	24 hr	75	Batch	−860	Giessauf and Gamse (2000)
	Candida rugosa	0.1	30 min	50	Semicontinuous	84	Fadiloglu and Erkmen (2002)

(*continued*)

Table 6.1 *(continued)*

Enzyme	Source of enzyme	Pressure (MPa)	Time	Temp. (°C)	System	Loss of activity (%)	References
	Commercial (68% water)	20	2 h	35	Batch	12	Kamihira *et al.* (1987)
	Commercial (62% water)	20	2 h	35	Batch	22	Kamihira *et al.* (1987)
	Pseudomonas	15	24 h	75	Batch	36.1	Giessauf *et al.* (1999)
	Candida cylindracea	15	24 h	75	Batch	36.1	Giessauf *et al.* (1999)
Lipase	Commercial (5–7% water)	20	1 h	35	Batch	4	Taniguchi *et al.* (1987)
	Commercial (62–68% water)	20	2 hr	35	Batch	12 to 22	Kamihira *et al.* (1987)
	Candida rugosa	1 bar 200 mL/min CO_2	30 min	45	Semicontinuous	88	Fadiloglu *et al.* (2002)
α-galactosidase	*Lactobacillus plantarum*	7	10 min	30	Batch	93.4	Hong and Pyun (2001)
	Lactobacillus plantarum	7	60 min	30	Batch	93.9	Hong and Pyun (2001)
β-galactosidase	Commercial (6% water)	20	2 h	35	Batch	2	Taniguchi *et al.* (1987)
	E. coli (ATCC 11775)	10	13 min	35	Continuous	100	Kobayashi *et al.* (2009)
	Lactobacillus plantarum	7	10 min	30	Batch	94.1	Hong and Pyun (2001)
	Lactobacillus plantarum	7	60 min	30	Batch	94.3	Hong and Pyun (2001)
α-amylase	Commercial (68% water)	20	2hr	35	Batch	−21	Kamihira *et al.* (1987)
	Commercial (62% water)	20	2hr	35	Batch	−35	Kamihira *et al.* (1987)
	Commercial (6% water)	20	2 h	35	Batch	6	Taniguchi *et al.* (1987)
	Bacillus subtilis	10 MPa 15.6 g/min CO_2	14.4 min	45	Continuous	83.9	Yoshimura *et al.* (2002)
Glocose oxidase	Commercial (6% water)	20	2 h	35	Batch	3	Taniguchi *et al.* (1987)
α-glucosidase	*E. coli* (ATCC 11775)	10	13 min	35	Continuous	100	Kobayashi *et al.* (2009)
	Lactobacillus plantarum	7	10 min	30	Batch	93.5	Hong and Pyun (2001)
	Lactobacillus plantarum	7	60 min	30	Batch	93.2	Hong and Pyun (2001)
Gluco-amylase	Commercial (5–7% water)	20	1 h	35	Batch	−2	Taniguchi *et al.* (1987)
Catalase	Commercial (5–7% water)	20	1 h	35	Batch	10	Taniguchi *et al.* (1987)
cystine arylamidase	*Lactobacillus plantarum*	7	10 min	30	Batch	91.2	Hong and Pyun (2001)
cystine arylamidase	*Lactobacillus plantarum*	7	60 min	30	Batch	92.8	Hong and Pyun (2001)

Glucose isomerase	Commercial (5–7% water)	20	1 h	35	Batch	–2	Taniguchi *et al.* (1987)
Pectinesterase	Orange juice	26.9	145 min	56	Batch	100	Balaban *et al.* (1991b)
	Orange juice	107	10 min	24	Continuous	46.3	Kincal *et al.* (2006)
	Mandarin juice	41.4	7 min	45	Continuous	50.7	Lim *et al.* (2006)
	Orange juice	30	10 min	55	Continuous	90.7	Zhou *et al.* (2009a)
	Apple juice	20	20 min	65	Continuous	82	Niu *et al.* (2010)
Polyphenol oxidase	Spiny lobster	5.8	1 min	43	Batch	98	Chen *et al.* (1992)
	Brown shrimp	5.8	1 min	43	Batch	78	Chen *et al.* (1992)
	Potato	5.8	30 min	43	Batch	91	Chen *et al.* (1992)
	Spiny lobster	0.1	30 min	33	Batch	98.5	Chen *et al.* (1993)
	Muscadine grape juice	27.6 MPa 7.5% CO_2	6.25 min	30	Continuous	35	Del Pozo-Insfran *et al.* (2007)
	Carrot juice	4.9	10 min	5	Batch	61	Park *et al.* (2002)
	Apple juice	30	60 min	55	Batch	60	Gui *et al.* (2007)
	Watermelon juice	30	30 min	50	Batch	95.8	Liu *et al.* (2011)
	Red beet extract	37.5	60 min	55	Batch	95	Liu *et al.* (2008)
	Red raspberry	30	60 min	55	Batch	36.6	Liu *et al.* (2010)
	Tea leaves extract	7	1 h	30	Batch	50	Primo *et al.* (2007)
	Apple juice	20	20 min	65	Batch	100	Niu *et al.* (2010)
	Hami melon juice	35	5–10 min	55	Batch	74.7	Chen *et al.* (2009)
Tyrosinase	mushroom	8	20 min	55	Batch	40	Hu *et al.* (2010)
Lipoxygenase	Soybean	10.3	15 min	50	Batch	100	Tedjo *et al.* (2000)
	Hami melon juice	35	5–10 min	55	Batch	99	Chen *et al.* (2009)
	Glycine max	50	30 min	30	Batch	98.2	Liao *et al.* (2009)
	Carrot juice	2.94	10 min	5	Batch	>70	Park *et al.* (2002)
Peroxidase	Horseradish	62.1	15 min	55	Batch	100	Tedjo *et al.* (2000)
	Hami melon juice	35	5–10 min	55	Batch	62	Chen *et al.* (2009)
	Horseradish	30	60 min	55	Batch	88	Gui *et al.* (2006b)
	Tea leaves extract	25	6 h	50	Batch	100	Primo *et al.* (2007)

6.2 Effects of gas bubbling

Most globular proteins such as enzymes have a strong tendency to accumulate at interfaces (Norde 1992). Studies on protein adsorption report irreversible protein denaturation on hydrophobic surfaces (Schmidt *et al.* 1990). Caussette *et al.* (1999) studied chicken egg-white lysozyme inactivation at gas–liquid interfaces. A solution of phosphate buffer 0.01 M at a given pH was pre-equilibrated at the assay temperature (70–75°C). Lysozyme was then solubilized in this preheated buffer solution and a 60-ml volume of the enzymatic solution (final concentration 0.05 g/L). Bubbling nitrogen at atmospheric pressure in a lysozyme aqueous solution strongly enhanced enzyme inactivation. At a 150-ml/min flow rate, the half-life of lysozyme was 12 min while at the same conditions without bubbling no activity loss is observed after 8 h. Inactivation changed linearly with the interfacial area which showed that the inactivation was an interfacial mechanism. The inactivation induced by nitrogen bubbling strongly depended on temperature and pH acting on the adsorption process. This study pointed out the potential of enzyme inactivation using gas–liquid interfaces.

6.3 Alteration of the protein structure

Proteins may be altered when exposed to DPCD. Details of changes in some enzymes will be discussed when the specific enzyme is mentioned below. For an example of protein alteration, Ishikawa *et al.* (2000a) investigated the conformational changes in α-helical poly-L-glutamic acid caused by microbubbling supercritical CO_2 with circular dichroism spectra. After microbubbling using a micropore filter at 35°C and 30 MPa for 30 min, α-helix content decreased to 37%, while without the filter it was reduced only to 68%. The α-helix structure was significantly decomposed by DPCD. No important changes were observed in heating, autoclaving or pH-lowering.

Rezaei and Temelli (2000) mention that different enzymes exhibit different activity patterns in supercritical media, possibly because of different immobilization supports (Yu *et al.* 1992) and different structures. They give the example that *C. cylindracea* lipase supported on Celite 545 was inactivated at a higher rate than *M. miehei* lipase supported on macroporous anionic resin beads (Yu *et al.* 1992). At 13.6 MPa and 40°C, the residual activity of the former was 75% and that of the latter was 85–90% after 7 and 6 d, respectively (Yu *et al.* 1992). Chymotrypsin and trypsin are examples to show the effect of structure on enzyme stability. These enzymes contain disulfide bridges and undergo partial inactivation during slow (~5–10 min) depressurization from 10 MPa after exposure to DPCD

(Randolph *et al.* 1991). However, penicillin amidase, which does not have cysteine, appears to be the least stable during depressurization (Randolph *et al.* 1991). Giessauf *et al.* (1999) reported that hydrolases (crude preparations of lipases and esterases) with disulfide bridges had a lower degree of inactivation compared to an enzyme without cystine after several pressurization–depressurization steps at 15 MPa and 75°C for 24 h.

In this chapter, the effects of DPCD on several enzymes important in the food quality are discussed.

6.4 Studies with multiple enzymes

Ishikawa *et al.* (2000a) studied the effect of pH on the inactivation of the enzymes acid protease from *Aspergillus niger* (E.C. 3.4.23.6), alkaline protease from *Bacillus subtilis* (E.C. 3.4.21.14), papain from *Carica papaya* (E.C. 3.4.22.4), and glucoamylase from Rhizopus delemar (E.C. 3.2.1.3) dissolved in McIlvaine buffer by microbubble supercritical (SC) CO_2 treatment in the range of pH 3 to pH 6. Ethanol was added to McIlvaine buffer at concentrations of 2.5, 5.0, 7.5, 10 and 15% (v/v). Each enzyme preparation was dissolved in the buffer solution at the concentration of 50–100 mg/mL. A semibatch DPCD system was used where CO_2 was fed continuously to the system. Inactivation was observed at pH less than 4. With gaseous and liquid CO_2 treatments little or no significant inactivation occurred. Microbubbling of SC–CO_2 at lower pH was effective at 40–50°C. Microbubbling of SC–$C0_2$ at pH 3 completely inactivated enzymes at temperatures 25° lower than that of the thermal treatment (65–75°C). The degree of inactivation in acid protease increased with increasing ethanol concentration.

Mun *et al.* (2011) treated *Pseudomonas aeruginosa* in neutral phosphate-buffered saline in a multibatch system at 37°C and 10 MPa for 5, 10 and 15 min with vigorous mixing. Several enzymes in the organism were evaluated using the API ZYM kit (BioMerieux, France), which monitors 19 different enzyme activities (alkaline phosphatase, esterase, esterase lipase, lipase, leucine arylamidase, valine arylamidase, cystine aryl-amidase, trypsin, α-chymotrypsine, acid phosphatase, naphtol-ASBI-phosphogydrolase,α-galactosidase,β-galactosidase,β-glucuronidase, α-glucosidase, β-glucosidase, N-acetyl-β-glucosaminidase, α-mannosi-dase and and α-fucosidase) from a complex sample. The fraction of insoluble protein in case of the DPCD treatment was significantly increased, possibly as a result from the transformation of the soluble proteins inside the cells into their insoluble forms. A considerable reduction was observed for most of the enzyme activities except for esterase. For instance, all of the activities of five enzymes, lipase, leucine arylamidase, valine arylamidase, trypsin and acid phosphatase, disappeared, possibly due to pH change after the DPCD treatment.

Hong and Pyun (2001) treated the cells of *Lactobacillus plantarum* with DPCD at 7 MPa at 30°C for 10 min or 60 min. Enzymatic activities of the cells with and without CO_2 treatment were determined by the use of APIZYM kit (BioMérieux, Marcy-l'Etoile,France), to monitor 20 different enzyme activities from a complex sample which had not been purified. The enzymes that were significantly reduced in activity were α-glucosidase: 6.8±1.2%, 6.5±2.0; β-glucosidase: 5.9±1.0, 5.7±1.4; N-acetyl-b-glucosaminidase : 9.7±1.8, 7.2±0.4; α-galactosidase: 6.6±1.7, 6.1±1.4; and cystine arylamidase: 8.8±2.2, 8.7±2.3 of residual activity for 10 and 60 min exposure, respectively.

Kim *et al.* (2007a) investigated the effect of DPCD (35°C and 10 MPa) on the cellular enzymes of *E. coli*. The enzymatic activities of *E. coli* cells with and without DPCD treatment were determined by the use of an APIZYM kit which permits the monitoring of 19 different enzymatic activities from a complex sample not requiring enzyme purification steps. A total of six enzymes, alkaline phosphatase, leucine arylamidase, acid phospatase, naphtol-AS-BI-phosphohydrolase, β-glucuronidase and α-glucosidase, were identified using the assay kit in *E. coli* O157:H7. Most of these enzymes showed an initial increase after 10 min then exhibit a continuous decrease in enzyme activity with increasing DPCD treatment time. For the generic *E. coli*, esterase activity was shown in addition to the six enzymes detected in O157:H7. The decreasing pattern of the enzyme activity in generic *E. coli* was similar to that of *E. coli* O157:H7.

Tisi (2004) studied the effects of DPCD on milk. A prototype continuous DPCD equipment designed by Praxair (Chicago, IL) was used. Enzyme activity was assayed by acid degree value and protein breakdown. Temperatures between 15°C and 40°C and pressures between 7 MPa and 62 MPa were investigated.

Because of milk's susceptibility to pH change, the process' effect on casein proteins was also studied. At higher temperatures (40°C) and CO_2 concentrations, the process was shown to significantly reduce the proteolytic and lipolytic end products. However, these conditions also changed the structure of the underlying casein protein.

6.5 Effects on specific enzymes

6.5.1 Alpha-amylase

Kamihira *et al.* (1987) treated dry α-amylase AD-C (Amano Pharmaceutical Co.) preparations with DPCD at 20 MPa, 35°C, 2 h in preparations with *E. coli* (68% water) or baker's yeast (62% water). The activity of the enzyme increased: 121% of original in *E. coli* preparation, 135% of initial activity in Baker's yeast preparation.

In the Yoshimura *et al.* (2002) study, α-amylase from *Bacillus subtilis* (optimum pH, 5.8 to 6.8; E.C.3.2.1.1; Nagase Biochemicals Ltd., Osaka,

Japan) was dissolved in McIlvaine buffer (0.05 M citric acid and 0.1 M disodium hydrogen phosphate, pH 7.0) in a concentration of 0.1 mg/mL. This buffer system had a buffer capacity similar to commercial liquid foods (e.g. natural juices, fruit drinks and Japanese sake). A continuous DPCD system with microbubbles was used for enzyme inactivation. Treatment conditions were 10 MPa and 15.6 g/min of CO_2 flow rate. Treatment temperature was set at 5°C intervals from 35 to 50°C. Treatment time (average residence time) was calculated by dividing the inner volume of the treatment vessel against the sample flow rate. D values were calculated for enzyme inactivation for both heat treated and DPCD treated samples. The activation energies from Arrhenius analysis were 111 ± 8.5 kJ/mol for continuous treatment with DPCD and 149 ± 9.2 kJ/mol for heat treatment.

6.5.2 Acid protease

Yoshimura *et al.* (2002) dissolved acid protease from *Aspergillus niger* (optimum pH, 2.5; E.C.3.4.23.6; Hankyu Bioindustry Co., Ltd., Osaka, Japan) in McIlvaine buffer (0.05 M citric acid and 0.1 M disodium hydrogen phosphate, pH 3.0). Treatment conditions in a continuous DPCD system with microbubbles were 10 MPa and 15.6 g/min of CO_2 flow rate. Treatment temperature was set at 5°C intervals from 35°C to 50°C. Treatment time (average residence time) was calculated by dividing the inner volume of the treatment vessel against the sample flow rate. D values were calculated for enzyme inactivation for both heat-treated and DPCD-treated samples. The activation energies from Arrhenius analysis were 64.1 ± 7.5 kJ/mol for continuous treatment with DPCD, and 294 ± 15.9 kJ/mol for heat treatment.

Ishikawa *et al.* (1996) investigated the conformational changes of enzyme-inactivated DPCD microbubble method by measuring the circular dichroic (CD) spectra at far-ultraviolet (UV) range (250–200 nm). Batch system pressure ranged from 8 to 30 MPa and the temperature from 35°C to 50°C. The CO_2 was fed at 4.0 g/min about 10 min until the pressure reached the experimental level, and then feeding was stopped. At the end of the treatment, the vessel was slowly depressurized by releasing the pressure over a period of 5 min. At 35°C and 25 MPa for 30 min, the residual activity of acid protease from *Aspergillus niger* (EC 3.4.23.6), dissolved in deionized water to a concentration of 10 μg/mL, was 25.3%. The decay of negative ellipticity of acid protease as a result of DPCD treatment was 37.6%. The plot of the residual activity against the residual α-helix content gave a linear relationship for acid protease.

6.5.3 Alkaline protease

Ishikawa *et al.* (1996) investigated the conformational changes of enzyme-inactivated DPCD microbubble method by measuring the circular dichroic (CD) spectra at far-ultraviolet (UV) range (250–200 nm). Batch DPCD system pressure ranged from 8 to 30 MPa and the temperature from 35°C to 50°C. The CO_2 was fed at 4.0 g/min about 10 min until the pressure reached the experimental level, and then feeding was stopped. At the end of the treatment, the vessel was slowly depressurized by releasing the pressure over a period of 5 min. At 35°C and 25 MPa for 30 min, the residual activity of alkaline protease from *Bacillus subtilis* (EC 3.4.21.14), dissolved in deionized water to a concentration of 10 μg/mL, was 0%. The decay of negative ellipticity of alkaline protease as a result of DPCD treatment was 31.3%.

6.5.4 Gluco-amylase

Ishikawa *et al.* (1996) investigated the conformational changes of enzymes inactivated DPCD microbubble method by measuring the circular dichroic (CD) spectra at far ultraviolet (UV) range (250–200 nm). Batch DPCD system pressure ranged from 8 to 30 MPa and temperature from 35°C to 50°C. The CO_2 was fed at 4.0 g/min about 10 min until the pressure reached the experimental level, and then feeding was stopped. At the end of the treatment, the vessel was slowly depressurized by releasing the pressure over a period of 5 min. At 35°C and 25 MPa for 30 min, the residual activity of glucoamylase from *Rhizopus delemar* (EC 3.2.1.3), dissolved in deionized water to a concentration of 10 μg/mL, was 17.5%. The decay of negative ellipticity of glucoamylase as a result of DPCD treatment was 12.4%. The residual α-helix content of glucoamylase treated with DPCD decreased from 12.4% to 5.1% during 10 days at 20°C, and the residual enzyme activity also decreased from 17.5% to 0%. The plot of the residual activity against the residual α-helix content gave a linear relationship for glucoamylase. The α-helix structure of glucoamylase remained 85.7% after the heat treatment at 80°C for 30 min, and increased to 97.8% during the initial 4 days, and then no additional change was observed.

6.5.5 Lipase

Kamihira *et al.* (1987) treated dry lipase OF (Meito Sangyo Co.) preparations with DPCD at 20 MPa, 35°C, 2 h in preparations with *E. coli* (68% water) or baker's yeast (62% water). The activity of the enzyme decreased: 88% of original in *E. coli* preparation, 78% of initial activity in Baker's yeast preparation.

Ishikawa *et al.* (1996) investigated the conformational changes of enzyme-inactivated DPCD microbubble method by measuring the circular dichroic (CD) spectra at far-ultraviolet (UV) range (250–200 nm). Batch system pressure ranged from 8 to 30 MPa and the temperature from 35°C to 50°C. The CO_2 was fed at 4.0 g/min about 10 min until the pressure reached the experimental level, and then feeding was stopped. At the end of the treatment, the vessel was slowly depressurized by releasing the pressure over a period of 5 min. At 35°C and 25 MPa for 30 min, the residual activity of lipase from *Rhizopus japonicus* (EC 3.1.1.3), dissolved in deionized water to a concentration of 10 μg/mL, was 0%. The decay of negative ellipticity of lipase as a result of DPCD treatment was 62.9%.

Fadıloğlu and Erkmen (2002) exposed lipase to CO_2 (under atmospheric pressure) treatment at different pH and temperature conditions, and showed a decrease in enzyme activity. Inactivation was found to be highest as the temperature increased from 20°C to 50°C at initial pH 7.15. About 84% of the initial activity of the enzyme was lost at initial pH 7.15 and 50°C. Faster drop in pH and enzyme activity occurred within 5 min. Without CO_2, no significant change ($P > 0.05$) was observed in activity at the highest temperature (50°C). The pH drop and loss in enzyme activity were higher at initial pH 8.20. No changes in enzyme activity and pH were observed when nitrogen gas was applied instead of CO_2. Studies using gel electrophoresis showed no differences in protein patterns between CO_2-treated and untreated control. After 6 weeks of frozen storage, no changes in pH and activity were observed for CO_2-treated lipase solution.

Giessauf and Gamse (2000) treated crude porcine pancreatic lipase powders with DPCD at 75°C and 15 MPa, and observed a time-dependent increase of enzyme activity compared to before treatment. After 24 h of incubation in DPCD, a maximum activity of 860% was measured, compared to the untreated enzyme. DPCD treatment caused a weight loss of about 4%. 0.47% of the weight loss was an oily extract during depressurization. In the extract, free fatty acids were found (mainly palmitic, stearic and oleic acids). No chemical modification on the protein reactive groups were found as studied by determination of reactive free amino groups, determination of free –SH groups and by determination of the carbonyl content (as an oxidation marker). No significant difference in the protein content between the untreated and DPCD processed lipase powders were found. Tryptophan fluorescence emission spectra showed no change of the emission maximum indicating no larger protein conformational change. DPCD-treated lipase showed no loss of its increased activity after storage in a freezer for 6 months.

Rezaei and Temelli (2000) used lipase from *Mucor miehei* immobilized on macroporous anionic resin (Lipozyme IM) during enzyme-catalyzed hydrolysis of canola oil in DPCD. There was a drop in triglyceride conversion over a 24-h reaction time at 38 MPa and 55°C, which may be an indication of loss of enzyme. This is consistent with the results of Marty *et al.* (1992), who used *M. miehei* lipase immobilized on a macroporous anionic resin and reported a 10% loss in the enzyme activity in both DPCD and n-hexane after 6 d of exposure to 13 MPa and 40°C. Giessauf *et al.* (1999) reported that lipase from *Pseudomonas sp.* and lipase from *Candida cylindracea* lost only 36.1% of its activity after 30 pressurization–depressurization steps in 24 h in DPCD at 15 MPa and 75°C.

Oliveira *et al.* (2006) investigated the influence of temperature, pressure, exposure times and depressurization rate on the activity of two commercial immobilized lipases (*Mucor miehei* (Lipozyme IM) immobilized on a macroporous anion exchange resin, and *Candida antarctica* (Novozym 435) immobilized on a macroporous anionic resin) submitted to DPCD, at temperatures from 35°C to 75°C, pressure range of 1–28 MPa, exposure times from 1 to 6 h and different decompression rates. Activity losses were observed for Lipozyme IM treated with DPCD. For Novozym 435, treatment in DPCD also led to activity losses. The decompression rates were not a relevant variable. An exploratory study was conducted to investigate the effect of compression–expansion cycles on enzyme activity. Thermogravimetric analyses showed that the thermal profiles of Novozym 435 treated in carbon dioxide were similar to the untreated enzyme. Scanning electron micrographs of Novozym 435 illustrated that the material submitted to DPCD presented morphological alterations when compared to the untreated enzyme.

6.5.6 Pectinesterase (PE)

Balaban *et al.* (1991a) studied PE inactivation in orange juice by DPCD. The pH of orange juice must be lowered to 2.4 for substantial PE inactivation. DPCD lowered pH only to 3.1. Therefore, only the pH-lowering effect was not sufficient to explain enzyme inactivation. Balaban *et al.* (1993) applied the following treatments to orange juice and observed the decrease of PE activity. Untreated control had a decrease in PE activity of 8% after 20 days storage. Supercritical CO_2 treatment (31 MPa, 40°C, 45 min) showed 31% reduction; juice acidified with HCl to pH=3.1 and pressurized with N_2 (24 MPa, 40°C, 45 min) had a 36% reduction; juice buffered to pH=3.8 with citrate buffer, then treated with supercritical CO_2 (31 MPa, 40°C, 45 min) reduced PE by 23%; juice pressurized with N_2 (20.6 MPa, 55°C, 1 h) showed an increase in PE activity. These results suggest that the buffered-juice PE activity decreased only by the molecular effect of CO_2, while the unbuffered CO_2 combined the effects of pH-lowering and CO_2 effects. Pressurized N_2 did not lower activity. The effect of DPCD process time, temperature and pressure on pectinesterase (PE) activity was determined. PE could be inactivated with supercritical CO_2 below temperatures necessary for thermal inactivation. Higher pressure and temperature and longer treatment time resulted in more inactivation. Inactivation kinetics showed activation energy was significantly reduced at DPCD treatment at 31 MPa (97.4 KJ/mole), compared to identical treatments at atmospheric pressure (166.6 KJ/mole). D values ranged from 2673 min at atmospheric pressure and 40°C to 10 min at 31 MPa and 60°C. z value at atmospheric pressure was 8.8°C, and at 31 MPa 5.2°C.

Kincal *et al.* (2006) used a continuous DPCD system to treat orange juice. Treatment pressures were 38, 72 and 107 MPa, and CO_2–juice (w/w) ratios were from 0.40 to 1.18. Residence time was kept at 10 min. PE was not completely inactivated: the highest PE inactivation (46.3%) was obtained when the pressure was 107 MPa and no heat was applied. The average temperature of the juice in the holding tube was 24°C and ranged between 21°C and 30°C. The CO_2–juice ratio was 0.43. Cloud increased between 446% and 846% after treatments.

Lim *et al.* (2006) treated mandarin juice in a continuous DPCD system. Temperatures were set to 25°C, 35°C and 45°C, residence times 5, 7 and 9 mins, and pressures 13.8, 27.6 and 41.4 MPa. Also, CO_2–juice ratios of 2%, 7% and 12% were used. Maximum PE inactivation was 50.7% at 41.4 MPa, 7 mins residence time, CO_2 ratio of 7% and temperature at 45°C. The maximum cloud increase was 38.4% at 27.6 MPa, 7 mins residence time, 2% CO_2 ratio, and 45°C.

6.5.7 Pectin methyl esterase (PME)

Corwin and Shellhammer (2002) added CO_2 at approximately 0.2 molar % to solutions processed at 500–800 MPa in high hydrostatic pressure treatment in order to further inactivate pectin methylesterase (PME), using the juice of Valencia oranges purchased from a local supermarket. In the inactivation of PME, pressure was a significant factor at 25°C and 50°C ($p = 0.0001$). Carbon dioxide was a significant factor in further inactivating PME beyond that which pressure would achieve alone at 25°C ($p = 0.0022$) and at 50°C ($p = 0.0008$). The interaction between pressure and CO_2 was found to be significant ($p = 0.0371$) at 25°C but insignificant at 50°C. At 25°C and 800 MPa for 1 min treatment, residual activity with CO_2 was 7%, while at 50°C it was 11%.

Zhou *et al.* (2009a) studied the influence of DPCD on the activity and structure of pectin methylesterase (PME) from orange. The pressures were 8–30 MPa, temperature 55°C and time 10 min. DPCD caused significant inactivation on PME, the lowest residual activity was about 9.3% at 30 MPa. The SDS-PAGE electrophoretic behaviour of DPCD-treated PME was not altered, while changes in the secondary and tertiary structures were found. The β-structure fraction in the secondary structure decreased, and the fluorescence intensity increased as HPCD pressures were elevated. After 7-day storage at 4°C, no alteration of its activity and no reversion of its β-structure fraction were observed, while its fluorescence intensity further decreased.

Zhi *et al.* (2008) worked on the inactivation of apple pectin methylesterase (PME) with DPCD combined with temperatures (35–55°C). DPCD increased the susceptibility of apple PME to temperature, and pressure had a noticeable effect on apple PME activity. A labile and stable fraction of apple PME was found and the inactivation kinetics of apple PME by DPCD were adequately described by a two-fraction model. The kinetic rate constants kL and kS of labile and stable fractions were 0.890 and 0.039 min-1, and the decimal reduction times DL and DS are 2.59 and 58.70 min at 30 MPa and 55°C. Z_T representing temperature increase needed for a 90% reduction of the D value and the activation energy E_a of the labile fraction at 30 MPa were 22.32°C and 86.88 kJ/mol, its Z_P representing pressure increase needed for a 90% reduction of the D value and the activation volume V_a at 55°C were 21.75 MPa and –288.38 cm^3/mol. The residual activity of apple PME after DPCD exhibited no reduction or reactivation for 4 weeks at 4°C.

Zhou *et al.* (2009b) studied the inactivation of pectin methylesterases (PMEs) from carrot and peach in buffer by DPCD at 55°C. The two PMEs were effectively inactivated by DPCD, their residual activity (RA) decreasing with increasing pressure. The RA of the two PMEs exhibited a fast decrease first and reached a constant after a prolonged treatment time; their inactivation kinetics was adequately modelled by a fractional-conversion model. The nonzero RA of the two PMEs was 6–7%, with increasing pressures the kinetic rate constant, k, increased and the decimal reduction time, D, decreased for the HPCD-labile fraction of the two PMEs. The labile fraction of carrot PME was more susceptible to HPCD than that of peach PME; the activation volume, V_a, and Z_P (the temperature increase needed for a 90% reduction of D) were –1079.37 cm^3/mol and 5.80 MPa for carrot PME, and –130.51 cm^3/mol and 48.31 MPa for peach PME.

In cloudy apple juice from apple slices treated by DPCD (25–65°C, time 20 min, pressure 20 MPa) and mild heat (MH, 65°C), Niu *et al.* (2010b) found that residual activity (RA) of pectin methylesterase (PME) treated with HPCD was significantly lower than MH, and its minimum level was 18%. The authors pointed out that the turbidity of the juice was not well related to RA of PME.

Plaza (2010) treated guava puree in a continuous DPCD system and observed an apparent increase in PE activity, with a reduction of cloud (between 60% and 91%) in all treated samples. The initial PE activity was quite low, in the order of 5×10^{-4} units/g. The pH of the puree was 3.7.

6.5.8 Polyphenol oxidase (PPO)

Zemel *et al.* (1990) inactivated apple polyphenoloxidase (PPO) to 98%, 63% and 34% of the initial activity after 20 min at adjusted pH values of 2.00, 2.25 and 2.50, respectively. Higher pH values resulted in slower inactivation. The juice was acidified with HCl, and then readjusted to original with NaOH. Sensory analysis of the juice indicated that the acidified samples had a lighter but less preferred colour and a disagreeable salty taste. No aroma difference was recognized.

Chen *et al.* (1992a) used a pH control and measured the activity of lobster PPO that was kept at pH = 5.3, which was the same pH of samples achieved by DPCD treatment. The flow rate of CO_2 at 1 atm was 10 mL/min for up to 30 mins. The pK_a value at equilibrium among dissolved CO_2, H+ and HCO_3^- in the solution is 6.1 (Montgomery and

Swenson 1969). Using the Henderson–Hasselbalch equation, it was determined that the molar ratio of the concentration of HCO_3^- to that of CO_2 in the solution was reduced from 251 to 0·16 when the pH was dropped from 8.5 to 5.3. Although the pH control sample retained 35% of its activity at 35°C after 30 min, DPCD-treated enzyme lost all its activity after 1 min at the same temperature. Therefore CO_2 was suggested to have a unique role in the inactivation of PPO. Bubbling N_2 under the same conditions did not cause inactivation, in contrast the PPO activity increased, presumably because volume decrease due to evaporation during heating. The authors suggested that the loss in enzyme activity of CO_2-treated PPO was not due to the physical effect of the gas bubbling. DPCD can change electrophoretic protein patterns, isoelectric profiles and secondary structures/protein patterns of PPOs (Chen *et al.* 1992a), but not by CO_2 at atmospheric pressure (Chen *et al.* 1993a). Chen *et al.* (1992a) obtained the circular dichroism spectra of untreated and treated lobster, brown shrimp and potato PPOs, and showed that DPCD caused conformational changes in the secondary structures (α-helix, β-sheet, β-turn and random coil) of the enzymes. They also showed that purified Florida spiny lobster, brown shrimp and potato polyphenol oxidases (PPOs) exhibited a time-related decline in activity following treatment at 43°C with high-pressure CO_2 at 5.8 MPa. Kinetic studies showed crustacean PPOs were more vulnerable than potato PP0 to high-pressure CO_2 treatment.

Chen *et al.* (2009c) treated Hami melon juice with DPCD. For each experiment, 50 mL of juice was placed in a 200 mL uncapped plastic tube and then placed in the DPCD vessel preheated to the experimental temperature (35°C, 45°C, 55°C and 65°C) and then exposed to 8, 15, 22, 30 and 35 MPa pressure for 5–10 min until the pressure reached the experimental level. At the end of treatment, the vessel was slowly depressurized over a period of 15 min. Unpressurized samples were used as control. The residual activity of PPO after treatment at 35 MPa was 25.26%, and stayed substantially constant after 4 weeks of storage at 4°C while untreated juice PPO went from 100% at week 0 to 62% at week 4.

Gui *et al.* (2007) investigated the inactivation of polyphenol oxidase (PPO) in cloudy apple juice exposed to DPCD treatment. Higher pressure, higher temperature and longer treatment time caused more inactivation of PPO. The maximum reduction of PPO activity reached more than 60% at 30 MPa and 55°C for 60 min. The experimental data followed first-order reaction kinetics; the kinetic rate constant k and the decimal reduction time D were closely related to the pressure and temperature of DPCD treatment. Higher pressures or higher temperatures resulted in lower D values (higher k); the D value of PPO was lowered to 145 min when treated by the combination of 30 MPa and 55°C. Activation energy of 18.00 kJ/mol was significantly reduced by DPCD treatment at 30 MPa, as compared to activation energy of

72.0 kJ/mol for identical treatment at atmospheric pressure. Pressure and temperature sensitivity of kinetic parameters were studied. Z_P at 55°C was 66.7 MPa and Z_T at 30 MPa was 108°C.

Gui *et al.* (2006b) studied the effects of DPCD treatment of 8, 15, 22 and 30 MPa for 60 min at 55°C on polyphenol oxidase (PPO) activity, colour and browning degree in cloudy apple juice during storage at 4°C for 4 weeks. The DPCD treatment had significant effects on inactivation of PPO and the minimum residual activity of PPO was 38.50% at 30 MPa. The restoration of PPO residual activity after DPCD treatment was also observed, which was dependent on the pressure level. A greater reduction of lightness *L* and a minor increase of redness *a* of cloudy apple juices after DPCD treatment occurred. Moreover, the total colour difference (ΔE), which was significantly less than that of untreated sample, was decreased by increasing the pressure. The changes of lightness *L* and browning degree *A* during storage were well fitted to a first-order kinetic model. The rate constants of k_L and k_A of samples subjected to DPCD treatment reduced from 4.75×10^{-2} to 0.42×10^{-2} and from 37.19×10^{-2} to 8.02×10^{-2}, respectively, when pressure increased from 0 MPa (untreated sample) to 30 MPa.

Liu *et al.* (2011) investigated the inactivation of polyphenol oxidase (PPO) in watermelon juice with a batch mode DPCD system. The maximum reduction of PPO activity reduced by HPCD treatment was 95.8% at 30 MPa and 50°C for 30 min, while the control treatment at 50°C for 30 min resulted in 50.9% reduction. The DPCD inactivation kinetics for PPO was described by a two-fraction model: a labile and a stable fraction in PPO from watermelon juice. The kinetic rate constants k_L and k_S of labile and stable fractions were 1.976 and 0.041 min^{-1} by DPCD.

Liu *et al.* (2008) measured the inactivation of polyphenoloxidase (PPO) in red beet extract with DPCD at 37.5 MPa, 55°C, 60 min in a batch system. The treatment reduced the initial activity by 95%. Compared with the thermal treatment (55°C), the decimal reduction time (D) of PPO was reduced from 161.29 min to 38.31 min, with an activation energy reduction from 57.06 kJ/mol to 53.58 kJ/mol. First-order kinetics were adequate to describe the inactivation, ($R^2 > 0.84$, $p < 0.05$) with D values declining with the rise of pressure and temperature.

Liu *et al.* (2010) simulated the inactivation kinetics of PPO from frozen red raspberry (*Rubus idaeus* L.) by a first-order reaction after treatment with DPCD. The minimum of PPO residual rate was 36.6% under 30 MPa and 55°C for 60 min in a batch HPCD treatment, while that by thermal treatment at 55°C for 60 min was 66.8%. The activation energy for inactivation of PPO was reduced from 98.9 (thermal) to 14.6 kJ/mol after HPCD treatment.

The effect of DPCD on polyphenol oxidase (PPO) activity in red muscadine grape (*Vitis rotundifolia* cv. Noble) juice under different processing pressures (27.6, 38.3 and 48.3 MPa) and CO_2 concentrations (0%, 7.5% and

15%) at a constant processing temperature (30°C) and residence time (6.25 min) was determined in a continuous system (del Pozo-Insfran *et al.* 2007). Processing muscadine grape juice at 27.6 MPa and 0% CO_2 resulted in a 40% decrease on initial PPO activity and this was not significantly improved when processing pressures were increased to 38.3 or 48.3 MPa. When CO_2 levels were increased from 0% to 7.5%, an additional 35% reduction in PPO activity was observed at 38.3 MPa, but no additional inactivation occurred when CO_2 levels were increased to 15%. This is probably due to the saturation concentration of CO_2 in the juice at the given treatment conditions: an increase from 7.5% to 15% did not achieve more solubility and therefore had no effect. Juices retained low concentrations of residual dissolved CO_2 (6.25 and 13.2 mM for juices at 7.5% and 15% CO_2, independent of pressure).

The effects of temperature, exposure time, solvent reduced density, pressure and depressurization rate on the activity of PPO in the crude extract of mate tea leaves (*Ilex paraguariensis* St. Hill) treated in a batch DPCD system were evaluated by Primo *et al.* (2007). In all assays, 40 g of mate tea leaves were homogenized with 90 mL of 3% (w/v) pirocatechol and polyvinylpyrrolidone K90 (PVP K90) in 0.05 M sodium phosphate buffer (pH 7.5). The suspension was filtered through five gauze layers and centrifuged for 30 min at 11,000 × g and 4°C. The supernatant was used in the experiments as crude enzymatic extract. A temperature of 30°C, pressure of 7 MPa, exposure time of 1 h, depressurization rate of 10 $kgm^{-3}min^{-1}$ and carbon dioxide–reduced density of 0.60 led to a PPO activity loss of 50%.

Niu *et al.* (2010b) compared cloudy apple juices from apple slices treated by DPCD and mild heat (MH). Temperatures were from 25°C to 65°C, time 20 min, and pressure 20 MPa. Polyphenol oxidase (PPO) was completely inactivated by DPCD. The minimum residual activity (RA) by MH at 65°C was 38.6%. L value of cloudy apple juice from DPCD-treated apple slices was significantly greater than that from MH-treated apple slices. The b value and browning degree (BD) were lower.

6.5.9 *Tyrosinase*

Corwin and Shellhammer (2002) added CO_2 at approximately 0.2 molar % to solutions processed at 500 to 800 MPa in high hydrostatic pressure treatment in order to further inactivate tyrosinase from mushroom (EC 1.12.18.1, Fluka 93898). It was dissolved at a concentration of 25 mg/L in phosphate buffer (0.1 M, pH 6.5). An interaction was found between CO_2 and pressure at 50°C for PPO. Activity of PPO was decreased by CO_2 at all pressure treatments. At 25°C and 800 MPa for 1 min treatment, residual activity with CO_2 was 52%, while at 50°C it was 22%.

The effects of DPCD treatments on aggregation and homogenization, surface charge, secondary and tertiary structure and activity of mushroom tyrosinase in an aqueous system were investigated by Hu *et al.* (2010)

using a number of methods including dynamic light scattering (DLS), zeta potential measurement, circular dichroism (CD) spectropolarimeter, and spectrofluorometer. With a treatment time of 20 min, three treatment temperatures (35°C, 45°C and 55°C) and four pressures (5, 8, 12 and 15 MPa) had been selected. The aggregation and homogenization of the globular protein particles were induced by DPCD as suggested by the particle size distribution (PSD) patterns that were closely related to the pressure and temperature. The surface charge of the tyrosinase decreased following the DPCD treatments, and its variation tendency showed a favourable consistency with that of its PSD patterns. The α-helix conformation in secondary structure and fluorescence intensity reflecting tertiary structure also decreased, and the λ_{max} red-shifted with increasing pressure. The results also indicated that DPCD could enhance inactivation effect of the temperature on the tyrosinase with its lowest residual activity being about 60% under the condition of 8 MPa, 55°C, and 20 min treatment time. The loss in the activity of the tyrosinase was correlated to its aggregation and homogenization effect induced by DPCD, which led to the change of surface charge as well as secondary and tertiary structures.

6.5.10 Lipoxygenase

Chen *et al.* (2009c) treated Hami melon juice with DPCD. For each experiment, 50 mL of juice was placed in a 200 mL plastic tube without the cap and then placed in the DPCD vessel preheated to the experimental temperature (35°C, 45°C, 55°C and 65°C) and then exposed to 8, 15, 22, 30 and 35 MPa pressure for 5–10 min until the pressure reached the experimental level. At the end of treatment, the vessel was slowly depressurized over a period of 15 min. Unpressurized samples were used as control. The residual activity of LOX after treatment at 35 MPa was 0.02%, and was reduced to 0% after 4 weeks of storage at 4°C while untreated juice LOX went from 100% at week 0 to 60% at week 4.

Liao *et al.* (2009) reported on the alterations of molecular properties of lipoxygenase (LOX, EC 1.13.11.12) from *Glycine max* (soybean) induced by DPCD, (pressures 10, 30, 50 MPa, temperatures 30°C and 55°C, treatment time 30 min) using transmission electronic microscope (TEM), SDS-PAGE and native-PAGE, far-UV CD and fluorescence spectroscopy. Compared to the control, the residual activity of LOX was reduced significantly to 26.39%, 9.74% and 1.88% corresponding to 10, 30 and 50 MPa at 30°C, respectively. The aggregation of LOX molecules was suggested by an increase in the particle size of LOX after DPCD treatment. The absence of LOX band was observed after exposure to DPCD at 55°C for 30 min on the native-PAGE due to the aggregation, but the electrophoretic behaviour of LOX was not altered on the SDS-PAGE. The CD and fluorescence spectra of DPCD-treated LOX were noticeably changed, its α-helix relative content decreased sharply to less than 10%, and its intrinsic relative fluorescence intensity (RFI) decreased linearly with increasing pressures.

Tedjo *et al.* (2000) showed that the application of DPCD (35.2 MPa, 40°C, 15 min) achieved approximately 35% inactivation in 30% sucrose solution. Total inactivation of LOX (10.3 MPa, 50°C and 15 min) could be achieved through DPCD treatment in unbuffered solution. Increasing the concentration of sucrose and buffering (pH range 4 to 9) of enzyme solutions resulted in increased resistance of the enzyme to DPCD treatment.

Chikere *et al.* (2000) measured the activity of immobilized soybean lipoxygenase-1 (LOX-1) in aqueous and DPCD media for the production of 13S-hydroperoxyoctadecadenoic acid (13S-HPODE). In DPCD, the optimal conditions were 33°C and 25 MPa. Addition of 5% (v/v) water to the reaction medium increased LOX-1 activity by approximately 10%. Above 40°C substantial loss in LOX-1 activity was observed probably due to protein denaturation. A higher space–time yield of 5×10^{-3} Ms^{-1}mg^{-1} LOX-1 for 13SHPODE was obtained in DPCD compared to only 5×10^{-5} Ms^{-1}mg^{-1} LOX-1 in aqueous medium. The stability of immobilized LOX-1 was only significantly affected by the pressurization and depressurization steps during reactions in DPCD. The authors noted that the inactivation of immobilized LOX-1 in DPCD following the depressurization steps was not critical for reactions since these batch operations have only one such step.

6.5.11 Peroxidase

Tedjo *et al.* (2000) reported that the application of DPCD (62.1 MPa, 55°C, 15 min) achieved approximately 65% peroxidase (POD) inactivation in 30% sucrose solutions. Total inactivation of POD (62.1 MPa, 55°C and 15 min) could be achieved through DPCD treatment in unbuffered solution. Increasing the concentration of sucrose and buffering (pH range 4–9) of enzyme solutions resulted in increased resistance of the enzyme to DPCD treatment.

Chen *et al.* (2009c) treated Hami melon juice with DPCD. For each experiment, 50 mL of juice was placed in a 200 mL plastic tube without the cap and then placed in the DPCD vessel preheated to the experimental temperature (35°C, 45°C, 55°C and 65°C) and then exposed to 8, 15, 22, 30 and 35 MPa pressure for 5–10 min until the pressure reached the experimental level. At the end of treatment, the vessel was slowly depressurized over a period of 15 min. Unpressurized samples were used as control. The residual activity of POD after treatment at 35 MPa was 38.46%, and stayed relatively constant after 4 weeks of storage at 4°C while untreated juice POD went from 100% at week 0 to 92% at week 4.

Gui *et al.* (2006a) reported on the influence of DPCD at 55°C on inactivation of horseradish peroxidase (HRP) in buffer solution, pH 5.6. Its structural change was analysed by far UV-circular dichroism (CD) and tryptophan fluorescence spectroscopy. DPCD treatment had significant effects on the residual activity of HRP, the lowest residual activity was only 12% at 30 MPa. HRP's secondary and tertiary structures were changed. The α-helix relative content in the secondary structure decreased and the intrinsic relative fluorescence intensity (RFI) increased as the pressure of DPCD treatment was increased. The HRP's inactivation closely corresponded to the loss of α-helix relative content and the increase of RFI. After a 7-day storage at 4°C, the restoration of residual activity and the reversion of the α-helix relative content were observed while RFI resumed, except for the 30 MPa treatment.

Fricks *et al.* (2006) assessed the influence of compressed CO_2 on the specific activity of peroxidase (POD) from radish (*Raphanus sativus* L.). The protein concentrate, in phosphate buffer, was obtained through the precipitation of proteins with $(NH_4)_2SO_4$. The peroxidase activity was determined using guaiacol as substrate. A semifactorial experimental design was adopted to evaluate the effects of temperature (30–50°C), exposure time (1–6 h), carbon dioxide density ($\rho_R = 0.6-1.8$, resulting in pressures from 7 to 25 MPa) and depressurization rate (10–200 kg/m^3/min) on the peroxidase activity. Samples of lyophilized protein concentrate were also submitted to high pressure, and the effect of water content on the enzymatic activity in compressed CO_2 was studied. The water content in lyophilized samples was determined by thermogravimetric analysis. The results showed that the treatment of the enzymatic concentrate with CO_2 at high pressure could be a promising route to increase the POD activity.

Liu *et al.* (2008) measured the inactivation of peroxidase (POD) in red beet extract with DPCD at 37.5 MPa, 55°C, 60 min in a batch system. The treatment reduced the initial activity by 86%.

Compared with the thermal treatment (55°C), the decimal reduction time (D) of POD was reduced from 555.56 min to 74.63 min, with an activation energy reduction from 92.54 kJ/mol to 68.63 kJ/mol. First-order kinetics were adequate to describe the inactivation, ($R^2 > 0.84$, $p < 0.05$) with D values declining with the rise of pressure and temperature.

The effects of temperature, exposure time, solvent reduced density, pressure and depressurization rate on the activity of peroxidase in the crude extract of mate tea leaves (*Ilex paraguariensis* St. Hill) treated in a batch DPCD system were evaluated by Primo *et al.* (2007). In all assays, 40 g of mate tea leaves were homogenized with 90 mL of 3% (w/v) pirocatechol and polyvinylpyrrolidone K90 (PVP K90) in 0.05 M sodium phosphate buffer (pH 7.5). The suspension was filtered through five gauze layers and centrifuged for 30 min at 11,000 × g and 4°C. The supernatant was used in the experiments as crude enzymatic extract. A temperature of 30°C, pressure of 7 MPa, exposure time of 1 h, depressurization rate of 10 kgm^{-3} min^{-1} and carbon dioxide reduced density of 0.60 led to an enhancement of around 25% in the peroxidase activity. On the other hand, a pressure of 25 MPa at 50°C for 6 h resulted in total inactivation of POD. The authors conclude that it is possible to 'tune' the activity of the enzyme by adjusting process parameters.

6.5.12 *Alkaline phosphatase*

The effects of temperature and CO_2 treatment on the inactivation of alkaline phosphatase (ALP) were studied by Fadıloğlu *et al.* (2004). ALP in the glycine–NaOH buffer exposed to CO_2 at atmospheric pressure and at different temperatures showed a decrease in enzyme activity. Inactivation was found to be higher as the temperature increased from 20°C to 50°C. After a 30 min treatment, residual ALP activity was found to be 84% and 19% at 20°C and 50°C, respectively (approximately 12% of activity was present at 50°C after 55 min of thermal treatment). Faster drop in pH and enzyme activity occurred within 5 min. The change in pH and enzyme activity dependant on CO_2 treatment was not observed in raw milk mainly due to strong buffering capacity of milk.

6.5.13 *Myrosinase*

Dunford and Temelli (1996) studied the effect of DPCD extraction conditions (temperature, pressure, extraction time and moisture content of the samples) on the myrosinase activity and glucosinolate hydrolysis in flaked and whole canola seeds. Combined effects of high temperature (75°C), pressure (62.1 MPa) and moisture (20% w/w) were necessary to achieve 90% enzyme inactivation in canola flakes in 3 h. Similar extraction conditions resulted in only 44% enzyme inactivation in whole canola seeds. Significant glucosinolate degradation was observed after 5 h of extraction of low moisture content (8.5% w/w) whole canola seeds. Glucosinolate degradation in canola flakes was minimal at all moisture levels studied. Therefore, myrosinase inactivation prior to SC–CO_2 extraction of canola flakes may not be necessary. The authors pointed out that the presence of enzyme activity in canola extracted at low moisture or temperature should be taken into account for further utilization of meal.

Yang *et al.* (2011) worked on the influence of temperature, pressure, and exposure times as they affect the activity of commercial myrosinase (MYR, thioglucoside glucohydrolase, EC 3.2.1.147) from *S. alba* seed, 374 units/g solid) treated with HPCD. At 22 MPa and 65°C for 5 min, only 1% of MYR activity was retained. Increase in pressure from 8 to 22 MPa at 55°C resulted in an increase of the first-order inactivation rate constant (k) from 0.015 to 0.024 min^{-1}, while the decimal reduction time (D) decreased from 157.2 to 96.1 min. The z value for pressure was calculated as 65.8 MPa. The far UV-CD spectra (190 to 250 nm) and tryptophan fluorescence spectra revealed that CO_2 dissolution plays an important role in MYR inactivation and its structural alteration.

6.5.14 Hydrolases

Bauer *et al.* (2000) studied the effects of dry and humid DPCD on hydrolases (EC 3.1.1.1 and EC 3.1.1.3). A crude and a purified preparation of esterase EP10 from *Burkholderia gladioli* were incubated in DPCD for long-term and repeated high-pressure treatments. Incubation for 24 h in DPCD at 15 MPa and 35°C of the crude preparation had no effect on enzyme activity, while incubation at 75°C led to a distinct loss of residual activity. After 30 pressurization and depressurization steps at 35°C and 15 MPa, the crude enzyme preparation showed an activity increase.

Using the purified enzyme preparation of esterase EP10 from *B. gladioli*, no significant effects could be observed. Fluorescence spectra indicated no conformational change before and after treatment with DPCD.

Treatment of a preparation of esterase from porcine liver with wet and dry DPCD at 20 MPa at various temperatures showed a significant denaturing influence of the dissolved water on residual activities of the enzyme at temperatures of more than 40°C.

Lipase from *Candida rugosa* and esterase from porcine liver were treated at 15 and 30 MPa at a constant temperature of 40°C and an incubation time of 22 h. During these treatments, different amounts of water were introduced into the DPCD. The results showed an increase of the water content in the treated enzyme preparation while the enzyme activity remained stable till the maximum amount of water soluble in this medium was injected into the DPCD.

6.6 Conclusions and suggestions

DPCD can be used either as a medium for enzyme-based reactions, and there are enzymes that will resist substantial CO_2 pressures, or it can be used as a means to inactivate enzymes. Much more needs to be done experimentally to understand the effects of the DPCD environment, including pressure, and that of the molecular CO_2 and its derivatives, the HCO_3^- and the CO_3^{-2} ions on the protein molecules, on their charges and their conformational state. The effects of the aqueous environment, whether resulting in acidification due to CO_2 or affecting mass transfer, need to be measured and their mechanisms understood. Finally there is an unknown territory where combinations of temperature, ultra-high pressure, pulsed electric fields, irradiation, other gases of modifiers and so on with DPCD are waiting to be explored.

7 The Kinetics of Microbial Inactivation by Carbon Dioxide under High Pressure

Maria G. Corradini and Micha Peleg

Abstract: Understanding the mechanisms through which carbon dioxide under high pressure inactivates microorganisms is very helpful in the development of the technology as a non-thermal alternative to heat preservation. But so is the knowledge of the inactivation kinetics. Although it is not yet possible to translate physical and chemical processes and events at the cellular level into a microbial population's response to the imposed stress, control of the kinetics is still essential for a successful treatment. Except for the different time scale, the survival of microbial populations exposed to CO_2 under high pressure follows patterns that are very similar to those observed in heat and chemical treatments. Therefore, the isobaric-isothermal survival curves of specific organisms can be described by the same kinds of mathematical models. These include the non linear Weibullian-Log logistic (WeLL) model, the traditional log-linear which is a special case, the Gompertz model for sigmoid survival curves and the special models previously proposed for extreme tailing ending in residual survival. At least in principle, all these models, regardless of type, can be transformed into rate equations. However, their utility in predicting the efficacy of dynamic processes will require determination of how their coefficients, which vary with the CO_2 pressure, are also affected by temperature.

Keywords: carbon dioxide; nonthermal processing; nonlinear microbial inactivation; Weibull model; kinetics; survival curves; mathematical modeling.

7.1 Introduction

Carbon dioxide has been increasingly considered as a means of nonthermal food preservation. The mechanisms by which pressurized CO_2 causes microbial inactivation have been extensively studied in recent years and they are addressed in other chapters of this book. Discussion of the various aspects of the technology and the science behind it can also be found in the two comprehensive reviews of Zhang *et al.* (2006b) and Garcia Gonzalez

Dense Phase Carbon Dioxide: Food and Pharmaceutical Applications, First Edition. Edited by Murat O. Balaban and Giovanna Ferrentino.

et al. (2007). Knowledge of what happens at the molecular and cellular level is key to understanding the lethal effect of high-pressure CO_2 on microorganisms. It is therefore essential to the proper application of the technology. But equally important is to understand the expression of these microscopic mechanisms at the population level. Once quantified, it can be used to establish the inactivation kinetics which, in turn, can be translated into effective process conditions (Peleg 2002).

Any effective food preservation process, be it thermal or nonthermal, is expected to be able to reduce microbial numbers by at least several orders of magnitudes. This is to make sure that even under the worst conceivable circumstances as far as the microbial contamination is concerned, the *probability* of encountering viable pathogen cells and spores will be negligibly small. The same is true for spoilage organism cells and spores in perishable foods. Although not totally eliminated, their number ought to be reduced to such a low level that the survivors will not be able to recover completely within the product's planned shelf life.

The basic information needed for any preservation process design is the targeted organism's *survival curves*, recorded under pertinent conditions. In our case, the CO_2 pressure is obviously the primary lethal agent but the food's composition (e.g. sugar concentration), temperature, and pH should also be considered. The survival curve is a record of how the targeted organism's count in a given food volume or mass changes with time during the exposure to the lethal agent. This is a clear and straightforward definition. However, one has to pay attention to the fact that only cells that have grown in an appropriate (i.e. optimal) medium and after a given time are actually counted. Thus, injured cells, which might repair the damage and become viable if given sufficient time, might not be included in the record. Or conversely, injured cells that could only recover under optimal growth conditions might be counted as viable survivors. The issue of injury has been amply mentioned in the literature (e.g. Singh *et al.* 1986; Kilimann *et al.* 2006; Miller *et al.* 2006) including its potential effect on the survival curve's interpretation (Corradini and Peleg 2007). Also, it is well known that different recovery methods may yield different counts (e.g. Feeherry *et al.* 1987). But, since what follows will primarily address the mathematical presentation and properties of microbial survival curves under high CO_2 pressure, these aspects of the survival curve will be only briefly addressed in this chapter. The focus will be on the number of survivors as reported, that is, regardless of whether some or may be even the majority of them were actually injured to at least a certain extent. Suffice it to add that had some of the counted cells recovered under optimal growth conditions been actually injured after the CO_2 exposure, many of them might not survive if left in the treated food. In such a case, the recorded survival curve would be an *underestimate* of the process's efficacy (Corradini and Peleg 2007). In the less likely but still possible scenario, the uncounted injured cells would be able to repair the damage and multiply in the food during its handling and storage. In such a case the situation will be reversed and the survival curve determined by a standard method would be an *overestimate* of the treatment adequacy. Which of the two scenarios is relevant to a particular food and CO_2 treatment, if at all, is a question that can be answered only by determining the kinds of injuries and their frequency experimentally, and then monitoring the product during its storage and distribution.

7.2 The survival curve

7.2.1 Primary models

Traditionally, microbial mortality during thermal processing at lethal temperatures has been assumed to follow the first-order kinetics (e.g. Jay 1992; Holdsworth 1997). The assumption has been extended to other microbial inactivation methods, including exposure to high CO_2 pressure. First-order kinetics entails that under isothermal and isobaric conditions the relationship between the survival ratio, $S(t)$, and time, t, must be log-linear. In our case, this translates into

$$\log_{10} S(t) = -\,k(P,T)t \tag{7.1}$$

where $S(t)$ is $N(t)/N_0$, the ratio between the momentary number of cells $N(t)$ per unit volume and the initial number N_0, and $k(P,T)$ the CO_2 pressure and temperature-dependent exponential inactivation rate constant (1/time). In other words, according to this model the plot of $\log_{10}S(t)$ versus t should always yield a straight line with a slope $-\,k(P,T)$.

Equation (7.1) is frequently written in the form of $\log_{10}S(t) = -t/D(P,T)$, where $D(P,T)$ is the pressure- and temperature-dependent 'D value'. (The 'D value' according to this model is the time needed to reduce the targeted microorganism's number by a factor of 10.)

Recent evidence suggests that the log-linear relationship between the survival ratio and time is the exception rather than the rule in heat processing (van Boekel 2002), and that there is no reason for its assumption for any method of inactivation, be it thermal or nonthermal (Peleg 2006). *By definition*, a survival curve is the cumulative temporal distribution of the mortality events, and as such it can assume a variety of shapes of which the log linear is just a special case (e.g. Peleg and Cole 1998; Peleg 2006). Nevertheless, the survival curve's *slope* has rate units (i.e. time reciprocal) and hence its relation to kinetics. In other words, the mortality event's temporal distribution and the inactivation kinetics are two sides of the same coin. Microbial mortality (and that of multicellular organisms) exposed to a hostile environment can be viewed as a 'failure phenomenon', equivalent to particulates breakage under mechanical stress. Consequently, the most likely distribution function that could describe it is the Rosin–Rammler distribution, better known as the Weibull

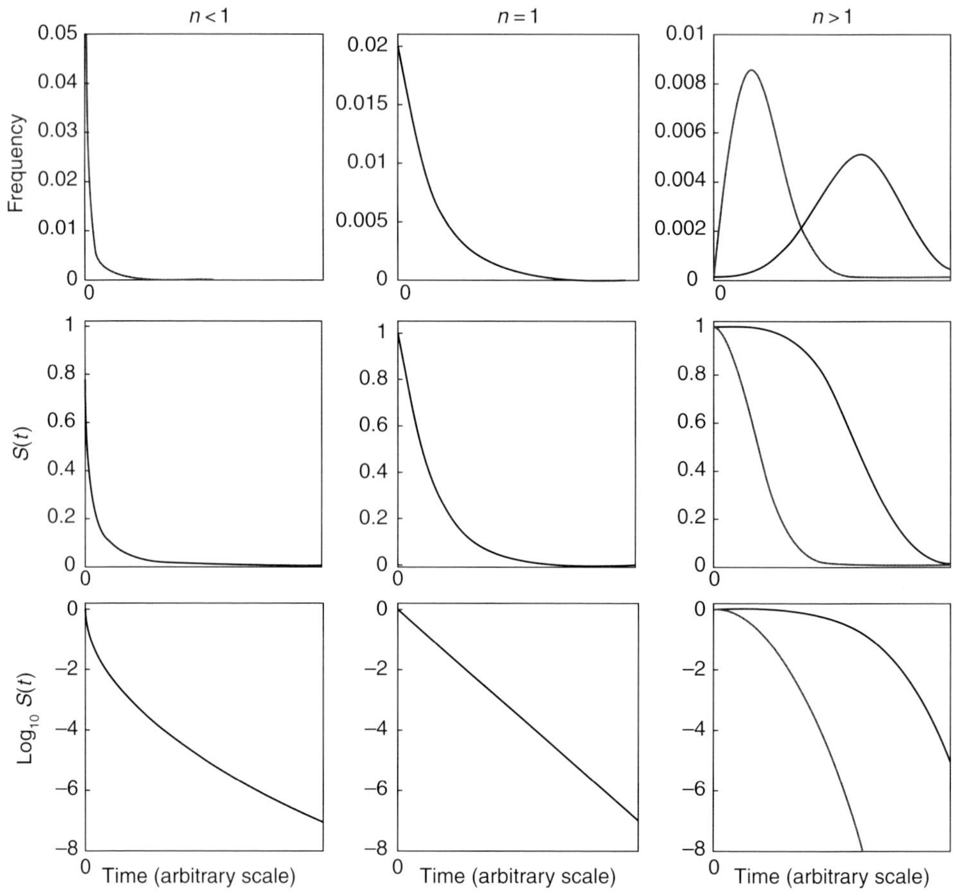

Figure 7.1 Weibullian inactivation patterns (Equation 7.2). Notice that the traditional log-linear inactivation model is just a special case where $n = 1$. Also notice that a 'shoulder' will always appear whenever the mortality events' temporal distribution has a spread much smaller than the mode.

distribution (e.g. van Boekel 2002; Peleg 2006). One way to write it is in the power law form (Peleg and Cole 1998):

$$\log_{10} S(t) = -b(P,T)t^{n(P,T)} \tag{7.2}$$

where $b(P,T)$ and $n(P,T)$ are pressure- and temperature-dependent coefficients.

According to this model – see Figure 7.1 – upward concavity ('tailing') is manifested in $n(P,T)<1$ and downward concavity ('damage accumulation') by $n(P,T)>1$. The traditional 'first-order kinetics' (Equation (7.1)) is just a special case of the model (Equation (7.2)) where $n(P,T)=1$ and $1/D(P,T)$ or $k(P,T)=b(P,T)$. Notice that a long 'shoulder' according to this model will always appear when the distribution's spread is considerably smaller than its mode, as shown in Figure 7.1. This will be true for all unimodal distributions, regardless of their symmetry or skewness, and their concavity direction (Peleg 2006).

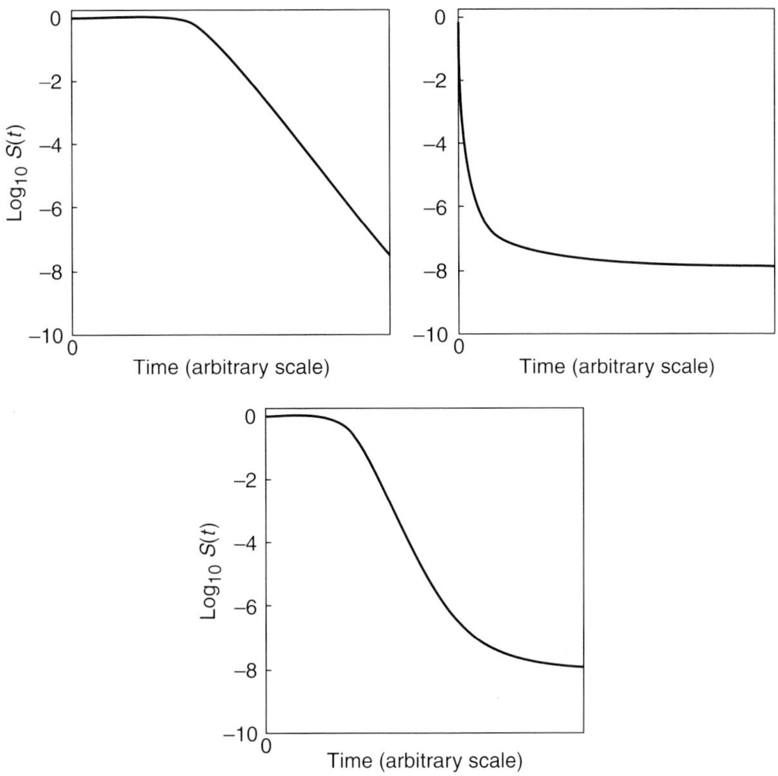

Figure 7.2 Non-Weibullian inactivation pattern generated with various models: top left with Equation (7.3), top right with Equation (7.4) or (7.5), and bottom with Equation (7.6).

In many instances, the power $n(T)$ of the Weibullian model (the 'shape factor' in its more common version) is only a weak function of temperature (van Boekel 2002) and the same can be true in nonthermal modes of inactivation (Peleg 2006). Consequently, $n(T)$, and so on can be assigned a constant value with little effect on the model's fit – as discussed in this chapter. This approach has been taken by Klotz *et al.* (2007) who fitted high-hydrostatic-pressure inactivation data with the power-law version of the model with a fixed $n(P) = 0.5$.

A survival curve that starts with a shoulder but whose continuation is a straight line can be described by the logistic distribution (Peleg 2006), which in the context of microbial survival can be written in the form:

$$\log_{10} S(t) = -\log_{10}\{1 + 10^{k'(P)[t - t_c(P)]}\} \tag{7.3}$$

where $t_c(P)$ is a pressure-dependent marker of the shoulder length in time units and $k'(P)$ the (approximate) slope of the semilogarithmic survival curve at $t >> t_c(P)$. (The reader will notice that according to Equation (7.3), when $t << t_c(P)$, $\log_{10} S(t) \sim 0$ and at $t >> t_c(P)$, $k'(P)[t - t_c(P)$ – see Figure 7.2.) For simplicity, we assume here that all the treatments are performed at the same temperature and therefore made the coefficients functions of the CO_2 pressure only. This kind of survival pattern may be difficult to discern if the experimental data has a scatter, in which case the Weibullian model (Equation (7.2)) might have a similar fit.

When pressurized CO_2 is used to suppress microbial growth, the possibility of encountering residual survival even after a long period of time cannot be ruled out. Since all the above mentioned models (Equations (7.1), (7.2), and (7.3)) imply that as $t \to \infty$, $\log_{10} S(t) \to -\infty$, they cannot describe such a survival pattern and therefore must be replaced. For a concave upward curve with a residual survival ratio, a candidate model would be

$$\log_{10} S(t) = -\frac{k_1(P)t}{k_2(P)+t} \tag{7.4}$$

or

$$\log_{10} S(t) = -\sum \frac{k_{1i}(P)t}{k_{2i}(P)+t} \tag{7.5}$$

where $k_1(P)$ and $k_2(P)$ or $k_{1i}(P)$ and $k_{2i}(P)$ are pressure- and temperature-dependent coefficients (Peleg 2006). According to these models, the asymptotic residual survival ratio at a given pressure and temperature is $-k_1(P)$ or $-\sum k_{1i}(P)$, respectively, while the logarithmic decay rate prior to its approach is $k_1(P)k_2(P)/[k_2(P)+t]^2$ in the first case (Equation (7.4)) or $\sum k_{1i}(P)k_{2i}(P)/[k_{2i}(P)+t]^2$ in the second (Equation (7.5)).

Sigmoid survival curves having a shoulder and ending in a residual ratio (Figure 7.2) can be described by the classic Gompertz model, for example. In our case all the treatments are performed at the same temperature, and can be written in the form:

$$\log_{10} S(t) = -A(P)\cdot\exp\{-\exp\{-B(P)\cdot[t-C(P)]\}\} \tag{7.6}$$

where $A(P)$, $B(P)$, and $C(P)$ are pressure- (and temperature-)dependent coefficients.

The apparent similarity of certain types of sigmoid survival curves to an inverted growth curve is only superficial. Thus, the fact that both can be described by the same mathematical expression preceded by an opposite sign does not imply that there is any analogy between growth and inactivation, as some investigators have assumed. In fact, there are numerous alternative three-parameter models that fit sigmoid survival curves, including those that can be encountered in high-pressure CO_2 treatments. Thus, unless supported by independent experimental evidence, one should be very cautious with the 'mechanistic interpretation' of the coefficients of any of the above model.

Extended exposure to chemical or heat stress can induce certain bacteria to adjust their metabolism in order to increase their survivability, a phenomenon known as 'adaptation'. To account for adaptation mathematically, one would have to render at least some of the model's coefficients a function of the stress history or rate (Corradini and Peleg 2009). Whether adaptation is a significant factor in pressurized CO_2 treatments is unknown to this chapter's authors. Similar to the role of injury, the proper approach to resolve the issue would be to isolate and examine the survivors before and after being treated under different conditions (see Chapter 12).

7.2.2 Secondary models – the effect of pressure alone

When exposed to an isobaric lethal CO_2 pressure at a given medium and under constant temperature, an organism's survival pattern solely depends on the pressure provided that no adaptation takes place and that the initial inoculum's size plays no significant role. Most of the published models on the effect of hydrostatic pressure on microbial lethality are analogous to those originally developed for heat inactivation, except that the temperature is replaced by pressure. Thus, the pressure dependence of the survival parameters is likewise described by what is known as *secondary models*. Where the first-order mortality kinetics had been invoked, it was frequently assumed that the 'D_p value' (the reciprocal of $k(P)$ in Equation (7.1)) has a log-linear dependence on pressure, characterized by a 'z_p value'. The 'z_p value' according to this assumption is the number of MPa's (or other pressure units) needed to change the 'D_p value' by a factor of 10. Alternatively, it has also been assumed that the pressure dependence of the exponential rate constant, $k(P)$ in Equation (7.1), follows an Arrhenius-type model with a corresponding 'energy of activation' (FDA 2000). Whenever the isobaric inactivation does not follow the assumed first-order kinetics, the meaning of either the 'D_p value' or $k(P)$ becomes obscure and the same can be said about the applicability of the secondary model, be it of the log-linear type in which the z_p is based or the Arrhenius type and the calculated 'energy of activation'. Both the log-linear model and the Arrhenius equation imply that there is no qualitative difference between high and low CO_2 pressures. However, at least for some organisms, there can be pressure levels that produce no measureable lethal effect at least on a practical time scale. For other deficiencies of the two traditional models, see Peleg (2006).

According to the Weibullian (Equation (7.2)) and other nonlinear inactivation models (Equations (7.3), (7.4), (7.5), and (7.6)), the isobaric (and isothermal) exponential inactivation rate is a function not only of pressure but also of *time* (Peleg 2003, 2006). Consequently, neither of the two traditional secondary models should be used to describe the pressure or temperature dependence of their coefficients. The pressure dependence of the rate parameter $b(P)$ can be described by the log-logistic model shown schematically in Figure 7.3 (Peleg 2006):

$$b(P) = \ln\left\{1 + \exp\left[k''\left(P - P_c\right)\right]\right\} \tag{7.7}$$

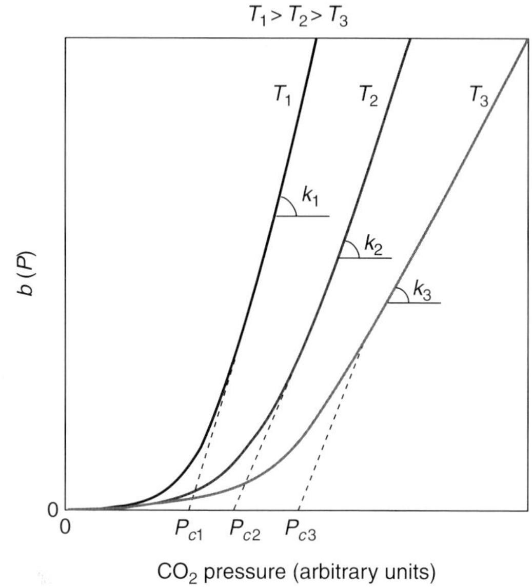

Figure 7.3 Schematic view of the log-logistic pressure dependence of *b*(*P*) expressed by Equation (7.7) as a model. Notice that increasing the treatment's temperature lowers the magnitude of $P_c(T)$ and increases that of *k*(*T*), the curve's slope at $P(T) >> P_c(T)$.

where $b(P)$ is the magnitude of b at pressure P and k'' and P_c are two survival parameters of the organism.

As before, when $P << P_c$, $b(P) \sim 0$ and when $P >> P_c$, $b(P) \sim k''(P-P_c)$. In the special case where the CO_2 is lethal to the targeted organism at any pressure, $P_c = 0$, and $b(P) = k''P$. In this case, the organism's resistance or sensitivity to the treatment will be characterized by two constants only, namely, k'' and n. Notice that the application of the log-logistic model does not require that $n(P)$ must be pressure independent (i.e. constant). This is because whenever $b(P) \sim 0$, $\log_{10} S(t)$ will remain ~0, regardless of the value of n on $n(P)$.

When a 'non-Weibullian' model is used to describe the survival pattern, as in the case where there is a residual survival ratio (Equations (7.4) or (7.5)) or when the survival curve has a sigmoid shape (Equation (7.6)), its survival parameters' pressure dependencies can be described by ad hoc empirical models. Some rules concerning their choice can be easily inferred. For example, it will be reasonable to assume that the residual survival level, $k_1(P)$ in Equation (7.4), will decrease as the CO_2 pressure increases. The inactivation rate will increase, which will be reflected in a decreasing $k_2(P)$. Similarly, if the survival pattern is sigmoid and described by the Gompertz model (Equation (7.6)), the CO_2 pressure increase will almost always lower the magnitude of $C(P)$ and increase that of $A(P)$. A pressure increase will also shorten the $t_c(P)$ of the logistic survival model (Equation (7.3)) and increase the magnitude of its $k(P)$. While these general trends are almost self-evident, their quantification has no established rules, except that the model choice should be guided by mathematical simplicity and the principle of parsimony (Ockham's razor), i.e., when several models can describe the same data set, the one with the smallest number of adjustable parameters should be selected.

The above is based on the assumption that increasing or decreasing the CO_2 pressure only affects the survival parameter's magnitude but does not change the general survival pattern. However, there is no theoretical reason to rule out the possibility that the pressure level can alter the inactivation pattern qualitatively as well as quantitatively. This could be manifested in the survival curve's concavity direction reversal, for example, or in the conversion of a sigmoid survival curve into one having a progressively increasing or decreasing slope, or vice versa. Certain secondary models can account for such transitions. Concavity inversion in Weibullian survival, for example, can be described by a variety of exponential or power-law expressions where, as the CO_2 pressure P monotonically increases, $n(P)<1$ (upward concavity) turns into $n(P)=1$ (log-linearity) and then to $n(P)>1$ (downward concavity).

7.2.3 The temperature effect and that of other auxiliary factors

Empirical evidence suggests that application of the pressurized CO_2 at an elevated temperature increases its lethality – as will be discussed in this chapter. The traditional approach to account for the combined effect of two or more factors, such as temperature and pH, pressure or chemical antimicrobial concentration, started with the assumption that either their effects are independent and additive or their interactions can be described in terms of a (generalized) linear model (e.g. Gänzle *et al.* 2007), in the form:

$$k(P,T)\ or\ D(P,T) = a_0 + a_1P + a_2T + a_3PT + \ldots \tag{7.8}$$

where k is the exponential inactivation constant, D is the decimal reduction parameter, and the a_i are constants determined by generalized linear regression.

Expressions such as Equation (7.8) have been commonly used in the 'response surface methodology'. However, one should remember that this technique was originally developed to identify optimal combinations, not to serve as a substitute for kinetic models. Therefore, our proposed approach is somewhat different (Peleg 2006). The principal lethal agent in our case is the pressurized CO_2 and the temperature plays only an 'auxiliary role'. This is because the process temperature is either ambient or only slightly elevated (i.e. it remains well below the level that would make it lethal by itself). Consequently, the temperature should not be treated as being on par with the CO_2 pressure, but as an enhancing factor. This requires that the inactivation model coefficients should be made temperature dependent, but not that the temperature be represented by an independent term. Consider the Weibullian log-logistic (WeLL) model, where for simplicity $n(P,T)$=constant=n. At the narrow temperature range of the CO_2 pressure application, it would be more prudent to modify the 'rate parameter' rather than to add a temperature term or terms to the survival model itself. This translates into converting $b(P)$into $b(P,T)$, writing it in the form (Peleg 2006):

$$b(P,T) = \ln\{1 + \exp\{k''(T)[P - P_c(T)]\}\} \tag{7.9}$$

where the survival parameters k'' and P_c become functions of temperature and are written as k''(T) and $P_c(T)$.

The same can be done for other factors such as pH, the presence of an antimicrobial, salt concentration and so on. Notice that determining the functions $k''(T)$ and $P_c(T)$, or their equivalent if several factors are involved, will require a larger number of experiments to cover the whole range of their pertinent combinations. At the time this chapter is written, finding a complete database of this kind is very difficult.

7.2.4 Dynamic treatments

Effective high-CO_2 pressure treatments usually involve exposure times from several minutes to one or more hours, during which both the pressure and temperature remain practically unchanged. Under such circumstances, the 'come up' and 'cooling' (or discharge) stages usually do not play a significant role and the process can be assumed as being truly isobaric and isothermal. It might be of interest, though, to investigate how non-isobaric and non-isothermal conditions might affect the process's lethality, at least theoretically. Consider a hypothetical organism whose isobaric and isothermal survival curves follow the WeLL model with a constant power $n(P,T)=n$ (Equation (7.2) and (7.9)). We'll assume that the function $b(P,T)$has been experimentally determined or estimated from published data on the same or similar organisms. We will also assume that under dynamic conditions (i.e. where the pressure, temperature or both are changing with time), *the momentary logarithmic inactivation rate is the isothermal–isobaric rate at the momentary pressure–temperature combination, at the time, that corresponds to the momentary survival ratio* (Peleg 2003, 2006). (A similar assumption has been experimentally verified for heat treatments, e.g. Mattick *et al.* 2001; Corradini and Peleg 2004; and chemical disinfection, e.g. Corradini and Peleg 2003.) In the context of high-pressure CO_2 inactivation, this translates into the following, in which the momentary isobaric–isothermal logarithmic rate is given by

$$\left.\frac{d\log_{10} S(t)}{dt}\right|_{\substack{P(t)\\T(t)}} = -b[P(t),T(t)]nt^{n-1} \tag{7.10}$$

where $P(t)$ and $T(t)$ are the momentary pressure and temperature at time t, respectively, and the time, t^*, which corresponds to the momentary survival ratio, $S(t)$, is

$$t^* = \left\{ \frac{-\log_{10} S(t)}{b[P(t),T(t)]} \right\}^{\frac{1}{n}} \tag{7.11}$$

Combining the two equations yields the dynamic rate equation:

$$\frac{d\log_{10} S(t)}{dt} = -b[P(t),T(t)]n \left\{ \frac{-\log_{10} S(t)}{b[P(t),T(t)]} \right\}^{\frac{n-1}{n}} \tag{7.12}$$

or, when $b(P,T)$ is defined by Equation (7.9),

$$\frac{d\log_{10} S(t)}{dt} = -\ln\left\{1+\exp\left\{k\left[T(t)\right]\cdot\left\{P(t)-P_c\left[T(t)\right]\right\}\right\}\right\}\cdot n\cdot \left(\frac{-\log_{10} S(t)}{\ln\left\{1+\exp\left\{k\left[T(t)\right]\cdot\left\{P(t)-P_c\left[T(t)\right]\right\}\right\}\right\}}\right)^{\frac{n-1}{n}} \tag{7.13}$$

Despite its cumbersome appearance, Equation (7.13) is an ordinary differential equation (ODE) and can be solved numerically for almost any conceivable pressure–temperature history. The solution is the survival curve, $\log_{10}S(t)$ versus time, which corresponds to the particular scenario. It can be computed and plotted with commercial mathematical software such as Mathematica® (Wolfram Research, Champaign, IL, US – the program used in this chapter). In principle at least, Equation (7.13) can also be solved with general-purpose software of the kind proposed for non-isothermal heat treatments (Peleg *et al.* 2005), except that the 'called values' of the coefficients would correspond to the momentary CO_2 pressure and temperature.

The main issue here is not the numerical solution, but as already stated, how to obtain sufficient experimental data to derive the relationship $b[P(t), T(t)]$ versus t. If this relationship could be derived from or assumed on the basis of published data, then one could easily generate hypothetical dynamic survival curves for analysis. Examples are shown in Figure 7.4. Notice that the assumption of $n(t)=n[P(t),T(t)]=n$, in the WeLL model, was made only to simplify the discussion. In a case where $n(t)$ does vary appreciably with pressure and/or temperature, all one would have to do is to replace the constant n in Equation (7.13) with the appropriate algebraic expression. (The modified Equation (7.13) will remain ODE and could be solved numerically almost as fast as the original model when $n(t)$ is a constant.) It ought to be added that all the stated assumptions can be tested and verified by comparing the model's predictions with actual non-isothermal and non-isobaric experimental survival data.

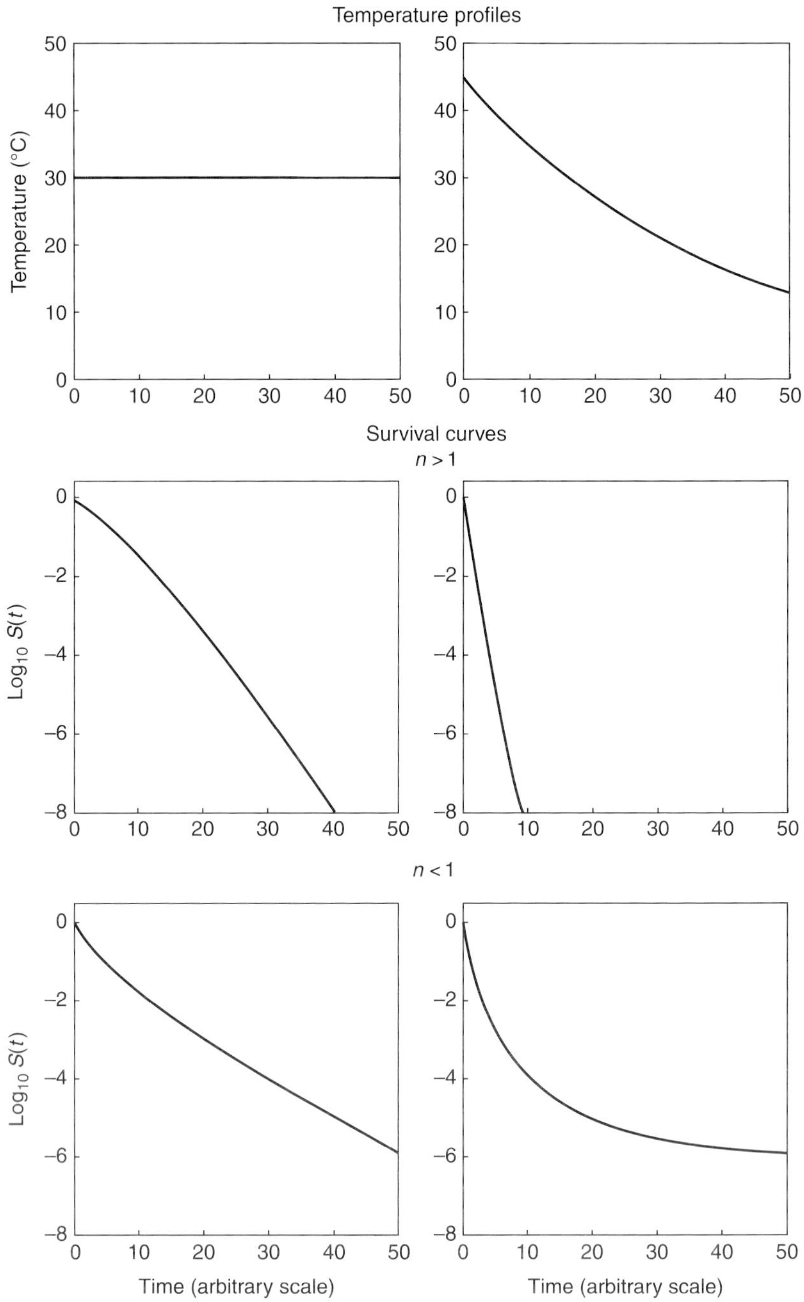

Figure 7.4 Examples of simulated isobaric and dynamic survival curves produced with Equations (7.2) and (7.12), respectively, as models.

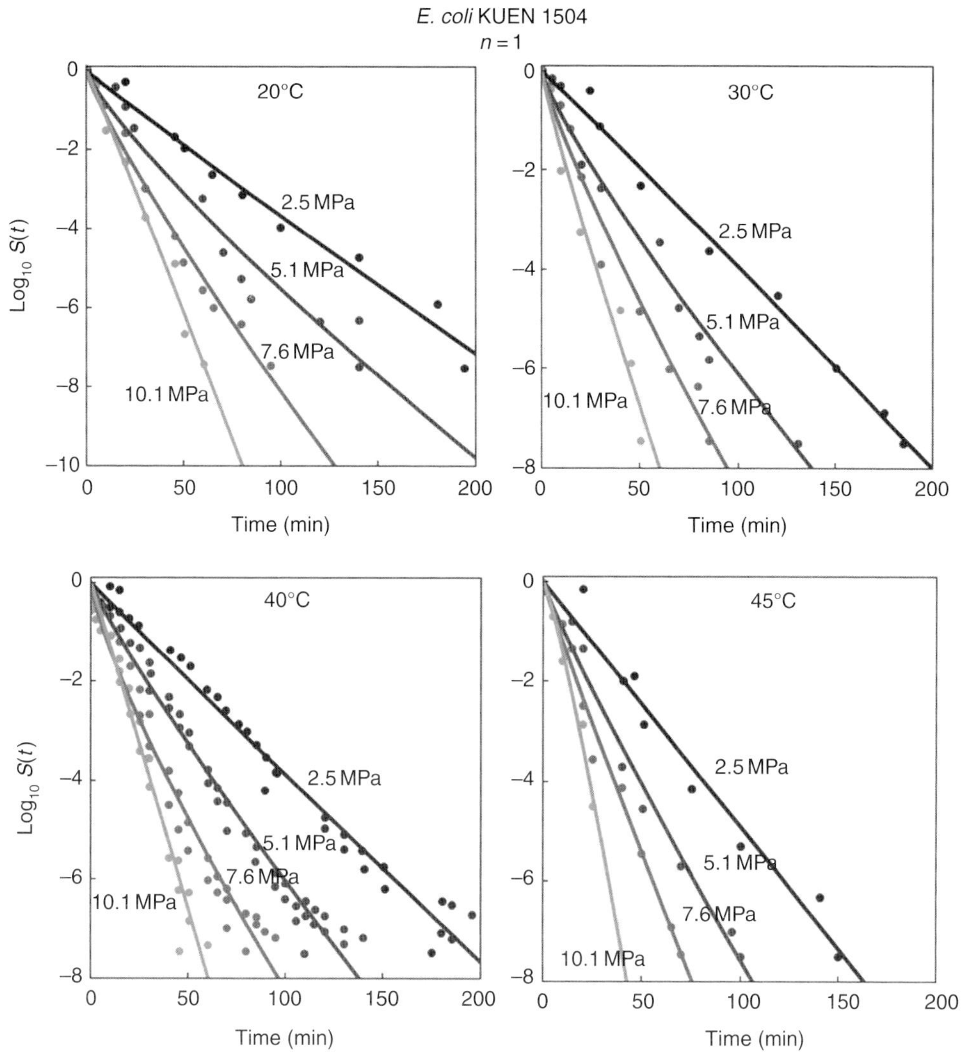

Figure 7.5 Log-linear isobaric and isothermal survival curves of *Escherichia coli* KUEN 1504 fitted with Equation (7.2) as a model. The linearized parameters are listed in Table 7.1. The original data are from Erkmen (2001b).

7.3 Application of the models to published experimental data

7.3.1 Primary model derivation

Examples of published survival curves fitted with Equations (7.1), (7.2), (7.3), (7.4), or (7.6) as a model are shown in Figures 7.5, 7.6, 7.7, 7.8, and 7.9. The corresponding regression parameters are listed in Tables 7.1, 7.2, 7.3, 7.4, and 7.5. The figures suggest that apart from the time scale, the effect of

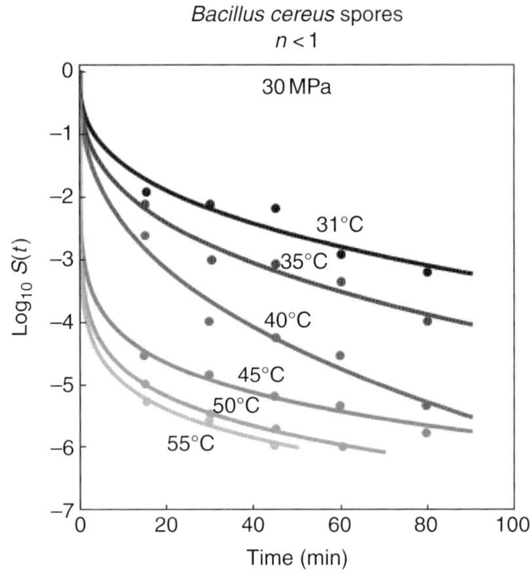

Figure 7.6 Weibullian isobaric (30 MPa) and isothermal survival curves of *Bacillus cereus* spores fitted with Equation (7.2) as a model with $n<1$. The regression parameters are listed in Table 7.2. The original data are from Ishikawa *et al.* (1997).

pressurized CO_2 on microorganisms is qualitatively similar to that of heat and other lethal agents. This is evident in the shapes of the survival curves which could be described by the same kinds of kinetic mathematical models (Peleg 2006). However, the magnitude of the survival parameter, although characteristic to the organism, need not be the same as in thermal, chemical, or other inactivation methods.

In most cases, the inactivation pattern is Weibullian with $n(P,T)=1$, that is, log-linear (Figure 7.5), $n(P,T)<1$ (Figure 7.6), and $n(P,T)>1$ (Figure 7.7). For all the cases of $n(P)>1$, the fixing of the curvature parameter at a representative value had hardly any effect, as expected – see Figure 7.7. Also the pressure or temperature dependence of the rate parameter, $b(T)$ or $b(P)$, could be adequately described by the log-logistic model as shown in Figure 7.10.

Notable exceptions to the Weibullian model are *Pichia awry* and wild yeasts, which under the reported treatments left residual survival that hardly varied with time (Figure 7.8), and *Lactobacillus plantarum*, which had sigmoidal survival curves (Figure 7.9). These organisms' survival curves are adequately described by Equations 7.4 or 7.6, respectively.

As expected, and regardless of the organism and model chosen, at least some of the survival parameters had strong CO_2 pressure dependence that was influenced by the treatment temperature. In almost all the cases examined, as shown in the figures, increasing the pressure did not alter the survival curve's overall shape, only its overall

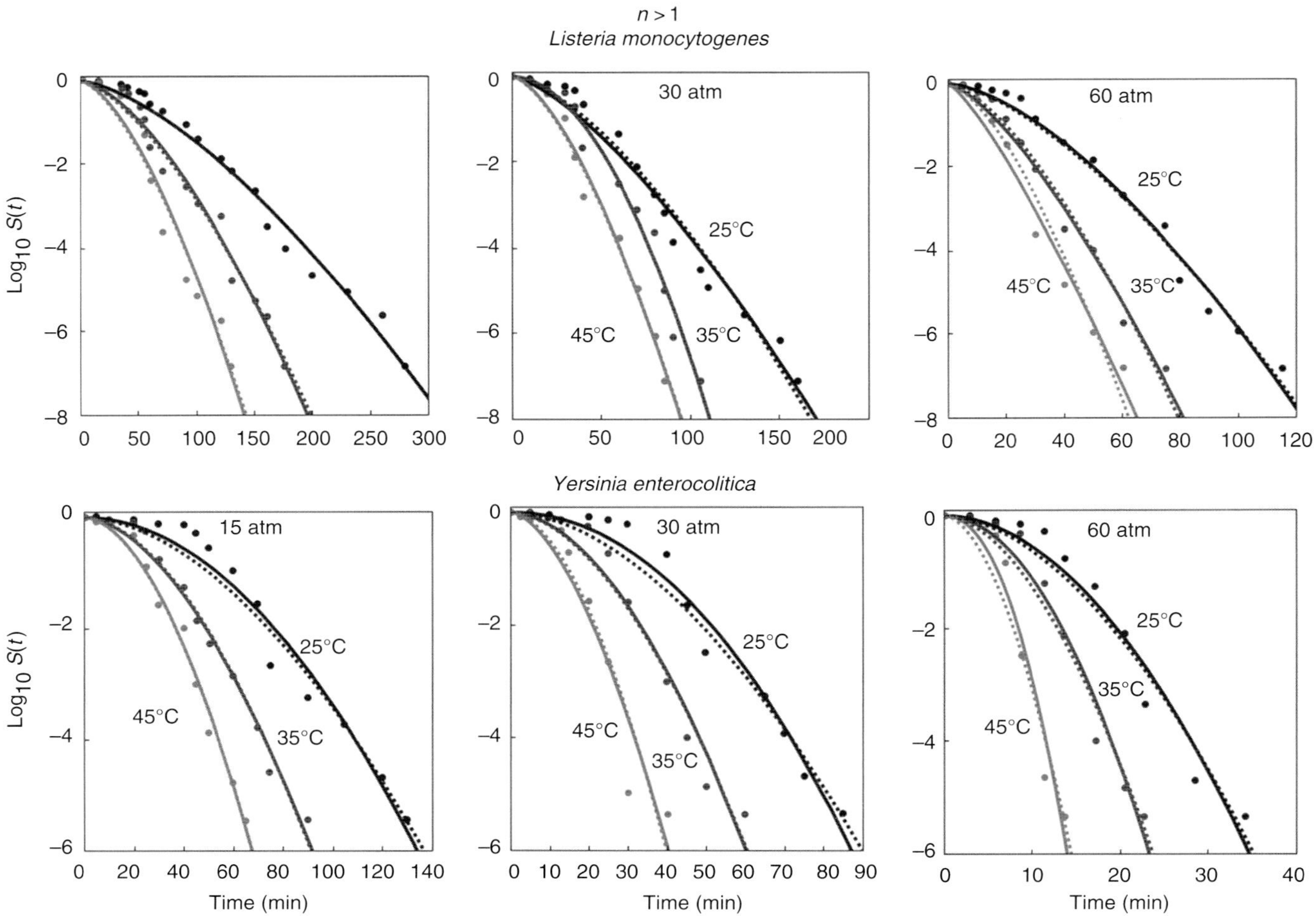

Figure 7.7 Weibullian isobaric and isothermal survival curves of *Listeria monocytogenes* and *Yersinia enterocolitica* fitted with Equation (7.2) as a model with $n(P) > 1$. The regression parameters are listed in Table 7.3. The original data are from Erkmen (2000c) and (2001a), respectively.

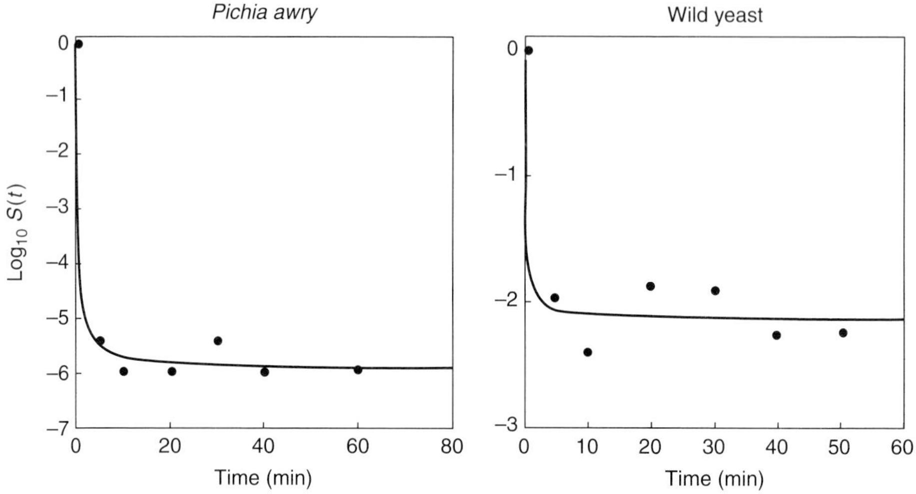

Figure 7.8 Non-Weibullian isobaric and isothermal survival curves of *Pichia awry* and wild yeast fitted with Equation (7.4) as a model. The regression parameters are listed in Table 7.4. The original data are from Parton *et al*. (2007). Notice the residual survival level.

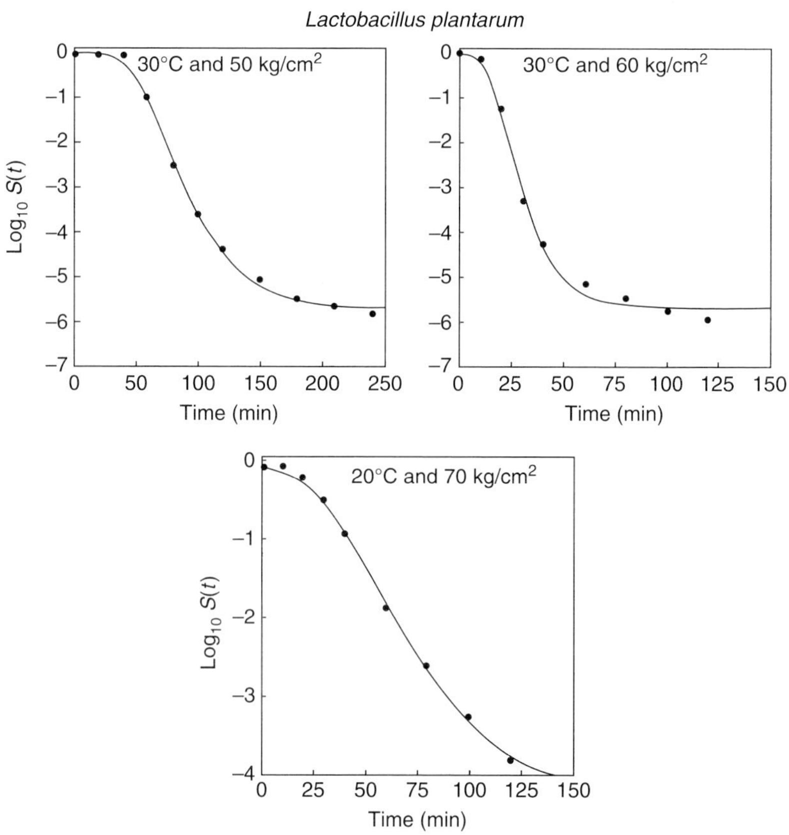

Figure 7.9 Sigmoidal isobaric and isothermal survival curves of *Lactobacillus plantarum* fitted with Equation (7.6) as a model. The regression parameters are listed in Table 7.5. The original data are from Hong *et al*. (1999).

Table 7.1 Parameters of inactivation of *E. coli* KUEN 1504 using Equation (7.2) as a model.

Temperature (°C)	CO_2 pressure (MPa)	$b(P,T)$ (min^{-n})	$n(P,T)$*	Mean square error (MSE)
20	2.5	0.04	1.0	0.140
	5.1	0.05	1.0	0.487
	7.6	0.09	1.0	0.167
	10.1	0.12	1.0	0.124
30	2.5	0.04	1.0	0.075
	5.1	0.06	1.0	0.212
	7.6	0.09	1.0	0.289
	10.1	0.14	1.0	0.290
40	2.5	0.04	1.0	0.109
	5.1	0.06	1.0	0.156
	7.6	0.09	1.0	0.358
	10.1	0.13	1.0	0.222
45	2.5	0.05	1.0	0.214
	5.1	0.08	1.0	0.177
	7.6	0.11	1.0	0.130
	10.1	0.18	1.0	0.138

*fitted with the log linear model

Table 7.2 Parameters of inactivation of *Bacillus cereus* at 30 MPa using Equation (7.2) as a model.

Temperature (°C)	$b(P,T)$ (min^{-n})	$n(P,T)$	Mean square error (MSE)
31	0.65	0.35	0.361
35	0.86	0.34	0.066
40	1.03	0.37	0.035
45	3.05	0.14	0.070
50	3.54	0.13	0.011
55	3.81	0.12	0.001

'steepness' and, where relevant, the residual survival level. The emergence of a sigmoidal survival curve (Figure 7.9) might indicate the pre-existence or formation of a resistant subpopulation within the treated cells. Such interpretation, however, needs to be verified by microscopic observations, biochemical assays and or testing the resistance of survivors after partial treatments (see the discussion of injury and adaptation in this chapter).

7.4 Concluding remarks

The available published results indicate that microbial survival under high pressure of carbon dioxide follows the same general patterns that are observed in thermal and chemical treatments. The main differences are in the time scale, specificity and level of inactivation. In most cases,

Table 7.3 Parameters of inactivation of *Listeria monocytogenes* and *Yersinia enterocolitica* using Equation (7.2) as a model.

Microorganism	CO_2 pressure MPa (atm)	Temperature (°C)	$b(P,T)$ (min^{-n})	$n(P,T)$	Mean square error (MSE)
Listeria monocytogenes	1.5 (15)	25	0.006 (0.001)	1.41 (1.50)	0.160 (1.85)
		35	0.001 (0.003)	1.92 (1.50)	0.143 (1.85)
		45	0.007 (0.005)	1.54 (1.50)	0.117 (1.85)
	3.04 (30)	25	0.0015 (0.004)	1.50 (1.50)	0.089 (1.85)
		35	0.002 (0.006)	1.58 (1.50)	0.120 (1.85)
		45	0.004 (0.009)	1.57 (1.50)	0.319 (1.85)
	6.08 (60)	25	0.005 (0.006)	1.55 (1.50)	0.086 (1.85)
		35	0.017 (0.011)	1.40 (1.50)	0.102 (1.85)
		45	0.043 (0.016)	1.25 (1.50)	0.122 (1.85)
Yersinia enterocolitica	1.5 (15)	25	0.0004 (0.0009)	1.97 (1.80)	0.121 (1.45)
		35	0.002 (0.002)	1.75 (1.80)	0.060 (1.45)
		45	0.003 (0.003)	1.82 (1.80)	0.058 (1.45)
	3.04 (30)	25	0.0006 (0.002)	2.07 (1.80)	0.104 (1.45)
		35	0.003 (0.004)	1.85 (1.80)	0.170 (1.45)
		45	0.012 (0.008)	1.68 (1.80)	0.308 (1.45)
	6.08 (60)	25	0.006 (0.01)	1.93 (1.80)	0.159 (1.45)
		35	0.011 (0.02)	2.00 (1.80)	0.154 (1.45)
		45	0.02 (0.05)	2.17 (1.80)	0.276 (1.45)

Table 7.4 Parameters of inactivation of *yeasts* using Equation (7.4) as a model.

	CO_2 pressure MPa (bar)	Temperature (°C)	$k_1(P,T)$ (–)	$k_2(P,T)$ (min)	Mean square error (MSE)
Pichia awry	8.5 (85)	40	5.93	0.38	0.065
Wild yeast	8.0 (80)	35	5.05	0.05	0.089

Table 7.5 Parameters of inactivation of *Lactobacillus plantarum* using Equation (7.6) as a model.

CO_2 pressure MPa (kg/cm^2)	Temperature (°C)	$A(P,T)$ (–)	$B(P,T)$ (min^{-1})	$C(P,T)$ (min)	Mean square error (MSE)
4.88 (50)	30	5.69	0.03	75.9	0.013
5.86 (60)	30	5.66	0.08	24.0	0.038
6.83 (70)	20	4.28	0.03	56.0	0.005

the organism's survival curve can be described by the Weibullian model, of which the traditionally assumed first-order mortality kinetics is just a special case. Like in other modes of inactivation, certain treatments, when applied to specific organisms, can exhibit an initial stage of relatively fast and massive mortality followed by a stage where a substantial

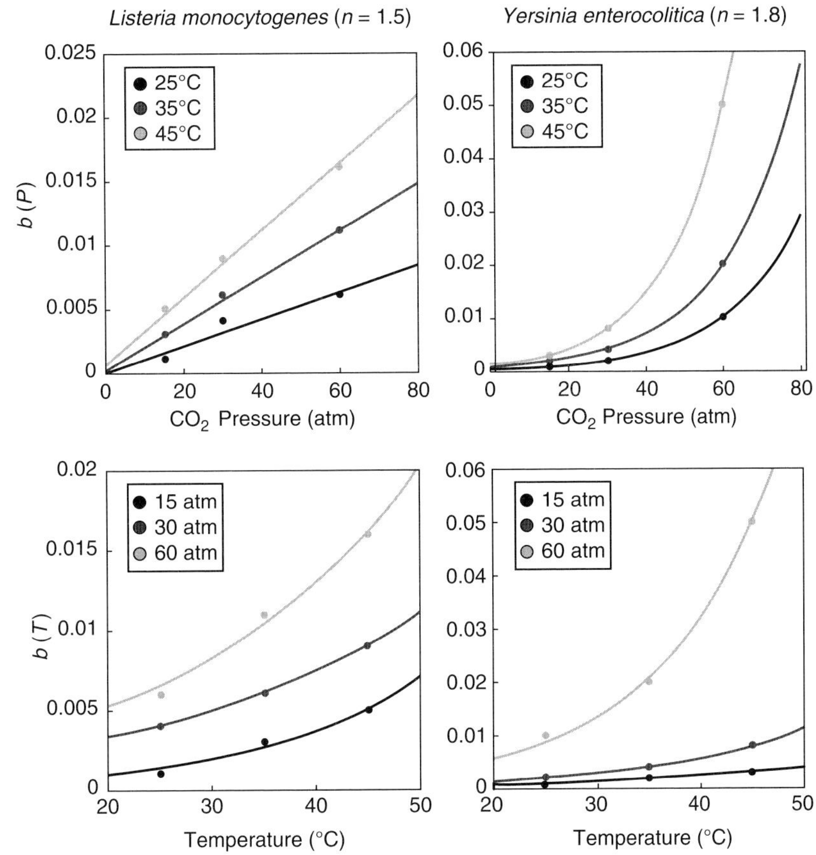

Figure 7.10 The log-logistic pressure and temperature dependencies of the Weibullian rate parameter, $b(P,T)$, of *Listeria monocytogenes* and *Yersinia enterocolitica*. The solid lines are the fit of Equation (7.7) to data calculated after fixing the value of $n(P)$ or $n(T)$ – see Table 7.3. Notice that the straight line (top-left plot) can be viewed as a special case of the model with $P_c = 0$.

number of survivors remain practically unaffected for a relatively long time. Certain microorganisms might exhibit a sigmoid survival curve of the Gompertz type, characterized by a short pause ('shoulder') followed by a stage of fast mortality, which may end in a constant number of survivors. These two 'non-Weibullian' patterns can also be described mathematically by kinetic models that have been originally proposed for thermal inactivation.

None of the models mentioned in this chapter has been derived from monitoring events at the cellular or molecular level. Therefore, they are purely phenomenological and are applicable only at the population level. Although able to adequately describe experimental results, they are not unique and their coefficients cannot be directly related to mechanisms that operate within the cells, unless supplemented by information obtained by independent assays. The discussed models, as explained, do not indicate whether adaptation, injury and recovery from injury play any role in shaping the observed survival pattern. Consequently, the survival curve's shape and the mathematical formula which describes it carry only a limited amount of information. What the models can do is to quantify the overall *manifestation* of biochemical and biophysical mechanisms tiggered by the pressurized CO_2 which determines the population's response to the stress induced. Such mechanisms, it should be reiterated, can be revealed only by direct analytical and microscopic methods, not by the mathematical properties of the survival model. In other words, mechanistic characterization of microbial mortality under high CO_2 pressure and mathematical characterization of survival patterns should be viewed as complementary endeavours and not as substitutes. Hopefully, future research would reveal whether the origin of certain survival patterns, those that are expressed in the survival curve's concavity direction for example, could be traced to specific inactivation mechanisms and/or the cells' genetic or physiological characteristics. The same can be said about the treatment's medium and the cells' history or growth stage. These too can affect both the survival pattern and what happens within the exposed cells. Only if the existence of a specific relationship between the cellular and population levels could be revealed, the survival curve's shape would become a tool to probe the mechanism of cell mortality when exposed to CO_2 under high pressure.

7.5 List of symbols

- a_i — Equation (7.8) constants determined by generalized linear regression
- $A(P,T)$ — pressure- and temperature-dependent coefficient in Equation (7.6)
- $b(P,T)$ — pressure- and temperature-dependent coefficient in Equation (7.2)
- $b(P)$ — magnitude of b at pressure P in Equation (7.7)
- $B(P)$ — pressure- and temperature-dependent coefficient in Equation (7.6)
- $C(P)$ — pressure- and temperature-dependent coefficient in Equation (7.6)
- $D(P,T)$ — decimal reduction parameter: time needed to reduce the targeted microorganism's number by a factor of 10
- D_p — reciprocal of $k(P)$ in Equation (7.1)
- $k(P,T)$ — CO_2 pressure- and temperature-dependent exponential inactivation rate constant; (1/s)
- $k'(P)$ — slope of the semilogarithmic survival curve at $t >> t_c(P)$ in Equation (7.3)

k''	survival parameter of the organism in Equation (7.7)
$k_1(P)$	pressure- and temperature-dependent coefficient
$k_2(P)$	pressure- and temperature-dependent coefficient
$k_{1i}(P)$	pressure- and temperature-dependent coefficient
$k_{2i}(P)$	pressure- and temperature-dependent coefficient
$n(P,T)$	pressure- and temperature-dependent coefficient in Equation (7.2)
P	pressure; bar
P_c	survival parameter of the organism Equation (7.7)
$S(t)$	survival ratio
t	time; min
t^*	time that corresponds to the momentary survival ratio in Equation (7.11); min
$t_c(P)$	pressure-dependent marker of the shoulder length in Equation (7.3); min
T	temperature; K
z_p	number of MPa's (or other pressure units) needed to change the 'D_p value' by a factor of 10

8 Applications of DPCD to Juices and Other Beverages

Murat O. Balaban and Xiaojun Liao

Abstract: Carbon dioxide can be used in the food industry because of its nontoxic, nonflammable, inexpensive, and generally regarded as safe (GRAS) status. A technology applied to foods is first and foremost concerned with the safety of the products. The literature in this area shows that DPCD treatment assures the microbial safety of the product, and this can be predicted to a satisfactory degree. The quality attributes of the processed foods are also taken into account, and these include physical (color, viscosity, particle size, Brix, etc.) and chemical (pH, flavor volatiles, and composition) properties. Enzyme activity, nutritional quality, and shelf life, including sensory properties, are also studied. Finally, technical and economic feasibility is also addressed regarding commercial implementation of the technology. This present chapter reviews the safety and quality aspects of applying DPCD technology to juices and beverages without going into the theory and mechanics of microbial or enzyme inactivation.

Keywords: juices; beverages; antimicrobial properties; quality attributes.

8.1 Introduction

Carbon dioxide (CO_2) is used in the food industry because of its nontoxic, nonflammable, inexpensive and generally regarded as safe (GRAS) status. Dense phase carbon dioxide (DPCD), a collective term for gaseous CO_2 under pressure, liquid CO_2 and supercritical CO_2, possesses antimicrobial and enzyme-inactivating properties. This chapter will briefly review the application of DPCD on juices and other beverages.

New technologies applied to foods are first and foremost of concern because of safety implications. Then, as data accumulate and microbial safety can be predicted to a satisfactory level, other concerns are addressed. These include quality attributes that involve physical (colour, viscosity, particle size, Brix etc) and chemical (pH, flavour volatiles and composition) properties. Enzyme activity, nutritional quality and shelf life including sensory properties are also studied. Finally, the technical and economic

Dense Phase Carbon Dioxide: Food and Pharmaceutical Applications, First Edition. Edited by Murat O. Balaban and Giovanna Ferrentino.

feasibility is addressed towards commercial implementation of the technology. This chapter will review the safety and quality involving the application of DPCD to juices and beverages, without going into the theory and mechanics of the microbial or enzyme inactivation, or other predictive areas, since these are addressed in other chapters of this book. The liquid foods discussed in this chapter are divided into juices and other beverages.

8.2 Juices processed with DPCD

8.2.1 *Orange juice*

Orange juice (OJ) is one of the most popular fruit juices in the world, with high nutritional value and desirable sensory attributes. Commercial orange juice is thermally processed to inactivate the enzyme pectinesterase (PE) that causes loss of cloud. Typically, pasteurization is done at 90°C for 10 sec and is also sufficient for microbial inactivation. However, this causes undesirable flavour and aroma changes. The effects of DPCD on OJ have been studied using static (batch) and continuous-flow DPCD systems.

8.2.1.1 Static DPCD equipment

Arreola *et al.* (1991) measured quality attributes (pH, Brix, cloud stability, total acidity, colour, ascorbic acid content and sensory attributes) of Valencia OJ treated with DPCD at 7–34 MPa, 35–60°C and 15–180 min in a batch system. They showed that there was no significant difference in pH or °Brix of the original juice and DPCD-treated juice ($p<0.01$). Ascorbic acid (AA) retention was significantly higher in DPCD-treated samples (71% to 95%) compared to temperature controls (samples held at the same experimental temperature as DPCD-treated samples for the same amount of time) (62% to 83%) ($p<0.01$) (Figure 8.1). Higher AA retention in DPCD samples was explained by the higher stability of AA under low pH provided during DPCD and also under the O_2-excluded environment. AA oxidizes easily when oxygen is present.

Cloud was enhanced in DPCD samples from 1.27 to 4.1 times (Figure 8.2), and was stable even in the presence of residual PE. This showed that PE inactivation is not the only reason for cloud enhancement and stability by DPCD. Shearing and homogenization through the expansion valve may cause size reduction in suspended particles in the juice. Cloud stability of DPCD samples (29 MPa, 50°C, 4 h) was retained after 66 days of refrigerated storage, whereas temperature control (50°C, 4 h) and room temperature control (25°C, 4 h) lost the cloud completely.

Instrumentally measured colour indicated higher brightness of DPCD-treated OJ compared to fresh untreated OJ. Sensory evaluations of untreated, DPCD and commercial but unpasteurized samples by 30 untrained panellists showed that overall acceptability, flavour and aroma

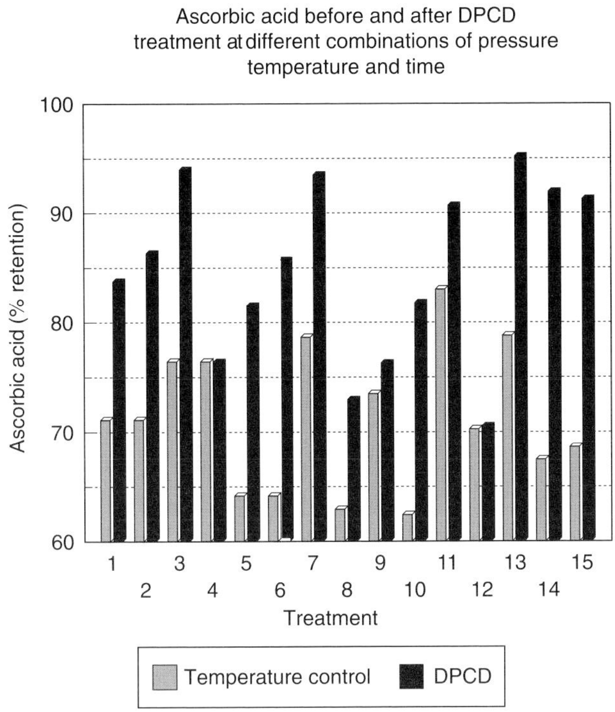

Treatment combinations			
#	P (atm)	T °C	t (min)
1	129.3	44	50
2	265.3	44	50
3	129.3	56	50
4	265.3	56	50
5	129.3	44	145
6	265.3	44	145
7	129.3	56	145
8	265.3	56	145
9	81.6	50	97.5
10	312.9	50	97.5
11	197.3	40	97.5
12	197.3	60	97.5
13	197.3	50	15
14	197.3	50	180
15	197.3	50	97.5

Figure 8.1 Retention of ascorbic acid after various DPCD treatments of fresh-squeezed orange juice. SCT: supercritical treatment. (Adapted from Arreola 1990.)

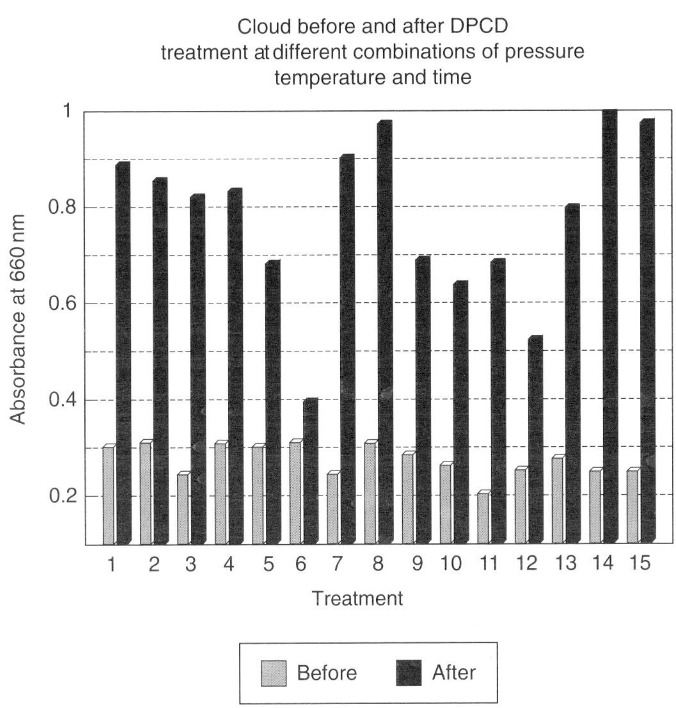

Treatment combinations			
#	P (atm)	T °C	t (min)
1	129.3	44	50
2	265.3	44	50
3	129.3	56	50
4	265.3	56	50
5	129.3	44	145
6	265.3	44	145
7	129.3	56	145
8	265.3	56	145
9	81.6	50	97.5
10	312.9	50	97.5
11	197.3	40	97.5
12	197.3	60	97.5
13	197.3	50	15
14	197.3	50	180
15	197.3	50	97.5

Figure 8.2 Cloud of orange juice before and after treatment with DPCD, at various conditions. (Adapted from Arreola 1990.)

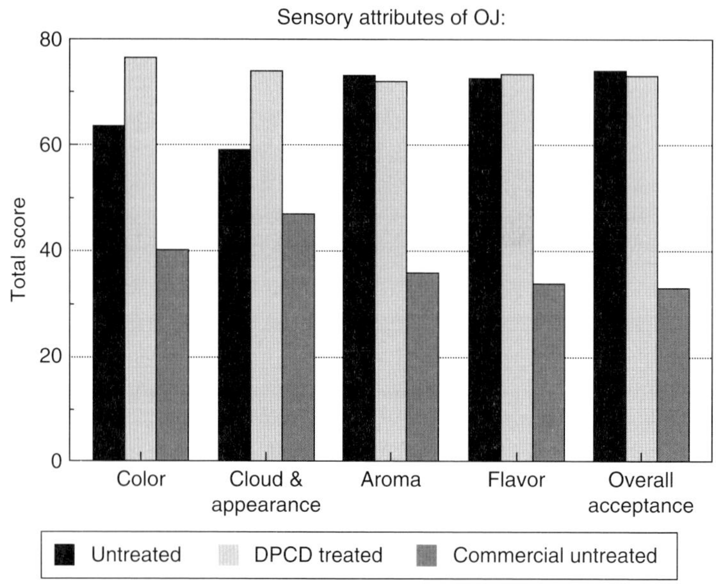

Figure 8.3 Various sensory attributes of DPCD-treated, control and commercial untreated orange juice samples. (Adapted from Arreola 1990.)

of fresh and DPCD-treated orange juice were not significantly different (Figure 8.3). Commercial untreated samples had significantly less acceptability, flavour and aroma. The colour, cloud and appearance of DPCD-treated OJ were liked more than those of fresh or commercial OJ.

Niu *et al.* (2010a) treated OJ in a DPCD process under 40 MPa at 55°C between 10 and 60 min. A control was thermally treated at 90°C for 60 s. Particle size, consistency coefficient and a* value tended to become smaller as the DPCD treatment time was extended. The proportion of the small particles increased after DPCD treatment, suggesting that the increase in cloud and stability of orange juice treated by DPCD is primarily due to homogenization by DPCD (Figure 8.4).

Zhou *et al.* (2009a) studied the effects of DPCD on the activity and structure of pectin methylesterase (PME) extracted from Valencia orange peel. The extract was dissolved in phosphate buffer and NaCl was added. This solution was treated with DPCD (8–30 MPa, 55°C and 10 min) and heat (55°C for 10 min). DPCD caused significant inactivation of PME, with the lowest residual activity about 9.3% at 30 MPa. The SDS-PAGE electrophoretic behaviour of HPCD-treated PME was not altered; however, changes in the secondary and tertiary structures were detected. The β-structure fraction in the secondary structure decreased as HPCD pressures increased. After 7-day storage at 4°C, no alteration of PME activity and no reversion of its β-structure fraction were observed.

Niu *et al.* (2010a) also studied the volatile components (ethyl butyrate, trans-2-hexenol, α-pinene, phellandrene, limonene, linalool, nonanal and

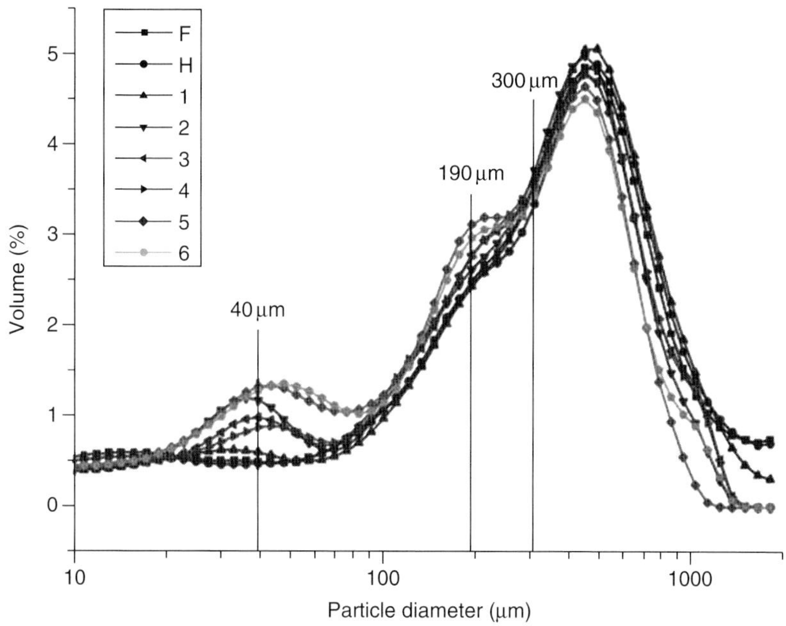

Figure 8.4 Changes in particle size distribution in DPCD-treated and thermally treated orange juice samples (Niu *et al.* 2010a). F: fresh-squeezed juice; H: thermally treated juice; and 1, 2, 3, 4, 5 and 6: treatment times representing 10, 20, 30, 40, 50 and 60 min of treatment with DPCD. (Reproduced with permission of Elsevier.)

citronellol) of the OJ treated in different ways. The relative contents of ethyl butyrate and trans-2-hexenol decreased linearly during 10–60 min of DPCD treatment, and their linear fits with higher correlation coefficients (R^2 = 0.968 and 0.804) suggested a good prediction for the DPCD process. The relative contents of nonanal and citronellol after DPCD treatment increased, higher than that after thermal treatment or even higher than that in fresh juice (Figure 8.5).

The authors found that the flow behaviour of the OJ fit the Herschel–Bulkley model. The consistency coefficient, K, and the flow behaviour index, n, were found to be constants and time-independent properties within the first 30 min and the second 30 min, respectively, and the consistency coefficient in the second period was about 28% lower than that in the first period. This may be significant in energy savings for commercial applications of the DPCD process.

8.2.1.2 Continuous-flow DPCD system

The sensory, nutritional and physical quality enhancement in orange juice by DPCD was reported by Kincal (2000) in a continuous-flow system with treatment conditions: pressures 38, 72 and 107 MPa, CO_2–juice ratios 0.40 to 1.18, and residence time 10 min. The cloud of orange juice was enhanced from 446% to 846% after DPCD treatments compared to controls. No significant changes were observed in pH and °Brix. However, %titratable

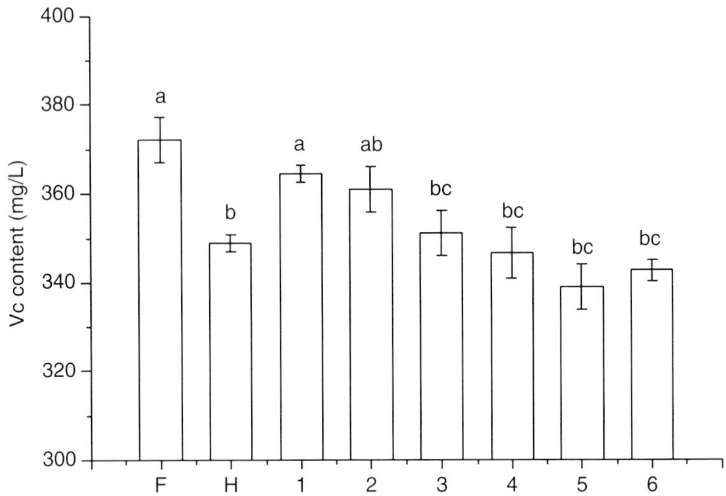

Figure 8.5 Volatiles content of the DPCD-treated, fresh-squeezed and thermally treated orange juice samples (Niu *et al.* 2010a). F: fresh-squeezed juice; H: thermally treated juice; and 1, 2, 3, 4, 5 and 6: treatment times representing 10, 20, 30, 40, 50 and 60 min of treatment with DPCD. Different letters in the same column indicate significant differences ($P < 0.05$). (Reproduced with permission of Elsevier.)

acidity (TA) of DPCD-treated juice was significantly higher than control juice ($p < 0.01$). Small but significant increases in L* and a* values were obtained after DPCD treatments. No significant differences were observed for sensory attributes between fresh OJ and DPCD-treated OJ after 2 weeks of refrigerated storage at 1.7°C.

Ho (2003) reported on Praxair's research with orange juice using a continuous DPCD system. *Listeria innocua* (ATCC 33090) was used as a surrogate in inoculated orange juice studies regarding the kinetics of DPCD processing. Results indicated that inactivation of bacteria by DPCD processing was primarily enhanced by an increase in CO_2 concentration. Processing pressure, to a lesser extent, also contributed to the effectiveness of the DPCD. No significant differences between the physical attributes (pH, Brix, and titratable acidity), nutritional content (vitamin C and folic acid) and aroma profile of the untreated and treated fresh-squeezed orange juice were detected. The bactericidal effects of the DPCD processing were not affected by pulpy and high-Brix juice.

8.2.2 Apple juice

Many studies exist regarding the various effects of DPCD on apple juice. For example, no visual changes or modifications in total soluble solid content of apple cider have been found (Van Ginneken *et al.* 2002). Gui *et al.* (2006b) reported a significant reduction of the browning degree in DPCD-treated cloudy apple juice processed at 55°C, 30 MPa for 60 min, and stored at 4°C. The sensory modification caused by DPCD treatment in apple juice was studied by Gasperi *et al.* (2009). Difference-from-control and ranking tests were performed on fresh untreated control and DPCD-treated juices. No significant differences were detected between the samples.

Niu *et al.* (2010b) reported on the quality of cloudy apple juice from apple slices treated by DPCD (temperatures 25°C to 65°C, time 20 min, pressure 20 MPa) and mild heat (MH). They stated that polyphenol oxidase (PPO) was completely inactivated by DPCD. However, the minimal residual activity (RA) by MH at 65°C was 38.6%. RA of pectin methylesterase (PME) with DPCD (18%) was significantly lower than that of MH. L value of cloudy apple juice from DPCD-treated apple slices was significantly higher than that from MH-treated apple slices, and the b value, browning degree (BD) and turbidity were lower. The authors found no differences in the a value, total soluble solids, pH and conductivity. After 7-day storage at 4°C, DPCD showed no change in BD but there was a significant turbidity loss. MH increased BD at 55°C and 65°C, and led to turbidity loss from 35°C to 65°C. The turbidity was not well related to RA of PME.

Ferrentino *et al.* (2009a) treated apple juice prepared from 'Annurca' apple puree with a DPCD batch system. The pH, °Brix, colour parameters and microbial load of the treated apple juice were compared with those of thermally processed juice. Thermal processes were performed at 35°C, 50°C, 65°C and 85°C at treatment times ranging between 10 and 140 min. Microbial inactivation kinetics showed that 5-log reduction of natural flora in apple juice was achieved at 85°C and 60 min for thermal treatment. The same decrease was observed at 16.0 MPa, 60°C and 40 min for HPCD treatment. Temperature was important for HPCD efficiency, with inactivation significantly enhanced when it increased from 35°C to 60°C. The effect of pressure was less significant. Also, 5-log reduction of natural flora in apple juice was obtained at lower temperatures by cyclic treatments of six compression and decompression steps. There were no significant differences between treated and untreated samples in °Brix ($\alpha = 0.05$). Significant differences were detected in pH values between the untreated and HPCD-treated samples ($\alpha = 0.05$). L* and b* values decreased significantly, and differences were detected in a* values between the untreated and the HPCD-treated samples ($\alpha = 0.05$). °Brix, pH and colour showed no statistically significant differences between the untreated and HPCD-treated samples in the first 2 weeks of storage at 4°C.

Alicyclobacillus acidoterrestris is of special interest to the fruit juice industry as common pasteurization techniques do not deactivate the spores. Bae *et al.* (2009) investigated the lethal effect of DPCD (temperature: 65°C, 70°C, pressure: 80, 100, 120 bar, time: 10–40 min) on *Alicyclobacillus acidoterrestris* spores (10^6–10^7 spores/ml) suspended in apple juice. *A. acidoterrestris* spores were completely inactivated by DPCD to undetectable levels above 65°C, 100 bar for 40 min and 70°C, 80 bar for 30 min. DPCD did not affect ($p > 0.05$) the pH and °Brix of apple juice. In electron microscopic observations, the surface and internal morphological changes and extraction of intracellular materials of the treated spores were observed.

Da Porto *et al.* (2010) reported on the effects of different operating conditions applied to a continuous DPCD system on antioxidant capacity and volatile compounds of apple juice compared with a heat pasteurization method. Only the microbial tests required by Italian Regulation to assure microbiological safety to 'fruit juices and vegetables' were performed, and DPCD-treated (15 MPa, 35°C, 15 min; 25 MPa, 35°C, 15 min), heat-pasteurized (75°C, 15 s) and untreated apple juice samples were considered. The DPCD processing carried out at 15 MPa proved as effective as that at 25 MPa in reducing microbial cells. Trolox equivalent of DPCD treated at 25 MPa resulted significantly ($P < 0.05$) lower than DPCD-treated samples at 15 MPa. Head space analysis of volatile compounds indicated the lowest decrease in apple aroma compounds (59% esters and 59% aldehydes) in DPCD treated at 15 MPa.

8.2.3 Mandarin juice

Mandarin cultivar 'Murcott' (*Citrus reticulata* Blanco) was processed in a continuous-flow DPCD system (Lim *et al.* 2006). Response surface methodology was used to investigate the effects of the processing parameters such as temperature, pressure, residence time and % (w/w) ratio of CO_2 to juice on total aerobic count (TAC), pectinesterase (PE) activity, cloud, °Brix, pH and titratable acidity (TA) of the juices. Maximum log reduction (3.47) of TAC was observed at 35°C, 41.1 MPa, 9 min of residence time, and 7% CO_2. PE was inactivated by 7~51%. Cloud was not only retained but also enhanced by 38%. Lightness and yellowness (L* and b*) increased, and redness (a*) decreased. The processing temperature and %CO_2 significantly affected DPCD processing of the juice in terms of pasteurization, PE inactivation, cloud increase and colour change. The °Brix, pH and TA before and after treatment were the same.

8.2.4 Grapefruit juice

Fresh-squeezed red-blush (*Citrus paradisi*) grapefruit juice was treated with continuous DPCD equipment to inactivate yeasts and moulds and total aerobic microorganisms (Ferrentino *et al.* 2009b). A central composite design was used with pressure (13.8, 24.1 and 34.5 MPa) and residence time (5, 7 and 9 min) as variables at constant temperature (40°C), and CO_2 level (5.7%) after experimentally measuring CO_2 solubility in the juice. Five log reduction for yeasts and moulds and total aerobic microorganisms occurred at 34.5 MPa and 7 min of treatment. A storage study was performed on the fresh juice that was DPCD treated at these conditions. °Brix, pH, titratable acidity (TA), pectinesterase (PE) inactivation, cloud, colour, hue tint and colour density, total phenolics, antioxidant capacity and ascorbic acid were measured after the treatment and during 6 wk storage at 4°C. During storage, the DPCD-treated juice showed no growth of total aerobic microorganisms and yeasts and moulds. Cloud increased (91%) while percentage of PE inactivation was partial (69.17%). No significant ($\alpha = 0.05$) differences were detected between treated and untreated samples for °Brix, pH and TA. Treated juice had higher lightness and redness and lower yellowness. No significant differences ($\alpha = 0.05$) were detected for the hue tint values while the colour density value was higher for the treated samples compared to the untreated. The treatment and the storage did not affect the total phenolic content of the juice. Slight differences were detected for the ascorbic acid content and the antioxidant capacity. The experimental results showed evidence that the DPCD treatment can maintain the physical and quality attributes of the juice, extending its shelf life and safety.

8.2.5 Watermelon juice

Watermelon (*Citrullus lanatus*) juice was extracted by a filter press (Lecky and Balaban 2004). The juice was treated by DPCD (pressure 10.3, 20.6 and 34.4 MPa; CO_2 ratio 5%, 10% and 15%; residence time 4, 5 and 6 min; temperature room T, 30°C and 40°C) and by thermal control. The effects on natural flora, pH, Brix, TA, lycopene and colour were measured. Taste panels were also conducted. The flavour profile of the treated samples was analysed by GC-MS and by GC-O.

It was determined on fresh watermelon juice that using 34.4 MPa, 40°C, 10% CO_2 in juice, and 5 min residence time, could reduce native aerobic microorganisms by 6 log cycles. In an acidified, sweetened and carbonated product (pH 4.3, °Brix 10.5), using the same processing parameters above could achieve a 4.5 log reduction. This level of treatment allowed the juice to be stored for 8 weeks at refrigerated conditions without major spoilage or unacceptable taste panel results.

During the 8-week storage study, °Brix, pH, titratable acidity, colour, lycopene, flavour and aroma of the HPCD-treated sample were evaluated and compared to a flash-pasteurized (74°C for 15 sec) sample. °Brix, pH and titratable acidity did not change after treatment or during storage. The colour data showed that HPCD-treated juice L*, a* and b* values were about 31, 27 and 24 respectively, and stayed stable over the total shelf life. The heat-pasteurized samples had average L*, a* and b* values of about 36, 19 and 18 respectively.

On week 0, the taste panel determined that the taste and aroma of the heat-treated and HPCD-treated samples had the same difference from control (untreated fresh juice); they could also be differentiated from the hidden control. However, on week 8, the taste panel could not differentiate the taste or aroma of the treated samples from the hidden control.

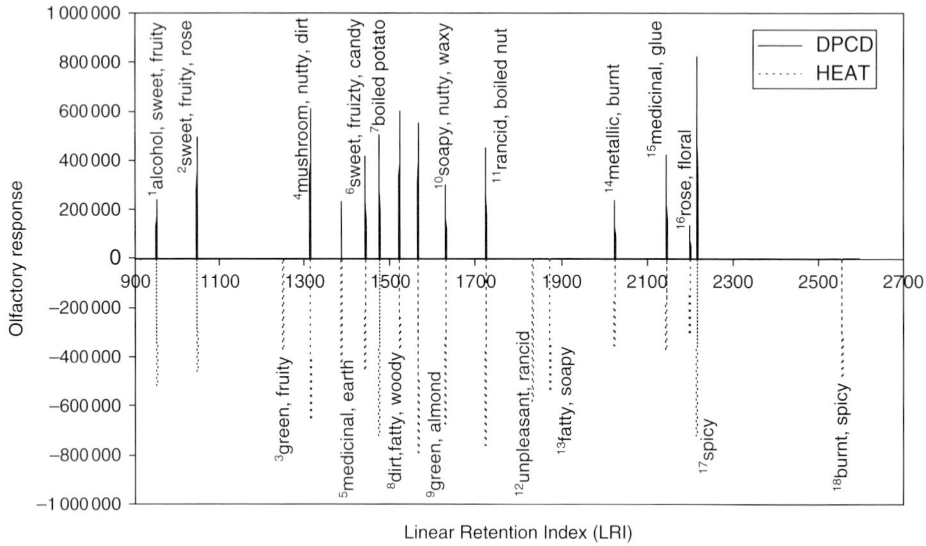

Figure 8.6 Aromagram representing the olfactory response of sniffers to coconut water flavour volatiles as they emerge from the olfactory port of a GC-O (Damar 2006).

Liu *et al.* (2011) used a semibatch DPCD system to treat watermelon juice to inactivate PPO. At 30 MPa and 50°C for 30 min, the PPO activity was reduced by 95.8%.

8.2.6 Coconut water

Damar *et al.* (2009) investigated the consumer likeability and flavour profile of a continuous-flow DPCD-processed (34.5 MPa, 25°C, 13% CO_2, 6 min) coconut water beverage in comparison to that of fresh-untreated and heat-processed (74°C, 15 sec) samples. Sensory panels that were conducted throughout 9 weeks of refrigerated storage (4°C) showed that DPCD-treated and fresh coconut water beverages were liked similarly, whereas heat-treated coconut water beverage was liked significantly less ($p > 0.05$). Aroma difference from control (untreated-fresh) scores of heat- and DPCD-treated samples was not significantly different, but taste difference from control scores was significantly different ($p > 0.05$). Heat-treated samples were rated with significantly higher off-flavour scores than fresh or DPCD-treated samples ($p > 0.05$). The gas chromatography–olfactory analysis of flavour compounds in DPCD and heat-processed coconut water beverages were conducted, and an aromagram was developed by taking average area of the olfactory responses with the aroma descriptors given by the sniffers. This aromagram showed that there were differences in the aroma profiles of DPCD- and heat-treated samples (Figure 8.6). Heat-treated samples had more aroma compounds that were described as green, fruity, nutty, rancid, unpleasant, fatty and burnt aromas. These aromas could be possibly developed by decomposition of aroma compounds during heating.

8.2.7 Guava puree

Guava (*Psidium guajava* L.) is a fruit with a unique tropical flavour. It is an excellent source of nutrients, phytochemicals and antioxidants, especially ascorbic acid. Guava puree is a common product that is pasteurized to extend its shelf life, but pasteurization has negative effects on sensory and nutritional quality. Plaza (2010) investigated if the use of DPCD would minimize or prevent undesirable changes in phytochemical composition compared to traditional heat pasteurization. Measurement and comparison of the chemical composition of the guava puree subjected to both treatments and an untreated guava puree (freshly thawed control) were conducted.

Microbial reduction was quantified as a function of pressure and residence time using 8% CO_2 and a temperature of 35°C. Optimum DPCD treatment conditions for microbial inactivation were 34.5 MPa for 6.9 min and 8% CO_2 at 35°C. Quality attributes, including pH, °Brix, % titratable acidity (%TA) and colour of DPCD-treated, freshly thawed and heat-pasteurized (90°C for 60 s) guava puree, were measured and compared throughout refrigerated storage (4°C for 14 weeks). DPCD did not cause a change in pH or °Brix but increased the titratable acidity and viscosity of the product. Pectinesterase (PE) was partially inactivated by DPCD treatment. DPCD-treated guava puree retained organic acids similar to fresh guava puree and served to protect polyphenolic and antioxidant levels throughout processing and storage. DPCD delayed the degradation of vitamin C during storage.

Flavour and aroma compounds in guava puree were identified using GC-MS. Flavour profiles showed that heat-treated guava puree had less aroma active compounds than DPCD-treated guava puree. Volatile compounds analysis showed a lower total peak area for the DPCD when compared to fresh and pasteurized (possibly due to stripping during the depressurization step) and differences in volatile composition were found for the three samples.

8.2.8 Grape juice

Grape juice is usually preserved by thermal pasteurization, and chemical preservatives are added in grape juice products and wine. However, these methods may have adverse effects on the nutritional and sensory qualities of grape juice. There are studies showing that DPCD can effectively inactivate yeasts that would cause spoilage of the juices and formation of off-flavours during fermentation of the wines.

The effects of DPCD processing parameters, including temperature (25°C and 35°C), CO_2 concentration (0, 85 and 170 g/kg) and pressure

Table 8.1 The effect of heat (75°C for 15 sec) or DP- CO_2 (D-1: 34.5 MPa, 8% CO_2; D-2: 34.5 MPa, 16% CO_2) pasteurization on the total anthocyanin, soluble phenolic and antioxidant content of unprocessed muscadine grape juice.

Treatment	Total anthocyanins (mg/L)	Soluble phenolics (mg/L)	Antioxidant capacity (μM TE/mL)
Unprocessed	1,105 a[1]	2,211 a	22.1 a
DP-1 (34.5 MPa, 8% CO_2)	1,077 a	2,213 a	20.7 a
DP-2 (34.5 MPa, 16% CO_2)	1,102 a	2,157 b	21.7 a
HTST (75°C, 15 sec)	866 b	1,859 c	18.2 b

[1] Values with similar letters within columns are not significantly different (LSD test, $P > 0.05$).
(Reproduced from del Pozo-Insfran *et al.*, 2006. Copyright 2006, with permission of American Chemical Society.)

(6.9, 27.6 and 48.3 MPa), on yeast inactivation and sensory properties of grape juice were investigated (Gunes *et al.* 2005). The DPCD process resulted in more than a 6 log reduction in yeast population. As the CO_2–juice ratio, temperature and pressure increased, the inactivation rate increased. CO_2 in the supercritical state was more effective in inactivating yeast than in the subcritical state. The process did not cause detectable flavour degradation.

Del Pozo-Insfran *et al.* (2006) pasteurized muscadine grape juice with DPCD (34.5 MPa, 8 or 16% CO_2) and heat treatments (75°C, 15 sec), and measured the amount of phytochemicals, antioxidant capacity and sensory attributes compared to unprocessed and heat-pasteurized grape juice. Results showed that thermal pasteurization decreased anthocyanins by 16%, soluble phenolics by 26% and antioxidant capacity by 10%, whereas DPCD-treated grape juice retained fresh-like levels of these (Table 8.1).

DPCD-treated grape juice retained higher amounts of total anthocyanins, antioxidant capacity (Figure 8.7) and total soluble phenolics (Figure 8.8) than thermally pasteurized juice at the end of 10 weeks refrigerated storage at 4°C. DPCD-treated grape juice, however, showed increase in the microbial counts starting from 5 weeks storage whereas there was no increase in microbial counts of the heat-pasteurized sample.

Sensory evaluation of DPCD- and heat-treated, and unprocessed control, grape juices were done by 60 untrained panellists. A continuous line scale from 0 = no difference to 10 = extremely different was used to rate the difference of each sample from the reference (fresh-control) for colour, flavour and aroma. Overall likeability of each sample was rated in a 9-point hedonic scale. Results showed that panellists liked DPCD-treated samples significantly more than heat-treated samples, with no significant difference between the DPCD and fresh sample. Flavour and aroma of DPCD-treated grape juice were not significantly different than those of fresh juice, whereas heat-pasteurized samples were significantly different

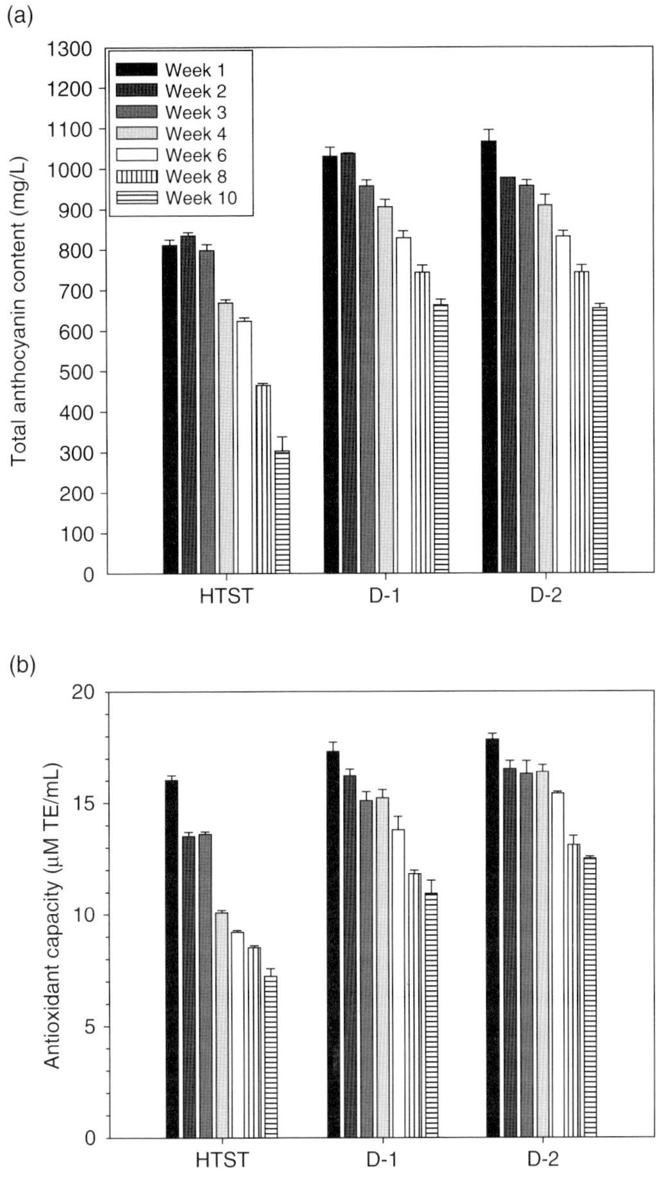

Figure 8.7 Changes in the anthocyanin content and antioxidant capacity of Muscadine grape juice treated with DPCD and thermally treated. HTST: 75°C, 15 sec, D-1: 34.5 MPa, 8% CO_2, D-2: 34.5 MPa, 16% CO_2. (Reproduced from del Pozo-Insfran *et al.*, 2006. Copyright 2006, with permission of American Chemical Society.)

than fresh grape juice. Colour of the fresh, heat-treated and DPCD-treated grape juice was not significantly different from each other. Therefore, DPCD treatment provided fresh-like quality in the first 5 weeks of refrigerated storage and offered advantages over thermal pasteurization.

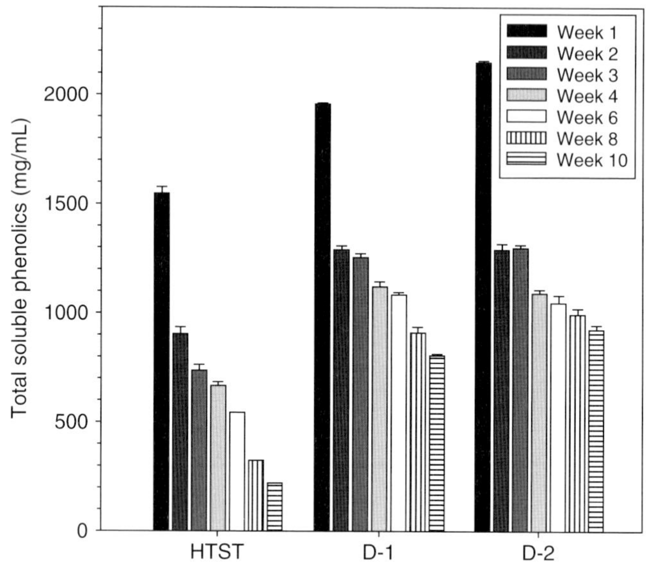

Figure 8.8 Changes in the total soluble phenolics of Muscadine grape juice treated with DPCD and thermally treated. HTST: 75°C, 15 sec, D-1: 34.5 MPa, 8% CO_2, D-2: 34.5 MPa, 16% CO_2. (Reproduced from del Pozo-Insfran *et al.*, 2006. Copyright 2006, with permission of American Chemical Society.)

8.2.9 Pear

Valverde *et al.* (2010) explored the use of DPCD to inactivate *Saccharomyces cerevisiae* in fresh-cut conference pears (*Pyrus communis*). Continuous CO_2 flow at different pressures (6–30 MPa), temperatures (25–55°C) and exposure times (10–90 min) were used. Heat-treated controls at similar temperatures and times were used. Inactivation of 5 $\log_{10}$ cycles of the yeast occurred at 55°C with HPCD while 70°C had to be reached in the case of heat. DPCD pressures and exposure times were relatively low (66 MPa and on the order of minutes, respectively) due to the direct contact between the CO_2 and the pear. The pH and °Brix were not affected by the DPCD treatment; however, the pears lost their texture and became darker due to a decrease in vitamin C and enzymatic browning. Peroxidase activity was partially reduced. The addition of an antioxidant did not help to prevent darkening.

8.2.10 Carrot

Zhou *et al.* (2009c) studied the inactivation of pectin methylesterases (PMEs) from carrot and peach in buffer by DPCD at 55°C. Both PMEs were inactivated by HPCD, and their residual activity (RA) decreased with increasing pressure. Their inactivation kinetics was described by a fractional-conversion model. The nonzero RA of the two PMEs was 6–7%. With increasing pressures the kinetic rate constant k increased, and the decimal reduction time D decreased for the HPCD-labile fraction of the PMEs. The labile fraction of carrot PME was more sensitive to HPCD than that of peach PME. The activation volume, V_a, and Z_P (the temperature increase needed for a 90% reduction of D) were 1079.37 cm^3/mol and 5.80 MPa for carrot PME, and 130.51 cm^3/mol and 48.31 MPa for peach PME.

8.2.11 Carrot juice

The effect of DPCD on the quality of carrot juice was also investigated by Zhou *et al.* (2009b). Carrot juice is one of the most popular vegetable juices (Marx *et al.* 2003), and it is preferably used as a natural source of provitamin A. To determine the effect of HPCD on the quality of carrot juice, the pressure used was 10, 20 and 30 MPa, the treatment time was 15, 30, 45 and 60 min, and the temperature was 25°C. The lightness (*L*-value) of HPCD-treated juices increased significantly compared to untreated juices, and the redness (*a*-value) exhibited an increase with increasing treatment time. However, the yellowness (*b*-value) of DPCD-treated juices did not change. The browning degree and pH of DPCD-treated juices decreased, the cloud and titratable acidity increased significantly, the UV-visible spectra of juices were lower, but the total soluble solid and the carotenoids of juices were stable. The particle size of juices treated by DPCD for 15, 30 and 45 min increased significantly, but for 60 min showed a noticeable decrease and was almost close to that of untreated juice. DPCD treatment did not alter the Newtonian flow behaviour of the carrot juice, but caused a significant increase in juice viscosity.

8.2.12 Peach

Zhou *et al.* (2010) continued their work on peach. When peach juice was treated with DPCD (30 MPa; temperature 55°C; treatment time, 30 s and 10, 40 and 60 min.), an acceleration of precipitation formation was observed. Particle size distribution (PSD), pH, ζ-potential (zeta potential: electro-kinetic potential in colloidal systems), protein and total phenols, PME activity, pectin and calcium, and viscosity in juice, all contributing to the precipitation formation, were studied. HPCD significantly altered juice PSD pattern (Figure 8.9), and the acceleration of precipitation formation confirmed this alteration. A significant decrease of protein and a declining trend of total phenols were obtained. Pectin and calcium contents were unchanged, and the difference in PME activity in juice was insignificant after HPCD. HPCD led to higher increase in juice viscosity, whereas pH and the absolute value of ζ-potential declined, suggesting that the pH and absolute value of ζ-potential declines caused the coagulation of proteins and decrease of particle charge, responsible for the acceleration of the precipitation formation.

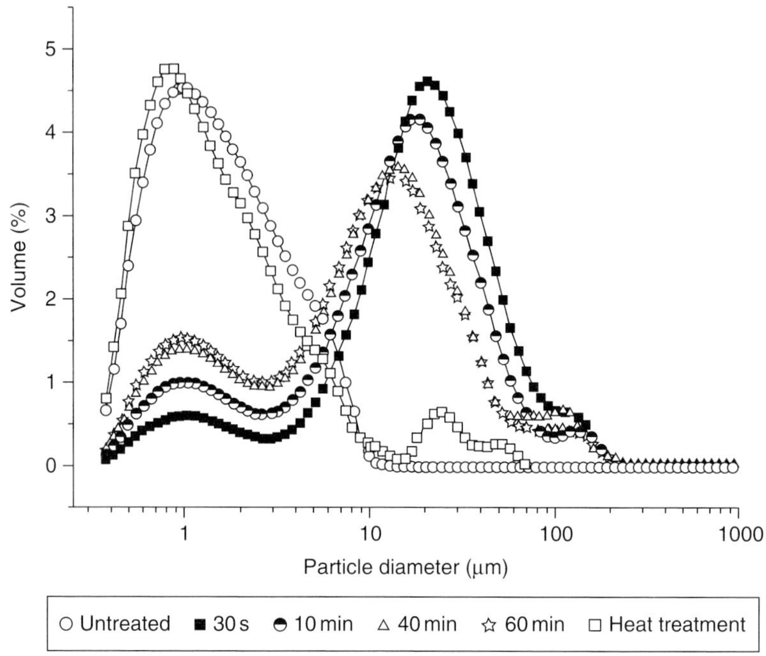

Figure 8.9 PSD patterns of peach juices treated by HPCD at 30 MPa and 55°C and heat at 90° for 1 min. (Reproduced from Zhou *et al.*, 2010. Copyright 2010 with permission of American Chemical Society.)

8.2.13 Kiwi

Spilimbergo and Ciola (2010) studied the microbial inactivation and quality parameters (pH, sugar content, titratable acidity, absorbance at 420 nm and turbidity) of peach and kiwi juices treated at 35°C with DPCD and supercritical nitrous oxide (SC-N_2O) as a function of pressure and treatment time. Total inactivation of naturally occurring microorganisms and a *Saccharomyces cerevisiae* strain (10^5 cfu/mL) was obtained after 15 min of DPCD–SC-N_2O treatment, 10 MPa and 35°C, for both juices. No significant changes in chemical–physical or sensory attributes between untreated and treated juice were detected. The results demonstrated the feasibility and the potential of DPCD–SC-N_2O treatment as an alternative low-temperature pasteurization process for peach and kiwi juices.

8.2.14 Melon

Chen *et al.* (2010) investigated the effects of DPCD treatment (8, 15, 22, 30 and 35 MPa for 5, 15, 30, 45 and 60 min at 35°C, 45°C, 55°C and 65°C) on microorganisms, enzymes and aroma compounds in hami melon (*Cucumis melo*) juice during storage at 4°C for 4 weeks. Colour, browning degree and vitamin C were also measured. The DPCD treatment had significant effects on inactivation of microorganisms and enzymes: higher pressures caused more inactivation. At 35 MPa, 55°C, 60 min, microorganisms were totally inactivated. Minimum residual activities of polyphenol oxidase (PPO), peroxidase (POD) and lipoxygenase (LOX) were 25.26%, 38.46% and 0.02%, respectively. Restoration of PPO, POD and LOX residual activity after DPCD treatment was also observed, depending on the pressure level. Aroma compounds were less affected after DPCD treatment, and the flavour of the melon juice was similar to that of the fresh juice after storage at 4°C for 4 weeks, without off-odours. The changes of lightness (L) and browning degree (A) during storage were well represented with a first-order kinetic model. Vitamin C concentration decreased by DPCD treatment, but this loss was lower than that of the untreated sample.

8.3 Other beverages processed with DPCD

8.3.1 *Beer*

Beer is a delicate and heat-labile beverage. Off-flavours are easily formed during heat pasteurization. Since freshness is top priority to brewers, the use of a nonthermal method in order to inhibit the growth of spoilage microorganisms is of great importance for quality. Sterile filtration is a nonthermal method that removes the spoilage bacteria and yeast without altering the colour and flavour. However, this method requires extensive monitoring and is labour intensive. Dagan and Balaban (2006) pasteurized beer with a continuous DPCD system (27.6 MPa, 21°C, 5% CO_2, 5 min) and evaluated the physical and sensory quality attributes such as foaming and foam stability, haze formation, aroma and flavour. Sensory tests showed that aroma and taste of DPCD-pasteurized beer were not significantly different from those of fresh beer after 1 month of storage (1.7°C) whereas those of heat-pasteurized beer (74°C, 30 sec) were significantly different ($p > 0.01$). Beer haze was significantly reduced by DPCD. Foam capacity and stability were significantly less after DPCD treatment, but not at levels detrimental to the quality.

8.3.2 *Kava kava*

Kava (*Piper methysticum*) is a traditional Polynesian beverage, popular in Hawaii and many South Pacific locations. It is known as a beverage that induces relaxation and sociability. The kavalactones, a collective name for δ-lactones and 5,6-dihydro-δ-lactones, are the pharmacologically active compounds in kava. Traditional, unpasteurized kava drinks have a shelf life of less than 3 days with refrigeration at 4°C. Hsieh *et al.* (2007) examined the effectiveness of DPCD on the bacteria reduction in a freshly prepared kava extract. A continuous DPCD system was used. Kava extract was prepared on site in Florida, with material shipped frozen from Hawaii. The treatment condition was set at a pressure of 34.5 MPa, 13% (v/v) CO_2, and a residence time of 7 min. The initial pH of the kava extract was 6.3 and was reduced to 5.5 after processing. The residual CO_2 in the liquid could be tasted. DPCD treatment also changed the mouth feel of the liquid.

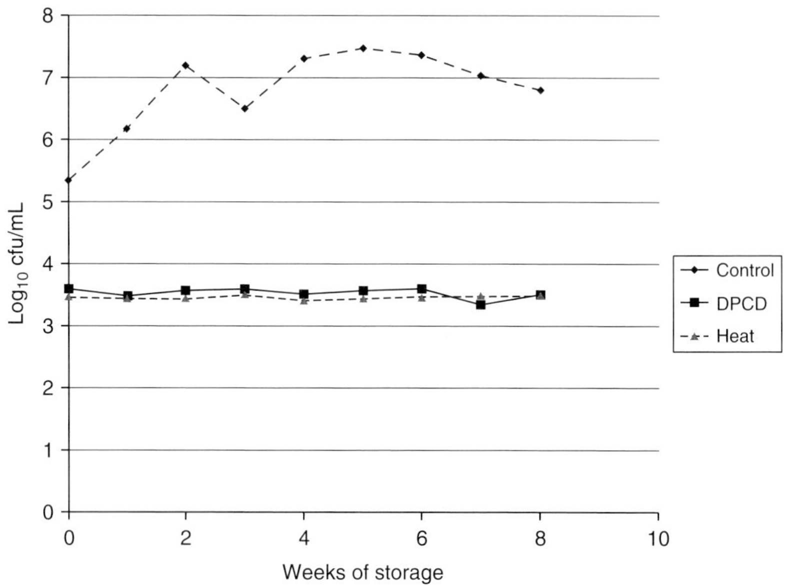

Figure 8.10 Changes in total aerobic microbial count in DPCD-treated, control and thermally treated kava beverage (Hsieh *et al.* 2007). Reproduced with permission.

The microbial counts for before and after treatment were approximately 4.0×10^5 CFU/mL and 1.0×10^2 CFU/mL, respectively. The DP-CO_2 system was able to produce an initial 3-log reduction. After 8 weeks refrigerated storage at 4°C, the microbial count was 4.0×10^3 CFU/mL (Figure 8.10). The study demonstrated that at the current setting, the bacterial reduction is not sufficient to produce a shelf-stable product. The high amount of starch suspension and type of bacteria species present in the kava could have resulted in a less-than-expected bacteria count reduction.

8.3.3 *Jamaica beverage*

Hibiscus sabdariffa (family Malvaceae) red calyces are rich in anthocyanins and other phenolic compounds. Fresh and dried hibiscus is used to prepare cold and hot beverages, and their preparation includes an extraction step followed by a pasteurization method (Ramirez-Rodrigues *et al.* 2009).

Solubility of CO_2 in a hibiscus beverage (5.06 g CO_2/mL at 31.0 MPa) and optimal processing conditions to inactivate yeasts and moulds were 30.8 MPa and 6.8 min (Rodrigues 2010). DPCD was a viable technology for processing hibiscus beverage since it extended its shelf life. Quality and colour attributes were maintained during storage. Lower losses of anthocyanins were observed in the DPCD hibiscus beverage as compared to a heat pasteurization process and no major changes in total phenolics content and antioxidant capacity occurred during storage. Changes in hibiscus aroma volatiles during storage did not affect panellists' overall likeability of the product.

8.4 Conclusions

The list of juices and beverages treated with DPCD, with varying levels of success in reducing microorganisms and in preserving quality attributes, is growing. Eventual commercial applications will be based on the accumulated data, as well as the results of technical and economic feasibility studies for each product, in the economic environment considered. Like other technologies, DPCD will find its niche where no other method will be satisfactory, and/or the economics will prove to be beneficial.

9 Use of Dense Phase Carbon Dioxide in Dairy Processing

Giovanna Ferrentino and Giovanna Ferrari

Abstract: Literature suggests that the application of DPCD technology for pasteurizing and protecting the quality of dairy products has not been extensively investigated yet. This may have been caused by the assumption that the concentrations of CO_2 required for a significant pasteurizing effect would negatively affect the organoleptic quality of the products. The present chapter will give first an overview on the effects of adding CO_2 to raw milk, then the effect of the treatment on microbial and enzymatic inactivation will be discussed, taking into account the concerns and positive implications of the process. A short review will also be presented on the application of CO_2 for the production of cottage cheese, yogurt, and fermented products. Considering that the precipitation of casein is a matter of concern during DPCD processing of milk, a description of CO_2 application to casein production will also be addressed.

Keywords: dairy products; organoleptic quality; casein precipitation.

9.1 Introduction

This chapter describes the application of dense phase carbon dioxide (DPCD) to milk and dairy products. The feasibility of using this process to induce microbial inactivation and improve the keeping quality of pasteurized milk has not been investigated in detail, probably on the assumption that the concentrations of CO_2 required to have a significant effect on shelf-life would detract from the sensory quality of the milk. The chapter will first give an overview on the effect of the addition of CO_2 to raw milk during thermal pasteurization considering that a significant amount of this gas is already dissolved in it. Then the effect of DPCD on the microbial and inactivation will be reported taking into account all the concerns and the positive implications of the process. A short review will be also reported on the application of CO_2 for the production of cottage cheese, yogurt and fermented products.

Dense Phase Carbon Dioxide: Food and Pharmaceutical Applications, First Edition. Edited by Murat O. Balaban and Giovanna Ferrentino.

Considering that the precipitation of casein is one of the results occurring during the DPCD process of milk, the description of the application of CO_2 to casein production will be also addressed.

9.2 Carbon dioxide in milk

Milk produced by animals contains a significant amount of CO_2, approximately 90 ppm (Vandenberg 1979). Some researchers (Van Slyke and Baker 1919; Van Slyke and Keeler 1920) suggested that a third of CO_2 in milk exists as carbonic acid and two thirds as bicarbonate. It has been demonstrated that the fraction of carbonate, bicarbonate and CO_2 ions varies in solution as a function of pH (Daniels *et al.* 1985). In milk at pH of about 6.3, approximately 88% of CO_2 exists as dissolved CO_2 2% as carbonic acid and the remaining 10% as bicarbonate. The CO_2 content of raw milk decreases when milk is exposed to air after milking, while O_2 and N_2 levels increase (Marshall 1902). Analysing raw milk samples received at a commercial milk plant, it has been shown that the CO_2 level is about 2.0 mM (Noll and Supplee 1941; Frayer 1941). This means that, prior to receipt at the processing plant, there is a significant loss of CO_2 from milk. Losses also occur during the pasteurization treatment due to the temperature and pressure changes. It has been demonstrated that the CO_2 content dropped from 1.6 mM in raw to 0.68 mM in processed milk (Smith 1964). These CO_2 losses have to be attributed to the aeration during pumping and no hermetic storage of milk.

9.3 Enzymes and microorganisms in milk

The composition of dairy products provides a favourable physical and chemical environment for the growth and propagation of a broad spectrum of microorganisms. Milk and dairy products are, in fact, excellent growth media for pathogenic and spoilage microorganisms. Microbiological deterioration of refrigerated raw and pasteurized milk, cottage cheese and similar products is often caused by the growth of psychrotrophic Gram-negative bacteria such as *Pseudomonas, Acinetobacter, Flavobacterium, Enterobacter, Klebsiella, Aerobacter, Escherichia, Serratia, Proteus, Aeromonas* and *Alcaligenes* or yeast and mould such as *Geotrichum, Scopulariopsis, Mucor, Alternaria* and *Penicillium*. The deterioration of milk results in a change in flavour and texture and is perceived as a visual spoilage. *Paenibacillus, Bacillus* and *Microbacterium* are the predominant spoilage microorganisms. They were found to be the most heat-resistant psychrotrophic Gram-positive rods identified in a study conducted by Fromm and Boor (2004) on pasteurized milk samples from three commercial dairy plants. These microorganisms produce extracellular protease and lipase activity, which reduce the functionality of milk proteins and produce undesirable aromas. Gram-positive microorganisms, particularly those producing lactic and acetic acids, can spoil dairy foods, but the numbers of microorganisms required are generally higher than for Gram-negative bacteria and the changes can be less noticeable.

The two broad enzyme types responsible for a majority of the degradation of milk quality are lipases and proteases (Ma *et al.* 2000). The principal lipase of concern, lipoprotein lipase, is associated with the casein micelles in the skim fraction of the milk (Fox 1992).

Lipases produce free fatty acids by cleaving the ester bond connecting a fatty acid to a glycerol molecule in a triglyceride. Free fatty acids have a pungent aroma, which results in a "rancid" note in the milk. While this may be considered an off-flavor in fluid milk, many dairy products (such as blue cheese) benefit from the characteristic "rancid" flavor.

To preserve the shelf life of milk, thermal processes such as ultra-high-temperature pasteurization have been developed. However, in many cases these processes alter the organoleptic properties and the food quality of the products.

Traditional pasteurization preserves milk not only because of the reduction in the number of spoilage microbes, but also because of its effective inactivation of lipoprotein lipase which is one of the more heat-labile enzymes and is inactivated at 60°C (Andrews *et al.* 1987).

Protease enzymes, in particular the protease plasmin, are another major category of endogenous milk enzymes responsible for the degradation in milk (Sousa *et al.* 2001). The plasminogen, which is a plasmin's precursor, is changed to plasmin by specific enzymes responsible for the conversion. The activation of plasmin is kept in check by other enzymes that inhibit the plasminogen activators (Fox 1992). Standard high temperature and short time (HTST) processing of milk (72°C, 16 seconds) destroys the inhibitors of the plasminogen activators, but not plasmin itself. By inactivating inhibitors, traditional thermal pasteurization increases the conversion of plasminogen into plasmin, effectively increasing the rate at which proteolysis occurs. Plasmin is extremely heat stable (Kaminoga *et al.* 1972). It would take very high temperatures to inactivate it and the activity remains even in ultra-high-temperature pasteurized milk (138°C, <2 seconds). However, processing milk at temperatures above this level causes severe organoleptic defects (Iwatsuki *et al.* 2000), so plasmin activity remains a problem even in thermally processed milks.

As an alternative to thermal pasteurization, some antimicrobial agents such as sorbic acid or nisin have been proposed but also discouraged due to labelling requirements and toxicity risks. Recent research showed the possibility of CO_2 to be added to liquid foods as an antimicrobial agent. Considering that CO_2 occurs naturally in raw milk, its direct addition has been proposed to dairy products, although the precise inactivation mechanism is not completely understood (Hotchkiss *et al.* 2006).

9.4 Application of carbon dioxide to milk

9.4.1 Carbon dioxide addition to raw milk

The effect of the addition of CO_2 to raw milk for bacterial reduction has been extensively studied in recent years. Research reported not only the possibility to add CO_2 to raw milk to prevent spoilage during its storage and transport to the processing plants, but also the possibility to use CO_2 at elevated pressures as an antimicrobial agent.

The first demonstration of the antimicrobial activity of CO_2 was evidenced by Hoffman (1906). The results showed that adding CO_2 at 5 MPa reduced the rate of increase of the microbial counts in milk. It was observed that milk kept under 1 MPa of CO_2 did not curdle even after 72 hours at ambient temperature while the untreated milk curdled within 24 hours.

More investigations documented the chemical and microbiological effects of the addition of low levels of CO_2 to raw milk. In particular, the ability of CO_2 to slow down the spoilage of raw milk was demonstrated to be of great interest during its storage and transport to the processing plants where the milk is pasteurized or sterilized (King and Mabbitt 1982; Mabbit 1982; Law and Mabbitt 1983). Researchers added CO_2 in concentration ranging from 10 to 40 mM to raw milk, sterilized skim milk inoculated with different types of bacteria and stored the samples at 4°C, 7°C and 10°C up to 6 days. They observed that the storage temperature had a great effect on CO_2 concentration added to the samples. In particular, when decreasing the temperature, CO_2 solubility increased and the microbial growth rate was reduced. They also demonstrated that the initial microbial quality of raw milk influenced the effect of CO_2, concluding that the lower the initial count in the untreated milk, the greater the CO_2 antimicrobial action (King and Mabbitt 1982). The same research group carried out experiments to observe if milk acidified with HCl and CO_2 to the same pH level showed similar behavior. The results were negative. While the addition of about 20–30 mM of CO_2 to milk had approximately 3 log units (cfu/ml) less than the untreated milk after 4 days of storage at 7°C, the milk treated with HCl did not show any difference in microbial counts when compared to the untreated sample stored at the same conditions. These results highlighted the antimicrobial effect of CO_2 not linked to the lowering of the pH or to the displacement of O_2. These findings supported the idea that CO_2 could be added to raw milk to inhibit microbiological deterioration during storage and subsequently easily removed without detrimental effects.

It is clear that the addition of CO_2 to milk inhibits the growth of microorganisms and retards the deterioration of raw milk stored under refrigeration, improving the overall quality of pasteurized milk. Many research studies reported the use of CO_2 on raw milk stored under refrigerated conditions while less work was focused on the potential effects of CO_2 on raw milk held at temperatures above normal refrigeration temperatures (7–10°C) or under changing temperature conditions experienced during transport. Some experimental studies reported that CO_2 had little effect on raw milk held at 20°C (Rashed *et al.* 1986).

Roberts and Torrey (1988) performed experiments on sterile milk inoculated with several common proteolytic psychrotrophic bacteria and investigated the effects of the addition of 20 to 30 mM of CO_2 on the microbial growth at 7°C. They showed that treatment with CO_2 decreased the growth rate of psychrotrophs and other groups of bacteria and that the refrigerated storage of raw milk could be extended from 1 to 3 days or longer depending upon initial microbiological quality. In addition, after the refrigerated storage CO_2 was easily removed from milk prior to processing by vacuum or gentle agitation. The results were was important considering that it was becoming common practice in the dairy industry to extend refrigerated storage of milk up to 5 days before pasteurization.

Some research reported the effect of the addition of CO_2 to the atmosphere surrounding the product to reduce the rate of growth of many food spoilage and pathogenic microorganisms (Farber 1991; Hanlin *et al.* 1995; Devlieghere *et al.* 1998; Devlieghere and Debevere 2000). The experimental results showed that the more sensitive microorganisms were the Gram-negative psychrotrophs, particularly *Pseudomonas* spp. while less inhibition effects were observed on the Gram-positive psychrotrophs, particularly *Lactobacillus* spp. In the action of adding CO_2 to the milk, it was also observed that factors such as microbial species, substrate and CO_2 concentration could influence microorganisms' life. However, it was difficult to explain how CO_2 could exert its antimicrobial action.

Recently several hypotheses have been developed to explain CO_2 action when it is added to dairy products for modified atmosphere packaging (MAP) to increase the shelf life of the products. One of the mechanisms by which CO_2 inhibits microorganisms is assumed to be the displacement of O_2. Another factor considered is the capability of CO_2 to dissolve in the food, forming carbonic acid and lowering the pH. Some experimental results demonstrated that the main variable controlling microbial growth in MAP is the dissolved CO_2 concentration (Devlieghere *et al.* 1998). A research study has been carried out by Hendricks and Hotchkiss (1997) to evaluate the effects of CO_2-modified atmosphere on the growth of *Pseudomonas fluorescens* and *Listeria monocytogenes*. The bacterial suspensions were incubated at 7°C under a continuous flow of O_2 and CO_2 gas mixture at different concentrations in order to better understand the direct effect of CO_2 on microbial growth. The results showed that CO_2 suppressed growth even when the amount of O_2 in the mixture was held constant to 20% and the media did not change its pH. This agrees with the findings of previous workers who concluded that CO_2 directly inhibits microbial growth in dairy products as opposed to the indirect effects of pH and O_2 displacement (King and Mabbitt 1982).

Another antimicrobial effect of CO_2 could be linked to its ability to concentrate in the lipid membrane of the bacterial cell, disrupting the physiochemical properties of the membranes. The lipophilic nature of CO_2 allows it to pass through membranes and concentrate inside the cell, lowering intracellular pH and affecting the enzymatic activity of the cell. Intracellular CO_2 could generate carboxylation and decarboxylation reactions resulting in a net energy expenditure and loss of ATP for the microbial cell.

Martin *et al.* (2003) reported the inhibition caused by CO_2 in concentration ranging from 0.6 to 61.4 mM on bacterial growth of natural microflora in raw milk and microbial populations of *Pseudomonas fluorescens, Bacillus cereus, Escherichia coli, Listeria monocytogenes, Enterococcus faecalis* and *Bacillus licheniformis* in inoculated sterile milk. The results on natural microflora showed that increasing CO_2 concentration from 0.6 to 61.4 mM, the time to reach the maximum growth increased as the growth rate decreased. CO_2 effect was greater on Gram-negative than Gram-positive bacteria. Microbial cells of *P. fluorescens* incubated with 0.4 mM of CO_2 showed a lag time of 3.3 hours while the lag time increased to 26.1 hours with 46.3 mM of CO_2. Similar effects were noticed on *L. monocytogenes* while for *B. cereus* no effects were observed on the lag phase but only on the growth rate by increasing CO_2 concentration.

Rajagopal *et al.* (2005) reported experiments on fresh raw milk treated with CO_2 at pressures ranging from $68 \cdot 10^{-3}$ to $689 \cdot 10^{-2}$ MPa, different temperatures (5°C, 6.1°C, 10°C and 20°C) and storage for up to 9 days. The treatment conditions were chosen to avoid protein precipitation. The results showed that the treatment significantly reduced natural microflora in raw milk also in terms of Gram-negative bacteria and *Lactobacillus*. Raw milk treated with CO_2 and held at 6.1°C for 4 days exhibited reduced bacterial growth rates at pressures of $68 \cdot 10^{-3}$, $172 \cdot 10^{-2}$, $344 \cdot 10^{-2}$ and $516 \cdot 10^{-2}$ MPa, while at $689 \cdot 10^{-2}$ MPa a significant loss of viability in standard plate counts was observed, reducing also Gram-negative bacteria and total *Lactobacillus* spp. At a storage temperature of 6.1°C, the time to reach 4.3 log units (cfu/ml) increased by 4 days as compared to the untreated control group. Coliform levels remained unchanged in the treated samples while they doubled in the control milk, and the level of thermoduric bacteria was significantly lower after 9 days of storage.

Few studies on the efficiency of CO_2 addition on the growth of spores have been carried out. The most significant was the research of Werner and Hotchkiss (2002) who observed the effect of the addition of 11.9 mM CO_2 on the growth of *Bacillus cereus* spores inoculated into sterile, homogenized whole milk. The research was important considering that spores of the organism are thermally resistant and can survive pasteurization and storage at refrigeration temperatures. In particular, psychrotrophic strains of *B. cereus* isolated from milk have been shown to grow at temperatures as low as 1°C (Coghill and Juffs 1979). During the experiments, 10^1 and 10^6 spores/ml were inoculated into sterile milk and stored at 6.1°C for 35 days. The results suggested that moderate levels of CO_2 did not enhance the outgrowth of *B. cereus* spores over long-term storage and did not increase the risk of foodborne illness due to the organism. Similar conclusions were also obtained in previous studies carried out on *Clostridium botulinum* spores inoculated in milk (Glass *et al.* 1999).

9.4.2 Carbon dioxide addition during thermal pasteurization of milk

Research was performed on the addition of CO_2 during the pasteurization process to increase microbial kill. During this operation it is very important to control the pasteurization temperatures and pressures and the initial CO_2 content in the milk to minimize buildup of deposits on the wall of the pasteurizer (Calvo and De Rafael 1995).

Some experimental studies also discussed the possibility to use the direct addition of CO_2 to pasteurized milk for shelf life extension or to add CO_2 before thermal inactivation of microorganisms in the pasteurization process. It has been reported that the addition of 1.81–3.18 mM CO_2 in full-fat pasteurized milk stored in paperboard cartons at 6°C for up to 14 days improved keeping quality. Sensory tests showed that the panellists found no differences between the untreated and CO_2-treated milk prior to 14 days (Duthie *et al.* 1985). Further results demonstrated that the addition of low levels of CO_2 in inoculated milk packaged in pouches inhibited the growth of psychrotrophic microorganisms and provided a moderate extension of shelf life (Chen *et al.* 1992b; Hotchkiss *et al.* 1998). Increasing CO_2 content from 0 to 21.5 mM, the microbial lag phase was extended and the growth rate reduced.

Ruas-Madiedo *et al.* (1996) reported experiments about the addition of CO_2 to 200 L batches of raw milk to lower the pH to 6.0. The milk was held at 4°C for 4 days, and subsequently residual CO_2 was removed with a vacuum system and the milk pasteurized. After the process no effect of CO_2 was observed on casein and whey proteins. Milk samples were evaluated organoleptically, microbiologically and chemically. The organic acid content of milk was not different, with the exception of lactic acid, which was slightly lower in CO_2-treated milk. The volatile organic compounds of the treated product were lower, probably due to the lower microbial activity. The experimental results showed that CO_2-treated milk had lower coliform, proteolytic and lipolytic psychrotrophic counts compared to the untreated raw milk after 4 days storage. No effects on fat, water-soluble vitamin and free monosaccharide content were detected.

Loss (2001) showed that increasing concentrations of dissolved CO_2 in raw milk between 1 and 36 mM linearly decreased the decimal reduction time of *P. fluorescens* in milk subsequently treated at 50°C.

Loss and Hotchkiss (2002) also demonstrated that CO_2 concentration ranging from 44 to 58 mM added to raw milk significantly reduced the number of surviving standard plate count microorganisms in milk heated over the range from 67°C to 93°C during the pasteurization process. Also in this study, a decrease in thermal survival rates for *Pseudomonas fluorescens* and *Bacillus cereus* spores was positively correlated with CO_2 concentration. It was also reported that better results could be obtained increasing temperature and CO_2 concentration and that the point at which CO_2 was injected into the pasteurization line could have also a positive or negative effect on microbial kill.

9.4.3 Effect of carbon dioxide addition on sensory properties of milk

The possible effects of the addition of CO_2 on the sensory properties of milk have also been investigated. Amigo *et al.* (1995) showed that a sensory evaluation of degassed and pasteurized milk resulted in no detectable differences between treated and untreated samples. They also reported that no degassed samples were perceived by the panellists as significantly different from the controls, due to the sensation associated with high levels of dissolved CO_2. An analysis was also performed to consider the possibility to introduce CO_2-treated milk in the market. In the United States, the Pasteurized Milk Ordinance Grade A regulations specify an upper microbial limit for raw milk prior to pasteurization of 5 log (cfu/ml) units (US Department of Health and Human Services 1999). Research showed that 5 log (cfu/ml) units were not reached in raw milk treated with CO_2 at $689 \cdot 10^{-2}$ MPa until the eighth day of storage while this limit was reached in the control milk in just 4 days, for both samples stored at 6.1°C. These data suggested that pressurized CO_2 might be an effective method for preserving raw milk with a higher shelf life and overall quality.

Gevaudan *et al.* (1996) analysed the milk treated by injection of CO_2 to study the effects on milk salts and on the buffering properties of skim milk in order to correlate them to the mineral balance. The results of the study suggested that no changes were observed in the concentrations of inorganic content such as phosphorous, calcium or magnesium in aqueous phase but that the buffering curves of the treated milk were different from those of the untreated milk showing that the solubilization of CO_2 induced the formation of a new salt system. Although CO_2 had a reversible effect on pH, an irreversible effect on inorganic colloidal calcium phosphate which was changed into other salt forms was detected.

A study about the influence of CO_2 addition to raw milk on microbial levels and also on some fat-soluble vitamin contents of raw and pasteurized milk was carried out by Ruas-Madiedo *et al.* (1998). The experiments were carried out by bubbling CO_2 until the pH of the milk ranged between 6.1 and 6.3. The milk was stored at 4°C for 4 days, degasified by vacuum, pasteurized and stored at 4°C for 7 days. The treated milk was analysed, and the results showed that CO_2 addition inhibited the growth of microorganisms without affecting the stability of vitamin A (retinol and β-carotene) and vitamin E (α-tocopherol). Vacuum degasification and the subsequent pasteurization on a pilot scale partially removed CO_2, making milk acceptable for consumption. The residual CO_2 which was found in the pasteurized milk extended the cold storage period by inhibiting bacterial survivors without detrimental effects on retinol, β-carotene and α-tocopherol.

An experimental study was also performed by Ma *et al.* (2001) to determine the effects of dissolved CO_2 in milk on the performance of four important routine testing methods: antibiotic residue test, freezing point test, infrared milk component analysis and alkaline phosphatase test. These tests are very important considering that the US Food and Drug Administration (FDA) requires that each truckload of milk needs to be tested for the presence of antibiotic residues (β-lactams) upon arrival at a dairy plant. Milk is also tested regularly for freezing point to detect water adulteration. For producer payment and dairy herd improvement programs, infrared analysis provides rapid and cost-effective measurement of milk fat, protein and lactose content. The test on alkaline phosphatase verifies that adequate pasteurization has been achieved. The milk carbonated with 1000 ppm of CO_2 was analysed and results showed that the addition did not affect the performance of antibiotic testing methods. Also a decrease of pH and freezing point was detected, but it was reversible upon removal of dissolved CO_2. The sample absorbance readings at the lactose reference wavelength region increased due to the addition of CO_2 causing the lactose signal to decrease and the fat signal to increase. The addition of CO_2 did not meaningfully affect the alkaline phosphatase test and would not change an analyst's ability to differentiate between pasteurized milk and pasteurized milk that was contaminated with raw or under-pasteurized milk.

9.4.4 Dense phase carbon dioxide process

DPCD has been investigated and recently applied to milk as an alternative and innovative process to replace thermal pasteurization.

Erkmen (1997, 2000d) studied the effect of DPCD on the inactivation of *E. coli* and *S. aureus* inoculated in whole and skim milk. The treatment was performed with a batch equipment and the object was to define the best process conditions in terms of pressure, temperature and treatment time to induce microbial inactivation. The results showed that a treatment carried out at 10 MPa, 30°C and 6 hours caused a decrease of 6.42 and 7.24 log cycles on *E. coli* inoculated in whole and skim milk, respectively. It was observed that microorganisms were inactivated more easily when they were suspended in skim milk than in whole milk, probably due to the protective effect of the fat globules on the cells, retarding the penetration of CO_2. The sterilization effects of CO_2 on *S. aureus* and aerobic bacteria were observed at 14.6 MPa for 5 hours and 9 MPa for 2 hours at 25°C in whole and skim milk, respectively.

Ho (2004) presented a preliminary study on DPCD processing and the combined treatment effects from DPCD processing and pasteurization using the Better Than Fresh™ System that was developed by Praxair Inc. (Burr Ridge, IL) and was test on whole milk, skim milk, and dairy-based beverages. Results showed that samples of the combined treatments (DPCD and thermal pasteurization) had a longer shelf life than those that were only pasteurized by heat, suggesting the possibility of using DPCD processing as a novel technology to achieve better sensory and nutritional quality while extending the shelf life of milk and dairy products.

Werner and Hotchkiss (2006) evaluated the effects of pressurized CO_2 on the reduction of bacterial population in raw skim milk. They investigated the possibility to use high-pressure CO_2 treatment as an alternative to thermal processes, which results in a range of undesirable texture and flavour changes in processed milk. The research compared the effectiveness of the process utilizing CO_2 at subcritical and supercritical phases carried out in a continuous-flow system to inactivate indigenous psychrotrophic vegetative cells of *Pseudomonas fluorescens* and spores in raw milk. Pressures between 10.3 and 48.3 MPa, temperatures of 15°C, 30°C, 35°C and 40°C and CO_2 concentrations of 3, 66 and 132 g/kg of milk were studied. A higher microbial lethality was achieved in raw milk treated with CO_2 in the supercritical phase than that obtained with subcritical CO_2, and the effects of supercritical CO_2 were more pronounced by increasing the pressure and temperature. At 30°C, no effects on total microbial count were observed even increasing the pressure up to 20.7 MPa with either 66 or 132 g/kg of CO_2. Increasing the temperature up to 35°C, with CO_2 in the supercritical state, there was a direct proportionality between the pressure and the lethality at CO_2 levels of 132 g/kg not only for psychrotrophic vegetative cells but also for *P. fluorescens*. The maximum reductions achieved were 5.36 log units for psychrotrophic vegetative cells and 5.02 log units for *P. fluorescens* at 35°C, 20.7 MPa and CO_2 concentration of 132 g/kg of milk. In all the treatment conditions tested, no effect on spore populations was detected.

9.5 Application of carbon dioxide for enzyme inactivation

DPCD treatment was also applied to investigate the effect on the stability of the alkaline phosphatase enzyme naturally present in milk. This enzyme is slightly more heat resistant than most pathogenic bacteria and the measurement of its destruction confirms proper pasteurization of the product. The activity of alkaline phosphatase in pasteurized milk generally indicates an inadequate pasteurization and its residual activity may be due to the contamination of pasteurized milk with raw milk or due to postprocess bacterial contamination. Fadıloğlu *et al.* (2004) reported the application of CO_2 under atmospheric pressure (CO_2 was bubbled through the solution at a flow rate of 200 ml/min for 30 min) to inactivate alkaline phosphatase enzyme in buffer and raw milk. The results showed that the activity of the enzyme sharply decreased after 5 minutes of treatment and then continued to decrease slowly. In a buffer with an initial pH of 9.6, the residual enzyme activity after a CO_2 treatment of 30 minutes was found to be 84%, 55%, 39% and 19% at 20°C, 30°C, 40°C and 50°C, respectively. No significant change in the enzyme activity was recorded in raw milk at any CO_2 process conditions tested. The phenomenon was attributed to the milk's strong buffering capacity which is the result of the properties of substances such as proteins, phosphates, CO_2 and citrate naturally occurring in it. These substances make the pH of milk very stable, protecting it from the effect of the DPCD treatment.

Studies to investigate the effect of CO_2 on lipase and lipolytic enzymes in raw whole milk were also carried out by Tisi (2004). During the experiments, a continuous high-pressure CO_2 equipment designed by Praxair Inc. (Chicago, IL) was used, capable of continuously treating milk with liquid CO_2 at pressures up to 69 MPa. The results showed that the in-line addition of CO_2 was able to significantly affect the chemistry of milk when processed at temperatures between 15°C and 40°C and at pressures between 7 and 62 MPa. Whole milk processed at 40°C with CO_2 showed the inactivation of lipases at temperatures lower than those used in thermal processing. However, after processing raw whole milk at 15°C, no inactivation of the lipolytic enzymes was detected. The inefficacy of CO_2 to inactivate lipase in the raw whole milk coupled with the increase in fat surface area in the homogenization step led to high levels of free fatty acid in the milk after 3 days. The treatment was accompanied by a change in the particle size distribution in both skim and whole milk. The altered particle size distribution also caused an inhibition of the proteolytic activity of the milk due to the change of its protein structure.

The effect of the addition of CO_2 to raw milk on proteolytic and lipolytic activity during 21 days of storage at 4°C was also investigated by Ma *et al.* (2003). The treatment was performed through a carbonation process adding approximately 500 ppm (pH 6.5 at 4°C), 1000 ppm (pH 6.3 at 4°C) and 1500 ppm (pH 6.2 at 4°C) of CO_2 to the milk. The experimental results showed that the addition of 1500 ppm of CO_2 delayed the growth of bacteria during 21 days of storage at 4°C. The proteolytic activity decreased by at least two mechanisms: the reduction of microbial proteases due to the reduced microbial growth, and the possible reduction in plasmin activity due to a lower milk pH. An effect was also observed on lipolytic activity due to a reduced microbial growth, while no effect of CO_2 addition or acidification on lipolysis induced by native milk lipase was reported.

9.6 Application of carbon dioxide to cottage cheese production

The use of CO_2 has been found to be commercially beneficial in the preservation of cottage cheese. Research has been carried out to observe the possible advantages of flushing CO_2 on cottage cheese sealed in flexible containers, or the direct addition of CO_2 into the cream dressing prior to mixing with the curd to form cream-style cottage cheese.

Kosikowski and Brown (1973) sealed the creamed cottage cheese in flexible containers and flushed CO_2 on the product before storing it at 4°C. The growth of psychrotrophs, yeasts and moulds was repressed. Fresh flavor was maintained for 73 days, but the cottage cheese had a "fizzy" taste due to the high levels of CO_2 added. More recently, other researchers demonstrated the efficiency of this technique for the preservation of cottage cheese (Rosenthal *et al.* 1991; Fedio *et al.* 1994). In particular it was reported that flushing with 100% CO_2 was preferred over other gas mixtures (Maniar *et al.* 1994). Moir *et al.* (1993) suggested that the addition of CO_2 throughout the cheese production before packaging was able to inhibit psychrotrophic microorganisms on the surface and within the volume of the cheese.

Some research groups tried to directly add CO_2 into the cream dressing before mixing with the curd to form cream-style cottage cheese (Chen and Hotchkiss 1991a, 1991b, 1993b). They inoculated the cottage cheese with 10^3 cfu/g of a mixture of three Gram-negative psychrotrophic spoilage bacteria and stored it in sealed glass containers at 4°C and 7°C for 80 days. Carbon dioxide was added by dissolution into the cream dressing prior to the addition of curd. The results showed no microbial growth including yeasts and moulds in the cheese containing CO_2 during storage. During the first 10 days of storage at 7°C or 17 days at 4°C, the cottage cheese packed without dissolved CO_2 had 10^4-fold more colony-forming units than that packed with dissolved CO_2. In addition the cheese containing CO_2 showed a fresh appearance during 70 days of storage at 4°C or 60 days of storage at 7°C.

The research showed that CO_2 addition was very important to avoid the alteration of the taste of the cottage cheese but the amount of CO_2 used should always be lower than that considered in the production of carbonated beverages or sodas. Chen *et al.* (1992b) demonstrated, using trained sensory panellists in triangle tests, that the lowest threshold CO_2 value not detected by the panellists in the treated milk was between 4.54 and 9.10 mM. It was also reported by Moir *et al.* (1993) that the injection of 10 mM CO_2 into cottage cheese cream dressing and package headspace could significantly increase the shelf life without affecting the pH and flavor of the product.

9.7 Application of carbon dioxide to yogurt and fermented products

The consumption of yogurt and fermented milk is increasing in industrialized countries due to the sensory aspects and perceived health benefits of these products. In yogurt, the survival of probiotic organisms is very important while the growth of yeasts and moulds needs to be avoided considering the development of off-flavors which are detrimental to shelf life of the product. It has been demonstrated by Tamine and Deeth (1980) that headspace flushing of yogurt packages with CO_2 could extend the shelf life by inhibiting the growth of spoilage microorganisms. Another possibility to beneficially impact the shelf life of fermented dairy products could be the direct incorporation of CO_2 in the product. Karagul-Yuceer *et al.* (2001) reported that the incorporation of CO_2 into yogurt had no significant effects on desirable and undesirable microorganisms. They hypothesized that the addition of CO_2 to the product could feasibly stimulate the growth of starter bacteria. *E. coli, L. monocytogenes* and *B. licheniformis* microbial growth was monitored. Dissolved CO_2 did not show a significant effect on *L. monocytogenes* growth which declined in both CO_2 treated and untreated product during storage. *B. licheniformis* showed no growth under any conditions while significant differences were detected for *E. coli* which decreased to nondetectable levels in the CO_2-treated yogurt during 60 days of storage.

In an earlier study, Karagul-Yuceer *et al.* (1999) reported no differences in the sensory properties and consumer acceptance between CO_2-added and regular yogurts.

The possibility to produce yogurt from raw milk modified with CO_2 during storage has been considered and experimental studies have been carried out to detect any improved microbial quality without affecting the finished product. First experimental results were reported by Calvo *et al.* (1999) showing that the yogurt made from CO_2 added to skim milk was not significantly different, including lactic acid production, from control yogurts made from untreated milk.

Gueimonde *et al.* (2002) determined the effect of CO_2 addition to raw milk on the chemical, microbiological and sensory properties of yogurt. The results demonstrated that the multiplication and acidification capacities of the lactic starter (*Streptococcus thermophilus* and *Lactobacillus delbrueckii* ssp. *bulgaricus*) and the evolution of organic acids were not affected by the CO_2 treatment of raw milk and by residual CO_2 present in the pasteurized milk. The authors concluded that milk acidified with CO_2 could be satisfactorily used in the manufacture of yogurt considering also that no differences on sensory properties were detected during the cold storage between yogurts made from CO_2-treated milk and the control milk.

Later, Gueimonde and de los Reyes-Gavilan (2004) found no significant differences in the evolution of organic acids between yogurts made with CO_2-treated and untreated milk.

Vinderola *et al.* (2000) also evaluated the possibility of using CO_2-modified raw milk in the manufacture of fermented milk beverages. They found that CO_2 modification decreased milk fermentation time for *Streptococcus thermophilus*/*Lactobacillus acidophilus* and *Streptococcus thermophilus/Lactobacillus acidophilus/Bifidobacterium bifidum* fermented milk products. No negative impact on sensory characteristics of the milks was noted. Noriega *et al.* (2003) examined *Streptococcus thermophilus/Lactobacillus acidophilus/Bifidobacterium bifidum* fermented milk inoculated with *B. cereus* cells and found significant inhibition of the growth of pathogens in CO_2-modified milk during incubation at 37°C. The authors stated the efficiency of the treatment preserving the milk from *B. cereus* contamination. They also observed a reduction of proteolysis and acid production in the CO_2-treated milk during storage at 4°C.

9.8 Application of carbon dioxide to casein production

9.8.1 *Casein properties*

Caseins are a group of proteins comprising four major fractions: α_{s1} casein, α_{s2} casein, β casein and κ casein. They account for nearly 80% of the proteins in cow's milk and cheese. Casein can be recovered from milk by acid precipitation (acid casein) or by enzymatic treatment (rennet casein). The industrial production of acid casein is based on the use of inorganic acids such as HCl or H_2SO_4, although lactic acid fermentation is a well-known option. During the process most of the acid casein is converted to sodium, potassium and magnesium or ammonium caseinate depending on the application. Casein molecules are associated into casein micelles forming aggregates with sizes ranging from 20 to 200 nm. Within the casein micelles, smaller structures can be identified, called 'submicelles', that are 10–20 nm in size and are held together by colloidal 'bridges' of calcium phosphate connecting phosphatized serine groups of the proteins. Besides calcium phosphate, other compounds are involved in these bridges, such as magnesium and citrate. Calcium is bound to the protein not only via these bridges but also by association with negatively charged groups, along with other cations. During precipitation, the calcium phosphate dissolves, breaking the bridge and losing the micelle structure. In addition, associated cations are exchanged for protons. In the industrial production of casein, the dissolution and the ion exchange must be complete. The final caseinate product does not contain residual calcium and phosphate which strongly influences the functional properties such as emulsification and gelling properties (Walstra and Jenness 1984). The stability of the colloidal protein depends on two main factors: the coulombic repulsion of the equally charged protein molecules, neutralized when the pH is decreased to the isoelectric point, and the steric hindrance, caused by the glycosyl oligomers bounded to the α casein and stuck out like hair into the solution. The precipitation of casein is a strong function of the temperature. The temperature used during the process is about 40–50°C. At these temperatures, firm particles can be produced with a high expulsion of water from the precipitate without co-precipitation of whey protein.

9.8.2 Casein production by high-pressure carbon dioxide

Although high-pressure CO_2 processing has the ability to kill bacteria and inactivate enzymes (Hong and Pyun 2001), its application in the dairy industry is difficult. This is due to the effects that CO_2 may have on casein proteins.

It is known that casein proteins precipitate out at the isoelectric point of pH 4.6 (Fox 1992). When CO_2 dissolves in aqueous solution, a portion reacts with water to form carbonic acid which, in turn, dissociates into carbonate and hydrogen ionic species lowering the pH of the solution. If enough CO_2 is incorporated into the milk to reduce the pH to 4.6, the casein will precipitate out.

Processes which utilize CO_2 for precipitation of casein have been developed for industrial manufacturing (Tomasula *et al.* 1997). For most other dairy applications, as in the pasteurization of milk, the precipitation of casein would be a negative consequence of the process.

In casein manufacture, obtained through the action of lactic acid producing bacteria or by the addition of a mineral acid such as HCl or H_2SO_4, the precipitation step is followed by a holding or acidulation stage for the agglomeration of the curd and stages for casein and whey separation, casein washing, drying and milling (Southward and Walker 1980). The whey may be treated using ultrafiltration or diafiltration to produce whey protein concentrates (Morr 1989).

While this method is effective, there are some advantages to use CO_2 instead of lactic or mineral acid for casein precipitation. First, CO_2 used as the acidifying agent can be removed and potentially recovered from the whey after processing, adding value to the uncontaminated sweet whey. Additionally, no acid remains to contaminate the curd, leaving a purer casein product.

Jordan *et al.* (1987) studied the ability of CO_2 to precipitate casein from skim milk with CO_2. At 3.5 MPa and 50°C, the process allowed the precipitation of the 99% of the casein in skim milk. Playing on pressure as a control parameter of the process, Jordan found out that the pressures required for precipitation decreased as the temperature increased. Although the lowest temperature investigated was 40°C, the trend of the data implied that if the process would be carried out at lower temperatures no precipitation occurred. The physical nature of the curd changed with the pressure used. Higher pressures resulted in a firm, rubbery curd, while more mild treatments produced a soft and more friable curd.

Tomasula *et al.* (1995) developed a batch system to determine the effects of pressure, temperature, residence time and weight of milk on casein precipitation by dissolving CO_2. Casein precipitated only when CO_2 was bubbled through the milk in the reactor. If CO_2 was added to the headspace of the reactor, very little precipitation of casein was observed. In this system, pressures between 2.8 and 5.5 MPa and temperatures of 38°C, 49°C and 60°C were tested. They showed that residence time was not a significant factor in casein precipitation, and once the isoelectric point was reached, all susceptible casein precipitated. To this extent they set the treatment time to 5 minutes. Curd appearance was strongly influenced by temperature. At 38°C and either 2.8 or 5.5 MPa, the curd was moist and had an appearance similar to that of cottage cheese. At 60°C and either pressure, the curd was stringy dry with a rubbery feel, with no discernible particles.

They also reported that solids, ash and calcium contents of casein precipitated with CO_2 were higher than those found in casein precipitated using mineral acids. The process conditions also influenced the mean particle size. At constant milk mass and temperature, the mean particle size decreased as pressure increased. The same trend was observed decreasing the pH while increasing the temperature, the mean particle size increased.

In this experimental study some considerations have been made to set up design criteria of casein precipitation processes utilizing CO_2 at high pressure. A critical factor to be considered in the design of the system was the position of the CO_2 sparger in the reactor to assure the optimal contact between milk and CO_2. They also noticed that it was not possible to operate with large volumes of milk. The instantaneous precipitation of casein induced great difficulties in sparging CO_2 in a large volume of milk due to the poor contact between milk and CO_2. The introduction of a stirrer did not improve the process. It was able to assure a good contact between milk and CO_2, but it affected the quality of the precipitate which was broken by the movement of the stirrer. The design of a continuous process would avoid this problem ensuring that the mix of the reactants did not interfere with the precipitate.

The experimental data also showed that a key point in the design of a continuous process could be the measurement of the amount of CO_2 solubilized into the matrix. A high-pressure pH probe (Innovative Sensors, Inc., Anaheim, CA) was inserted directly into a vessel port to control the pH of the sample up to 6.9 MPa and in turn CO_2 concentration. The measured pH in the sample was 5.4 at a pressure of 6.9 MPa and remained constant, increasing the pressure up to 10.0 MPa. A buffering effect due to the interactions between Ca^{2+}, PO_4^- and other ionic species in the milk and H^+ and HCO_3^- ionic species from the dissolution of CO_2 occurred during the pressurization of the system (Tomasula *et al.* 1995).

Considering that the amount of CO_2 was the key parameter in the process, the solubility of CO_2 in milk was also measured by Tomasula and Boswell (1999). They observed the same trend as the solubility of CO_2 in pure water. At 25°C the value of the solubility in milk was lower than that in pure water in the entire range of pressure. In particular, its value was the 77% of the solubility in water at low pressure, while at high pressure it was approximately 90%. At 38°C and pressures higher than 3.5 MPa, the solubility in milk reached that of CO_2 in water within ±5%. Although the pH value due to the dissolved CO_2 was 5.4, the curd produced with CO_2 was not as fibrous or sticky as that produced by HCl precipitation at the same pH. In addition, the whey of casein precipitated by CO_2 had the slightly yellow, opalescent appearance of acid whey even though pH was 5.8 and not 4.6.

Further studies on the precipitation of casein using CO_2 were carried out by Hofland *et al.* (1999). The authors measured the pH, casein yield and mineral content during the process. In particular, they investigated the functional properties of CO_2-precipitated casein and the mechanism of the process taking into account the high pressure, the high calcium concentration and the pH as factors causing precipitation. The experiments showed the trend of the pH in milk and water in the range of temperature between 5°C and 50°C and pressure up to 6.0 MPa. As stated previously (Tomasula *et al.* 1995, 1999a), considering the buffer capability of the components, the pH decreased much less in milk than in water. Although the isoelectric point was not reached during CO_2 solubilization, the precipitation of casein as flocks was visually observed at pH values ranging from 5.1 to 5.3. Calculations indicated that the milk should be diluted to a 4.46 wt.% water solution to obtain a pH close to the isoelectric point. Comparing the CO_2-assisted precipitation process with the conventional acid precipitation process with sulfuric acid, it was observed that the yield approached 100% even though the isoelectric point of 4.6 was not reached. For this reason the authors stated that the primary cause of the casein precipitation was the acidification of the solution. The analysis on the calcium and phosphate content of the whey showed a low calcium concentration for milk treated with CO_2 at 40°C and 5.5 MPa, probably due to the faster precipitation kinetics resulting in a compact and inaccessible precipitate with a negligible diffusion of the calcium phosphate in the whey.

Calvo and Balcones (2001) carried out experiments on skim milk in a batch system showing that while caseins were precipitated, whey proteins were not precipitated at the same process conditions.

The mildest treatment (2 MPa, 30 min and 40°C) precipitated 35% of the caseins, while the most severe treatment (5 MPa, 180 min 40°C) precipitated 87% by weight.

Raouche *et al.* (2007) studied the stability of casein micelles subjected to reversible acidification by carbonation. CO_2 was injected at 4°C and a pressure of 0.2 MPa with an increase of 0.1 MPa each 3 minutes until the pH dropped to 5.2. The milk was held in these conditions for 15 and 60 minutes. They showed that the milk subjected to drastic acidification had a significant destabilization of the micelles which seemed to return to their initial state when the pH rose from 5.2 to 6 after the treatment. The holding time (15 or 60 minutes) did not have any influence on the micelle composition. The calcium and protein content of the carbonated milk were the same as those of the nontreated one. CO_2-reversible acidification did not induce any change in the micelle average size demonstrating that before and after the treatment the mean diameter of the micelles was about 200 nm. They also demonstrated the stability of the carbonated casein micelles during a cold storage for 6 days, although a possible reorganization of the micelles and their surface could occur.

9.8.3 *Comparison between continuous and batch systems for casein production by carbon dioxide*

The optimum design and scale-up of the process are the key factors for the implementation of this technology. It was shown that batch processes are effective only for the recovery of small amounts of high-valued products, and that a large-scale batch production requires significant downtime for depressurizing, cleaning and refilling of the system. The economical production of food proteins is possible only if the design of a continuous process, consisting of a reactor-precipitator, and the setup of a system for continuous pressure reduction are considered.

Reactor-precipitator design requires a precise knowledge of the coagulation time of the protein and of the manner in which the reactants are in contact to optimize quality and yield. Coagulation time for casein precipitated by CO_2 has not been determined accurately.

Jordan *et al.* (1987) showed that in a 120 ml reactor the same amount of casein precipitated by CO_2 was obtained in a test carried out for 1 and 5 minutes. Tomasula *et al.* (1995), working with a 1 l batch reactor, showed that the coagulation time was at most 5 minutes. However, the difficulties in manipulating the pressurized reactors did not allow the determination of the exact coagulation times. Muller and Hayes (1962) showed that coagulation started about 0.1 seconds after mixing the acid with skim milk at 43–46°C. Jablonka *et al.* (1988) showed, in light-scattering experiments of batch and continuous reactors at different temperatures and pH, that coagulation time for casein precipitated by mineral acid was less than 1 second. Acid–milk mixing was identified as the rate-controlling step for the formation of precipitate and was critical for curd development.

Coagulation rate control is also an important factor in reactor-precipitator design because it affects curd quality. Muller and Hayes (1962) attributed the formation of a sticky curd or of an inconsistent one disintegrating during further processing to the precipitation at pH higher than

4.4 and higher temperatures. Methods for the regulation of the coagulation rate included the lowering of the mixture temperature and the use of pumps to ensure a constant flow of milk and acid. The mixture of milk and acid was gradually heated to the coagulation temperature and then held for curd agglomeration prior to casein–whey separation. Jablonka and Munro (1987) also found that particle size, solids content, cohesiveness, firmness and the curd calcium content increased with increasing pH and temperature.

For continuous recovery of CO_2-precipitated casein, it was necessary to keep the pressure constant in the reactor while removing the precipitate and whey. Another requirement for continuous recovery is maintaining the integrity of the casein curd. Particle size distribution is used as a measure of the quality of freshly precipitated casein (Jablonka and Munro 1987). While a standard particle size distribution of wet casein must is not defined, it is generally accepted that freshly precipitated casein be comprised of distinct curds. Larger curds are desirable because they are easily transported and handled, with minimal losses in the washing and drying phases encountered in casein manufacture. Small particles tend to be too soft, while too-large fused particles indicating heat damage are unacceptable because they tend to be tough and rubbery, thus very difficult to handle in subsequent operations.

Tomasula *et al.* (1997) suggested the design of a continuous process that uses high-pressure carbon dioxide instead of organic acids as a precipitant to isolate casein from milk. The experimental activity was performed in order to highlight the best equipment design to obtain casein of good quality and high yield. Two reactor-precipitators were designed and tested. One was a spray reactor in which milk and CO_2 were in contact by spraying milk into CO_2. The other was a tubular reactor in which a mixture of liquid CO_2 and milk was fed. They showed that the quality of casein depended on temperature and pressure process conditions. The experiments in the tubular reactor were carried out at 32°C, 38°C, 43°C and 49°C and at pressures of 5.52, 6.9 and 7.6 MPa. The experiments in the spray reactor were carried out at the same temperatures but at pressures of 4.13 and 5.52 MPa.

Casein produced at 32°C in both systems was watery. A firm product with distinctive particles was obtained for all pressures and temperatures in the range from 38°C to 49°C with the tubular reactor while, when using the spray reactor, the best product was obtained at 38°C or 43°C and 4.13 MPa. The results demonstrated the great potential of this technology as an alternative to processes utilizing strong acids or enzymes. The casein showed high calcium contents and a whey with a high pH. The possibility of a continuous process utilizing a pump to keep the pressure constant in the reactor-precipitator while allowing the removal of casein and whey without altering the system pressure and temperature conditions could be of great interest for industrial applications.

9.8.4 Economic comparison between high-pressure carbon dioxide and a conventional process for casein production

The experimental results discussed in the previous section highlighted the advantages of the use of high-pressure CO_2 for casein production. In particular using CO_2 as a precipitant allows eliminating it from the whey upon release of pressure while this is not possible with mineral acids or salts which remain in the curd. Moreover, it is not necessary to neutralize the acids in the whey because after the depressurization of the system the pH returned to its initial value. The product has a high calcium content (1.5%) which is about the same as the content of caseinate prepared by the addition of $Ca(OH)_2$ to casein.

Considering all these advantages, Tomasula *et al.* (1998) proposed a continuous process to produce casein using carbon dioxide as precipitant. An economic analysis was proposed in which costs were estimated for a continuous CO_2-precipitated casein-processing plant and compared to that one encountered in the precipitation steps for the production of mineral acid and lactic caseins.

The plant operating with mineral acid (Figure 9.1) processed 10^4 liters of milk/h with a caseinate output of 320 kg/h.

Casein manufacture started after the fat was skimmed from the milk and the milk was pasteurized. During the process, a series of precipitation steps with curd formation occurred followed by separation of the curd from the whey, curd-washing steps, curd pressing, drying, tempering, milling, sieving, blending, bagging and storage (Southward and Walker 1980). Casein was precipitated using a mineral acid, such as HCl or lactic acid producing bacteria. If the precipitation with HCl was considered, two HCl storage tanks were sized for a capacity of about 90 000 L. The tanks were fed with a 5 hp centrifugal pump. Also a dilution tank with a static mixer was used to assure a good contact between the acid and the milk.

The flow sheet of a proposed commercial process for casein precipitation using CO_2 was designed by Tomasula *et al.* (1998) (Figure 9.2).

The process took into account the recovery and recycle of the CO_2 but it could consider also the possibility that none of the CO_2 used in the precipitation was recovered. The process included a storage and refrigeration system for liquid CO_2 and a high-pressure pump to pump milk at 4°C from atmospheric pressure to the line pressure of 6.9 MPa. CO_2 at temperature of −18°C was pumped from the storage of 2.07 MPa to the line pressure. A second mixing chamber was designed with a static mixer ensuring good contact between the milk and CO_2 streams. The reactor was a double-pipe heat exchanger used to heat the milk from 3°C to 38°C with a residence time of 5 minutes. After precipitation of the curd, the casein, whey and CO_2 moved to a progressing cavity pump for the reduction to atmospheric pressure. At this point CO_2 could be recovered from the whey, compressed to 2.07 MPa, dried and cooled from 38°C to −14°C. To continue

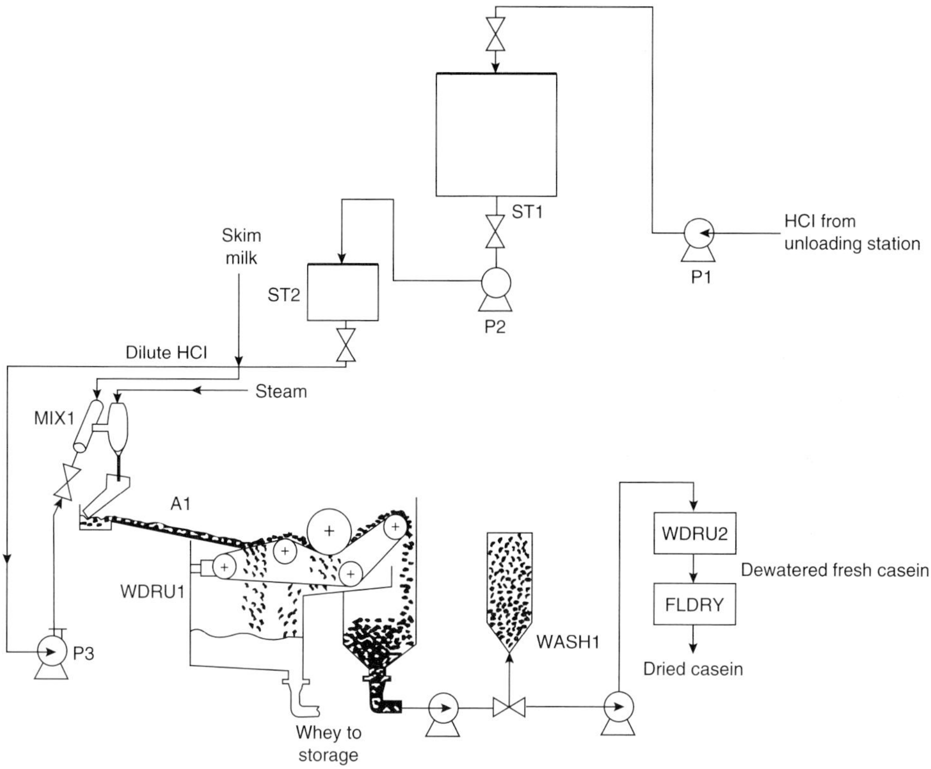

Figure 9.1 Schematic diagram of a casein-processing plant (Tomasula *et al.*, copyright 1998). P1: Pump; ST1: HCl storage tank; P2: pump; ST2: dilution tank; P3: pump; MIX1: acid–milk mixing chamber; A1: acidulation pipeline with steam injection; WDRU1: dewatering drum; WASH1: tower washer for casein; WDRU2: dewatering drum; and FLDRY: fluid-bed dryer. Reproduced with permission of Elsevier.

the process a makeup of CO_2 was taken into account to replace CO_2 that was not recovered. The whey was then separated from the casein and dried as in the conventional process.

The cost analysis of both processes showed that the recovery of CO_2 was an important factor. If CO_2 was not recycled, the process was not economical.

The analysis also confirmed a potential limitation of the use of high-pressure CO_2 in large-scale processes due to the availability of large sanitary pumps to reduce pressure and to remove casein from the process continuously. Pumps of high capacity were not available and needed to be made by special order. An alternative to this problem could be the design of smaller pumps arranged in parallel in order to replace one large pump, but this could increase the operating costs. Dewatering and washing costs were less because CO_2 did not bring water into the process with the precipitant. The higher costs of the high-pressure CO_2 process compared to the conventional process were linked to the operating costs due to the higher utility costs associated with operation of the high-pressure pump.

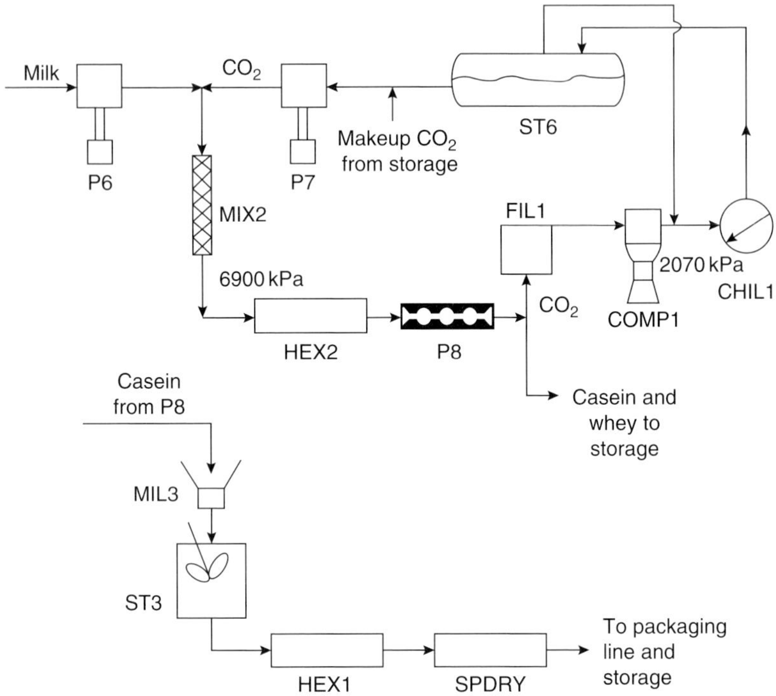

Figure 9.2 Schematic diagram of a proposed CO_2-precipitated casein-processing plant (Tomasula *et al.*, copyright 1998). Dewatering, and washing, and pressing of casein following P8, not shown in the figure. P6: High-pressure pump; ST6: CO_2 mix tank; P7: CO_2 pump; MIX2: mixing chamber; HEX2: heat exchanger; P8: progressing cavity pump; FIL1: CO_2 filter; COMP1: CO_2 compressor; CHIL1: CO_2 chiller; MIL3: cone mill; ST3: tank; HEX1: heat exchanger; and SPDRY: spray dryer. Reproduced with permission of Elsevier.

However, the calculations for return on investment, which are useful for comparing alternative innovative processes, showed that the DPCD process is competitive. It could be considered as an economical alternative to the use of acids and disproves the idea that processing with DPCD is significantly more expensive than processing with acids.

9.9 Conclusions

The relatively short shelf life and rapid loss of quality coupled with the desire to consolidate manufacturing into larger plants have stimulated the study of the application of the DPCD process to dairy products. This chapter discussed the feasibility of the process, also highlighting its limits and advantages on the quality of the treated products.

Additional research is needed to increase the efficiency of the process and contribute to a better understanding of the fundamental basis of the action of CO_2 applied to dairy products.

10 Particle Engineering by Dense Gas Technologies Applied to Pharmaceuticals

Neil R. Foster, Raffaella Mammucari, Luu Thai Danh, and Wen Hui Teoh

Abstract: Dense phase carbon dioxide is used in pharmaceutical processing since it is nontoxic, nonflammable, inexpensive, and easily accessible. For decades it has been utilized in food and chemical processing due to its ability to provide a high degree of purification, and to produce submicron particles with uniform particle size distribution. In recent years, DPCD technology has been explored and used in pharmaceutical processing to obtain ultra-high-purity active pharmaceutical ingredients and, more significantly, to modify their physical characteristics regarding particle size distribution, crystalline structure, and surface topography crucial for the drug's stability, bioavailability, ease of delivery, and clinical performance. The aim of this chapter is to describe the application of dense phase CO_2 as a solvent or antisolvent to pharmaceutical processes, presenting the experimental equipment, the main results obtained from the application of the technology, and the advantages of using pressurized CO_2.

Keywords: particle size distribution; micronization; microencapsulation; bioavailability.

10.1 Introduction

Dense gas technologies (DGT) have been utilized for decades in food and chemical processing due to their ability to provide high-grade purification and to produce submicron particles with uniform particle size distribution (PSD). In recent years, DGT have been explored and used in pharmaceutical processing to ultra-purify active pharmaceutical ingredients (APIs) and, more significantly, to modify the physical characteristics of the APIs. Physical modifications of an API such as its PSD, crystalline structure and surface topography are crucial to a drug's stability, bioavailability, ease of

---

Dense Phase Carbon Dioxide: Food and Pharmaceutical Applications, First Edition. Edited by Murat O. Balaban and Giovanna Ferrentino.

delivery and clinical performance. Hence, considerable research energy has been directed towards re-engineering the morphology of pharmaceutical compounds in dense gases.

The early investigations of solid precipitation in dense gases (DGs) began with the use of DGs as solvents. Pharmaceutical compounds were dissolved into DGs that were then rapidly expanded, thereby generating submicron particles. This concept of solid precipitation was then further developed into using DGs as antisolvents whereby DGs are used to dissolve in and expand organic solvents to precipitate solids contained in an organic solution. Micronisation plays an important role in the engineering of drug particles since it increases the surface area and, in many cases, results in increased drug bioavailability and dissolution rates (Kayrak *et al.* 2003; Perrut *et al.* 2005a).

The most commonly used DG is carbon dioxide since it is nontoxic, nonflammable, inexpensive and easily accessible. This circumvents the need for toxic organic solvents in the micronisation of drug particles (Johnston 1994; Mishima 2008).

The therapeutic effectiveness of APIs can be enhanced by maintaining a constant drug concentration in the body at some optimum, therapeutic level over the dose time. Controlled release delivery systems are beneficial in that optimum concentration of APIs can be maintained over a longer period of time. Reducing concentration fluctuations can increase the half-life and effectiveness of APIs in the body. Subsequently, smaller doses of drug are required which can reduce side effects.

The bioavailability of certain proteins, peptides, enzymes, hormones and genetic materials fall into this group of APIs. The bioavailability of drug molecules can be improved by controlled delivery. Controlled delivery systems are generally classified as matrix systems or reservoir systems. In a matrix system, the drug is encapsulated within a host matrix, usually a polymer, forming a drug–polymer composite. In a reservoir system, a drug phase is dispersed within a host phase where the rate of release is dependent on the diffusivity of the drug through the host wall. Biodegradable and bio-erodable polymers are generally used as these degrade into nontoxic materials over time. The rate of release depends on the diffusivity of the drug through the matrix, or the dissolution rate of the matrix. Drug encapsulation also allows for site targeting that limits the side effects of the drug involved. The size of the APIs affects the efficacy of the controlled released formulations, with smaller particles giving to higher absorption rates.

Conventional micronisation and encapsulation techniques are generally unable to produce the desired particle characteristics on the production level. Additional drawbacks of conventional processes include the extensive use of toxic solvents, chemical and thermal degradation of the APIs, and the presence of surface defects on particles.

Dense gas technologies can respond to the limits of conventional techniques by utilising the peculiar properties of fluids in the proximity of their critical points. A compound is supercritical if both the temperature and the pressure are above the critical values; a dense gas is a compound close to, but not necessarily above, the critical point. As a fluid approaches its critical point peculiar effects are observed. The isothermal compressibility and the heat capacity, at constant pressure, become infinitely large. The density and dielectric constant of the fluid increase sharply as the pressure increases through the critical point. Although the densities approach liquid-like values, the transport properties such as diffusivity and viscosity remain gas-like. The combined properties of liquid-like densities, gas-like diffusivities and viscosities and zero surface tension have been the catalyst for the development of DG-based processes. Additionally, the sharp changes in the physical properties of fluids in proximity of the critical point allow for their tuning by fine adjustments in the operating conditions conferring to the technology a distinctive versatility.

The advantages of DGT in regard to the micronisation and microencapsulation of APIs include moderate operating temperatures suitable for thermally labile compounds and the ability to control the morphology of products.

10.2 Dense gas as a solvent

Dense gases can be utilized as solvents as their solvation power is similar to that of liquids. The solvation power of a DG is directly related to its density and is manipulated through small temperature and pressure changes. Solutes tend to have higher solubilities in DGs as the pressure is increased. The liquid-like densities of SCFs and DGs allow them to act as solvents suitable for various pharmaceutical applications allowing solutes to dissolve and form DG-solute solutions.

10.2.1 Rapid expansion of supercritical solutions

The rapid expansion of supercritical solutions (RESS) method is an easy two-step process that utilizes the solvation power of DGs to re-engineer the morphology of pharmaceutical and neutraceutical compounds. It is particularly useful for the micronisation and encapsulation of compounds with low aqueous solubility and has been utilized to produce solvent-free micron and submicron particles, fine fibers and thin films (Lele and Shine 1992; Helfgen *et al.* 2001). In the RESS process, a pharmaceutical or neutraceutical compound is first dissolved into a DG which is then rapidly expanded through a fine throttling device (usually a nozzle or a capillary). The rapid expansion causes a significant decrease in the density and the solvation power of the DG, leading to supersaturation. The solutes consequently precipitate out of the DG solution. The high supersaturation ratios and the rapid mechanical perturbation occurring during expansion create uniform conditions within the nucleating media allowing for the formation of small particles with narrow particle size distributions (PSDs) (Tom *et al.* 1993). A schematic diagram of the RESS process is given in Figure 10.1.

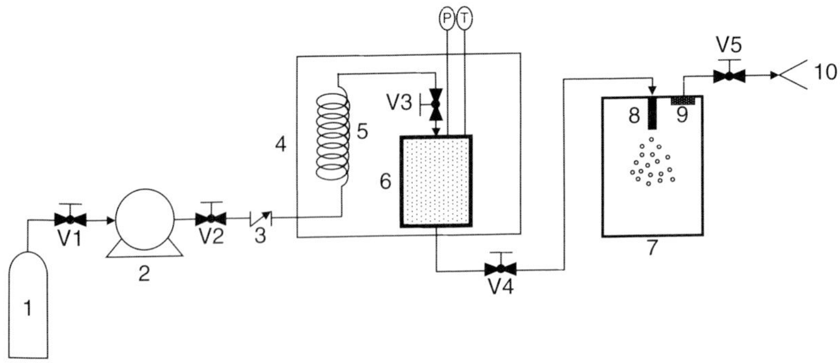

Figure 10.1 Schematic diagram of the RESS process.

The ability of pharmaceutical and neutraceutical compounds to solubilize in DGs is central to the RESS process. In many cases, DG carbon dioxide (DG CO_2) is used as the only solvent in the RESS process, circumventing the need for toxic organic solvents. Pharmaceutical compounds that have been micronized by the RESS process utilising DG CO_2 include nifedipine, lidocaine, griseofulvin, aspirin and lovastatin (Domingo *et al.* 1997; Frank and Ye 2000; Martin *et al.* 2000; Perrut *et al.* 2005a). However, many drugs show limited solubilities in DG CO_2 and the RESS processing of any compound with solubility below 10^{-3} kg/kg-DG has been shown to be commercially not viable (Mishima *et al.* 2000; Perrut and Clavier 2003; Calderone and Tallon 2008; Reverchon *et al.* 2009). Hence, organic solvents are used as the dense gas solvent or are added as a co-solvent to the DG CO_2 to increase drug solubilities in DG CO_2 (Perrut *et al.* 2005a, b; Mishima 2008). A number of studies have been conducted involving the use of dimethyl ether as the supercritical solvent and ethanol and trifluoromethane (CHF_3) as co-solvents in the micronisation of nifedipine, phenacetin and taxol (paclitaxel) (Perrut *et al.* 2005a, b; Yıldız *et al.* 2007). However, the use of organic solvents necessitates an extra washing step to purge the products of minuscule toxic residuals that remain. The extra processing requires longer production time and higher CO_2 volume, thus reducing the economic efficiency of the RESS process. A number of pharmaceuticals, neutraceuticals and DG solvents used in the RESS process are listed in Table 10.1.

The morphology and the particle size of drug produced by the RESS process are affected by the operating parameters such as the pre- and post-expansion temperatures and pressures, solute concentration, and length-to-diameter ratio of the nozzle or capillary (Lele and Shine 1992, 2002; Gosselin *et al.* 2003; Huang *et al.* 2005; Yildiz *et al.* 2007; Varshosaz *et al.* 2009). In most cases, particle size increases with increasing pre-expansion temperature, decreasing pre-expansion pressure, increasing nozzle diameter and lower concentration and solubility in DG (Turk 2009). The dependence of particle properties on the post-expansion temperatures and pressures are evident, although the results varied considerably.

Domingo *et al.* (1997) found that an increase in the post-expansion temperature had no effects on the crystal size of benzoic acid while Alessi *et al.* (1996) observed an increase in progesterone particle size as they increased the post-expansion temperature from 40°C to 60°C. Huang *et al.* (2005) found an increase in the post-expansion temperature commensurate with a decrease in the particle size of aspirin. Meanwhile, an increase in the post-expansion pressure was found to correspond with an increase in the particle size of progesterone and a change in the particle morphology (Alessi *et al.* 1996). In contrast, Huang *et al.* (2005) observed a decrease in particle size when they increased the post-expansion pressure while Liu and Nagahama (1996) found that a pressure decrease in the post-expansion chamber did not affect the particle morphology of naphthalene.

Generally, an increase in nozzle diameter increases the particle size of a drug compound while changing the nozzle length-to-diameter ratio (L/D) also changes the morphology of the product (Lele and Shine 1992). Su *et al.* (2009), Charpentier *et al.* (2008) and Alessi *et al.* (1996) concluded that a nozzle of smaller diameter yielded smaller particles while Liu and Nagahama (1996) observed no correlations between the particle size of naphthalene and the diameter of an orifice. Spraying distance between the tip of the nozzle and the stub also affects particle size and their degree of aggregation (Charoenchaitrakool *et al.* 2000; Yildiz *et al.* 2007). In addition, the type of phase behaviour and proximity to the phase boundaries during the RESS process affects the morphology of the particles while the effects of co-solvents on particle size were also noted by Lele and Shine (1992, 2002) and Yildiz *et al.* (2007).

In the microencapsulation of drugs via the RESS process, both the drug and polymer must be soluble in a DG. Previous studies have shown the successful encapsulation of naproxen in poly(L-lactic acid) between 90–115°C and 170–200 bar (Kim *et al.* 1996)and of 2,5-distyrylpyrazine in polyacrylate at 110–130°C and 275–483 bar (Pestov *et al.* 2003). However, the conduct of the RESS process at temperatures and pressures above 80°C and 200 bar is not economical due to the higher energy input required (Tandya *et al.* 2007). The RESS process is also limited by the low solubility of most polymers in DGs at temperatures and pressures below 80°C and 200 bar and the sensitivity of thermally labile drugs to high temperatures. Therefore, it is only practically viable when it is conducted at relatively low temperatures and pressures. Recent RESS applications had shown successful encapsulations of drugs at moderate temperatures and/or pressures. Rosekranz *et al.* (2008) encapsulated insulin in paraffin between 40–55°C and 80–120 bar while naproxen was encapsulated in poly(L-lactic acid) between 50–70°C and 200–400 bar with the use of a pneumatically operated nozzle in pulsed mode (Gadermann *et al.* 2009). A number of drugs encapsulated by the RESS process are listed in Table 10.1.

The RESS process is an attractive and easy method for the production of uniform submicron particles and composites. In many cases, the

Table 10.1 Pharmaceuticals and neutraceuticals re-engineered by the RESS process.

Substances	Solvent	References	Substances	Solvent	References
Micronization					
Acetaminophen[a]	DME	Calderone and Tallon, 2008	Racemic Ibuprofen + S-ibuprofen[d]	CO_2	Hermsdorf *et al.*, 2007
Aspirin	CO_2	Domingo *et al.*, 1997	Racemic ibuprofen	CO_2	Charoenchaitrakool *et al.*, 2000
Aspirin	CO_2	Huang *et al.*, 2005	Lidocaine	CO_2	Frank and Ye, 2000
Astaxanthin	CO_2	Quan *et al.*, 2009	Lidocaine + poloxamer 188 (Lutrol®, BASF)	DME	Perrut *et al.*, 2005b
Barbital	CO_2	Shinozaki *et al.*, 2006	Lovastatin	CO_2	Perrut *et al.*, 2005a
Beclomethasone-17, 21-dipropionate	CO_2	Charpentier *et al.*, 2008	Medroxyprogesterone acetate	CO_2	Alessi *et al.*, 1996
Benzoic acid[b]	CO_2	Domingo *et al.*, 1996	Mefenamic acid	CO_2	Hezave and Esmaeilzadeh, 2010
Benzoic acid	CO_2	Domingo *et al.*, 1997	Meloxicam[e]	CO_2	Chiou *et al.*, 2007
Benzoic acid	CO_2; CHF_3	Helfgen *et al.*, 2001	Nabumetone	CO_2	Su, Tang *et al.*, 2009
Benzoic acid	CO_2	Turk *et al.*, 2002	Naphthalene + Phenantrene	CO_2	Liu and Nagahama, 1996
Carbamazepine	CO_2	Gosselin *et al.*, 2003	Nifedipine	DME	Perrut *et al.*, 2005a
β-Carotene	Ethene; Ethene + toluene	Chang and Randolph, 1989	Nifedipine + poloxamer 188 (Lutrol®, BASF)	DME	Perrut *et al.*, 2005b
Cefuroxime axetil	CO_2	Varshosaz *et al.*, 2009	Phenacetin	CO_2; CHF_3	Loth and Hemgesberg, 1986
Chitin in hexafluoroiso propanol solution[c]	CO_2	Salinas-Harnalndez *et al.*, 2008	Phenylbutazone	CO_2	Moribe *et al.*, 2005
Cyclosporine	CO_2	Tandya *et al.*, 2006	Phytosterol	CO_2	Jiang *et al.*, 2003
Dihydro-artemisinin (DHA)	CO_2	Chingunpitak *et al.*, 2008	Progesterone	CO_2	Alessi *et al.*, 1996
Griseofulvin	CO_2	Martin *et al.*, 2000	Sulfathiazole + poloxamer 188 (Lutrol®, BASF)	DME	Perrut *et al.*, 2005b
Griseofulvin	CHF_3	Turk *et al.*, 2002	Salicylic acid	CO_2	Domingo *et al.*, 1997
Griseofulvin	CHF_3	Helfgen *et al.*, 2001	Salicylic acid	CO_2; CO_2 + EtOH	Yildiz *et al.*, 2007

Ibuprofen in phosphate buffer (pH 6.3)	CO_2	Foster *et al.*, 2003a	β-Sitosterol	CO_2	Turk *et al.*, 2002
Ibuprofen	CO_2	Kayrak *et al.*, 2003	Stigmasterol	CO_2	Ohgaki *et al.*, 1990
Ibuprofen[d]	CO_2	Hermsdorf *et al.*, 2007	Taxol (Paclitaxel)	CO_2; CO_2 + EtOH	Yildiz *et al.*, 2007
Ibuprofen	CO_2	Hezave and Esmaeilzadeh, 2010	Tolbutamide	CO_2	Shinozaki *et al.*, 2006
Encapsulation					
Bovine serum albumin in paraffin[f]	CO_2	Rosenkranz *et al.*, 2008	Naproxen in poly(D, L-lactic acid)	CO_2	Kim *et al.*, 1996
2,5-Distrylpyrazine in diethyl p-phenylenediacrylate	F-22	Pestov *et al.*, 2003	Naproxen in poly (D,L-lactic acid)[d]	CO_2	Gadermann *et al.*, 2009
Insulin in paraffin[f]	CO_2	Rosenkranz *et al.*, 2008	Phytosterol in poly (D,L-lactic acid)	CO_2	Turk *et al.*, 2006
Lovastatin in poly (D,L-lactic acid) and poly(hydroxyl acid)	CO_2	Tom *et al.*, 1993	Retinyl palmitate in poly(D,L-lactide)[g]	CO_2	Sane and Limtrakul, 2009
Lysozyme in poly (D,L-lactic acid)	CO_2	Whitaker *et al.*, 2005	Ribonuclease A in poly (D,L-lactic acid)	CO_2	Whitaker *et al.*, 2005

CHF_3 = trifluoromethane; F-22 = cholorodifluoromethane; DME = dimethyl ether; EtOH = ethanol.
[a] Expansion of a flash boiling solution–solution passed through a liquid–vapor phase transition during depressurization.
[b] Capillary or orifice replaced by porous sintered metal plate.
[c] RESS and RESOLV were conducted.
[d] Pneumatically controlled nozzle was used in pulsating mode.
[e] Filtration membrane added prior to expansion chamber.
[f] Combined high-pressure fluidized bed with RESS.
[g] Rapid expansion of supercritical solutions into an aqueous receiving solution (RESOLV).

micronisation of APIs using the RESS process resulted in higher dissolution rates, a result of increased surface area (Frank and Ye 2000; Martin *et al.* 2000; Jiang *et al.* 2003; Perrut *et al.* 2005a, b). However, the applicability of the RESS process is limited by the low solubility of polar drugs and polymers in DGs and the agglomeration of fine particles. These limitations are compounded by the plugging of nozzles and capillaries during expansion and the difficulties in scaling up such throttling devices. The production of a polymer encapsulated drug usually requires very high pre-expansion pressure to generate high yields. Hence, high energy input is required which reduces the economic return of the RESS process.

Depending on the requirement of a product, various modifications of the RESS process have been developed. These modifications include expanding a supercritical solution into aqueous solutions (RESOLV) to prevent agglomerations, adding a solid compound or a nonsolvent into the process to increase drug solubility, pulsating the nozzle to avoid plugging and regulating partial depressurisation of the dense gas solutions to allow for the recycling of a higher pressure DG to reduce energy usage.

10.2.2 Rapid expansion of supercritical solutions with a solid solvent

Domingo *et al.* first introduced the concept of adding a solid compound into a binary drug–DG solution to increase the drug solubility (Domingo *et al.* 1999). Thakur and Gupta (2006) further expanded this concept by introducing the rapid expansion of supercritical fluids with a solid co-solvent (RESS-SC), whereby a solid co-solvent is added into the drug–DG solution to increase the solubility of the drug in a ternary drug–DG–solid co-solvent solution. The introduction of the solid co-solvent also hinders particle growth, prevents agglomeration and results in nanoparticle formation. The solid co-solvent is later removed by sublimation. The particle sizes of drugs produced with the RESS-SC method were found to be smaller and the production rates were higher than with the RESS process (Thakur and Gupta 2006). Thakur and Gupta (2005 and 2006) stated that the solid co-solvent should have sufficiently high vapour pressures for easy removal by sublimation, behave as a solid at exit condition and not react with the drug processed. The co-solvent should also be nonflammable, nontoxic, soluble in the DG and inexpensive. Some of the drugs engineered under the RESS-SC process include phenytoin and griseofulvin in DG CO_2 with menthol as the solid co-solvent (Thakur and Gupta 2005, 2006).

10.2.3 Rapid expansion of supercritical solutions with a nonsolvent

Mishima *et al.* (2000) proposed the rapid expansion of supercritical fluids with a nonsolvent (RESS-N) to overcome the limitations of the RESS process in cases where insoluble polymers are used to encapsulate pharmaceutical compounds. In this process, co-solvents, which behave as nonsolvents for the polymers at ordinary conditions, were added to the DG to facilitate the formation of non-agglomerated polymeric microparticles. The addition of the nonsolvent can increase the solubility of the polymers at supercritical and near-supercritical conditions and prevent the swelling and the agglomeration of certain polymeric microparticles during expansion (Mishima *et al.* 2000). The RESS-N process was able to produce fairly monodispersed PSDs with particle sizes being regulated by the feed composition (Mishima *et al.* 2000; Matsuyama *et al.* 2003). The thickness of the coating on the drugs can also be controlled through the polymer-feed composition (Matsuyama *et al.* 2003). A number of studies were conducted on the encapsulation of *p*-acetamidophenol, acetylsalicylic acid, 1,3-dimethylxanthine, flavones and 3-hydroxyflavone in polyethylene glycols [PEG4000, PEG6000 and PEG20000], poly(methylmethacrylate) [molecular weight: 5000], poly(L-lactic acid) [molecular weight: 5000] and PEG–PPG–PEG triblock copolymer (molecular weight: 13 000; PEG/PPG = 0.85/0.15) (Matsuyama *et al.* 2003). Methanol, ethanol and 1-propanol are the more commonly used nonsolvents (Mishima *et al.* 2000; Matsuyama *et al.* 2003).

10.2.4 Particles from gas-saturated solutions

In order to overcome the limitations of the RESS process, Weidner *et al.* (1995) developed a new process called particles from gas-saturated solutions (PGSS). In the PGSS process, the dense gas (generally CO_2) is dissolved in a solid matrix leading to viscosity reduction and, in some cases, melting point depression. Due to the absorption of DG, the solid might either swell or melt at temperatures significantly lower than the normal melting or glass transition temperature. The concentration of DG increases with pressure forming a gas-saturated melt. Rapid depressurisation of this melt through a nozzle causes the dense gas to vapourize and lowers the temperature (Joule–Thompson effect). The result is the formation of particles (Weidner *et al.* 1995, 2003). The schematic diagram of the PGSS process is presented in Figure 10.2.

In pharmaceutical application, the PGSS process has been used mainly for microencapsulation of drugs in polymer matrix where high amounts of DG can be dissolved. The PGSS process offers a mechanism for controlling both macro- and microporosity via a single step process that can be used to control the characteristics of microencapsulated drug formulation (Tandya *et al.* 2007). Upon addition of a DG, the polymers can melt due to the reduction of the glass transition temperature, resulting in the formation of a viscous suspension that can be subsequently atomized thereby producing composite microparticles containing the suspended drug (Reverchon *et al.* 2009). The microencapsulation process is suitable for the production of thermally sensitive polymer composites, such as D, L-PLA or PLGA containing bioactive compounds, namely ribonuclease A, catalase and β-D-galactosidase (Howdle *et al.* 2001). The composites of

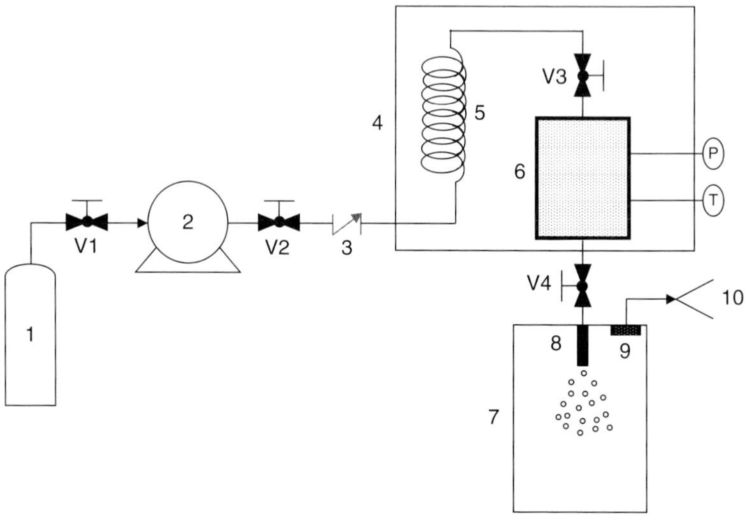

Figure 10.2 Schematic diagram of the PGSS process.

nifedipine–PEG and felodipine–PEG formed by the PGSS process had a dissolution rate much higher than that of raw materials (Kerc *et al.* 1999; Sencar-Bozic *et al.* 1997).

For particle production, the PGSS process has been successfully used for the micronisation of a limited number of pharmaceutical compounds. Nifedipine and felodipine were processed by the PGSS process to form microparticles with mean particle sizes between 15 to 42 μm (Kerc *et al.* 1999; Knez 2000). Similarly, micron-sized cyclosporine was successfully produced by PGSS with particle size less than 1 μm, and it exhibited a 97% reduction in size compared with the unprocessed cyclosporine (Tandya *et al.* 2006).

The particle characteristics can be controlled by manipulation of operating parameters, such as initial solute concentration, pre-expansion temperature and pressure, expansion temperature and pressure and nozzle geometry (Tandya *et al.* 2007).

10.3 Dense gases as antisolvents

Dense gas antisolvent methods were first developed by Gallagher *et al.* (1989) to process compounds with poor solubility in DGs. The techniques involve using a DG as an antisolvent to induce precipitation of dissolved solids from solutions of organic solvent. The processes are based on the solubility and high diffusivity of DGs into conventional solvents. The combined effects of high solubility and efficient mass transfer result in the fast expansion of conventional solutions containing the compound to be precipitated. The dramatic reduction of solution density significantly

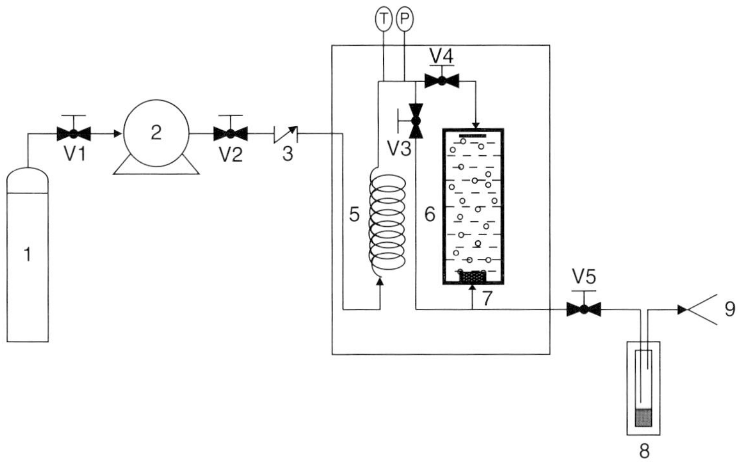

Figure 10.3 Schematic apparatus of the GAS process.

reduces the solvation power of the conventional solvent, triggering solute separation. The rapid mass transfer of the DG antisolvent into the conventional solution and the extraction of the conventional solvent into the DG phase determine high supersaturation levels and fast nucleation thus favouring the formation of small and uniform particles.

The applicability of DG antisolvent methods relies on three requirements: the (1) ability of the organic solvent to dissolve a large amount of gas, (2) low affinity of DGs for solutes and (3) mutual miscibility of DGs and organic phases. Different antisolvent processes have been developed on the basis of different mixing modes between DGs and organic solutions.

10.3.1 Gas antisolvent process

The gas antisolvent (GAS) process is a batch precipitation technique with a simple set-up as illustrated in Figure 10.3. In the GAS process, solutes previously dissolved in an organic solvent are loaded into a precipitation chamber. A DG gas is delivered into the chamber until the desired operating pressure is reached. In order to improve the mixing between the dense gas and organic solution, a stirrer is often inserted in the chamber. Due to the dissolution of the DG, the volume of the solution is expanded severalfold leading to a reduction of the solvent strength. The mixture becomes supersaturated and solutes precipitate in microparticles. After the precipitation, the mixture of DG and organic solvent is purged out of the system and the precipitates can be washed with pure DG to completely remove residual solvent (Tandya *et al.* 2007). The majority of the GAS investigations have been focused on micronisation and micro-encapsulation of various pharmaceutical compounds (Dehghani and Foster 2003; Tandya *et al.* 2007).

The GAS process has been used to produce microparticles of low- and high-molecular-weight drugs. For low-molecular-weight drugs, such as

Table 10.2 Pharmaceutical compounds re-engineered by the GAS process.

Solute	Solvents	Antisolvent	Observation	References
Micronization				
Lysozyme Insulin	DMSO, MeOH, EtOH, H_2O,	Ammonia, CO_2	0.01–1.8 μm	Thiering *et al.*, 2000a, b
Myoglobin	ethylacetate			
Lysozyme	DMSO	CO_2	0.05–1 μm	Muhrer and Mazzotti, 2003
Insulin	DMSO	CO_2	1–4 μm	Yeo *et al.*, 1993
	EtOH/H_2O	CO_2	< 1 μm	Tom *et al.*, 1993
Cu-indomethacin	DMF	CO_2	< 20 μm	Warwick *et al.*, 2002
Tobramycin	MeOH	CO_2	< 0.5 μm	Foster *et al.*, 2003b
Beclomethasone-17,21-dipropionate (BECD)	Ac, MeOH, EtOH	CO_2	2–44 μm	Bakhbakhi *et al.*, 2006
Paracetamol	Ac	CO_2	90–250 μm	Fusaro *et al.*, 2005
Itraconazole	DMF, THF, DCM	CO_2		Barrett *et al.*, 2008
Hydrocortisone	EtOH	CO_2	~2 μm	Corrigan and Crean, 2002
d-Glucosamine Hydrochloride	MeOH	CO_2	6–39 μm	Gimeno *et al.*, 2006
d-Glucosamine sulphate	MeOH	CO_2	2–45 μm	Gimeno *et al.*, 2006
Micro-encapsulation				
Insulin-PEG/PLA	DCM/DMSO	CO_2	387–685 nm	Elvassore *et al.*, 2001a
Insulin-PLA	DCM/DMSO	CO_2	0.5–2 μm	Elvassore *et al.*, 2001b
Carbamazepine-PEG	Ac, ethylacetate, DCM	CO_2	< 175 μm	Moneghini *et al.*, 2001 and 2003
Hydrocortisone-PVP	EtOH	CO_2		Corrigan and Crean, 2002

EtOH: ethanol, MeOH: methanol, Ac: acetone, DMSO: dimethyl sulfoxide, DMF: dimethyl formamide, DCM: dichloromethane, THF: tetrahydrofuran, PEG: polyethylene glycol, PLA: poly-*l*-lactide, PVP: polyvinylpyrrolidone.

Cu-indomethacin, tobramycin, beclomethansone-17,21-dipropionate and paracetamol, the mean particle sizes after GAS processing were in the range of 0.5–250 μm (Table 10.2). Protein powders (insulin, lysozyme and myogloblin) produced by the GAS process consisted of uniformly sized non-aggregated spherical particles between 0.05 and 2.0 μm (Table 10.2).

The GAS process has also been used to produce drug-loaded polymeric formulations for controlled release purposes. Co-precipitation of carbamazepine with polyethylene glycol (PEG), a hydrophilic polymer, dramatically increased the rate of dissolution of the drug (Moneghini *et al.* 2001; Sethia *et al.* 2002). The dissolution rate of a poorly water-soluble drug, theophylline, in composites with PEG produced by the GAS process was found to be 21 times greater than that of the original materials (Charoenchaitrakool *et al.* 2009). Microparticles of insulin coated with polylactide–glycoside with drug loading over 80% and an average size of 3 μm have been formed by using the GAS process (Richard 2003; Richard *et al.* 2001). The production of insulin-loaded poly-l-lactide (PLA) and insulin loaded in PLA–PEG using the GAS process resulted in the formation of nanospheres with smooth surface, compact internal structure and mean particle size between 400 and 700 nm (Elvassore *et al.* 2001a). Insulin coated with poly-l-lactide (PLA) presented a slow release profile as the PLA has low biodegradability and high hydrophobicity. The addition of high-molecular-weight PEG, a hydrophilic polymer, into insulin–PLA nanoparticles by GAS processing caused a burst in insulin release. On the other hand, the entrapment of low-molecular-weight PEG in polymeric networks allowed a slow and prolonged drug release. Consequently, the selection of polymers and their ratios added in drug formulations through the GAS process can be a suitable strategy to obtain the desired systems for drug delivery.

The organic solvent residues in the processed drugs may impose health risks to consumers. However, in GAS processing the concentration of residual organic solvent was generally found to be dependent on the mass of antisolvent used during the washing cycle, and the residual solvents can be easily removed within the production step (Thiering *et al.* 2000a, b; Warwick *et al.* 2000; Byrappa *et al.* 2008).

10.3.2 Aerosol solvent extraction system

The aerosol solvent extraction system (ASES) technique is a semicontinuous crystallisation process utilising the ability of DGs to extract organic solvents for the production of nanoparticles or microparticles. The technique is also referred to as *supercritical antisolvent* (SAS) and *precipitation by compressed antisolvent* (PCA). In the ASES process, a DG is first delivered into a precipitation chamber at a constant rate until the desired pressure is reached. When steady-state conditions (constant pressure, temperature and dense gas flow rate) are obtained, a liquid solution containing a drug or a mixture of drug and excipient is atomized through a nozzle into the compressed DG in the precipitation chamber. The DG antisolvent is fed simultaneously with the solutions, often through a concentric nozzle. The dissolution of the DG into the fine solution droplets is accompanied by a large volumetric expansion. The volumetric expansion leads to a reduction in the solvent power of the liquid, which causes a sharp rise in supersaturation within the solution. The phenomenon results in the formation of small uniform particles with a narrow particle size distribution (Byrappa *et al.* 2008). The mixture of DG and solvent exits the chamber and flows to a depressurisation tank where gas and liquid separate. After sufficient product is formed, the solution flux is stopped while pure DG is passed through the system to remove residual solvents. A schematic diagram of the ASES process is presented in Figure 10.4. The list of pharmaceutical compounds re-engineered by the ASES process is shown in Table 10.3.

Generally, the ASES process can be used to re-engineer pharmaceutical compounds with smaller particle sizes than those produced by the GAS process. The concomitant introduction of the solution and the DG

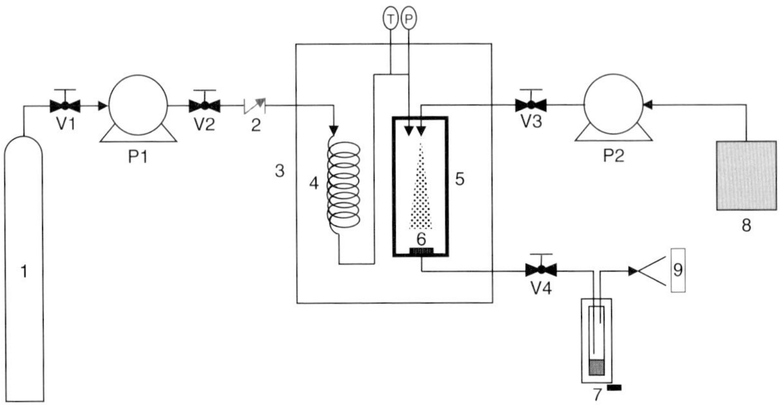

Figure 10.4 Schematic diagram of the ASES process.

antisolvent results in higher ratios of antisolvent–liquid solution compared to the GAS process. Solution atomisation, simultaneous introduction of solution and DG antisolvent produce more effective mixing than in the GAS process contributing to the formation of smaller particles. In the GAS process, the precipitation phenomena occur at a slower rate as antisolvent is slowly added to the solution, leading to the formation of larger particles with wider particle size distribution (Jung and Perrut 2001). It is clearly demonstrated that the ASES processing of various pharmaceutical compounds (d-glucosamine hydrochloride, d-glucosamine gulphate, lysozyme, insulin and Cu-indomethacin) produces smaller particles than the GAS process (Tables 10.2 and 10.3). On the other hand, the ASES process has a more complicated design, especially at the nozzle, and is generally operated at higher pressures than the GAS process. Operating pressures in the ASES process are such that the organic solvent and the DG form a single phase.

The ASES process has been successfully conducted on both organic and aqueous solutions generating micronized particles of various pharmaceutical compounds (Foster *et al.* 2003c, d; Bustami *et al.* 2001). Cefpodoxime proxetil (CPD), a poorly water soluble drug, was micronized through the ASES process using DG CO_2 as the antisolvent. Particles in the 0.1–0.6 μm range were produced that presented a higher dissolution rate than the raw material; 90% of the ASES processed CPD dissolved within the first 10 minutes in water, whilst the raw material required 60 min to produce the same result (Chu *et al.* 2009).

Particles of acetaminophen as small as 3 μm were produced from ethyl acetate solutions using DG CO_2 as the antisolvent. Changes in the operating conditions, specifically the solvent–antisolvent flux ratio and solution concentration, produced dramatic modifications of the particle morphology that varied between flakes, rods and spheres (Li *et al.* 2006). Micron-sized particles of d-glucosamine hydrochloride and d-glucosamine sulphate salts with very narrow particle size distributions have been successfully produced with high yields from methanol (Gimeno *et al.* 2006). Proteins (lysozyme, albumin, insulin and recombinant human deoxyribonuclease (rhDNase)) from aqueous solutions were precipitated as spherical particles ranging in size from 100 to 500 nm (Bustami *et al.* 2000). Particles of β-sitosterol less than 1 μm in size were produced from acetone by using the ASES process (Wenli *et al.* 2008).

The fine-particles produced by ASES can be suitable for pulmonary administration. One of the major prerequisites for a successful pulmonary drug application is that particles with aerodynamic diameter less than 5 μm are produced (Malcolmson and Embleton 1998; Timsina *et al.* 1994). Ipratropium bromide particles were formed by the ASES process using DMF as the organic solvent. The particles produced were between 2 and 3 μm in size, regular in shape and with low agglomeration; all useful characteristics for inhalation therapy (Kim and Shing 2008). Similarly, the particle size of various steroids (beclomethasone-17,21-dipropionate, betamethasone-17-valerate, dexamethasone-21-acetate, budesonide, lunisolide, fluticasone-17-pro-pionate, triamcinolone acetonide and prednisolone) processed by the ASES were less than 5 μm and consequently within the respirable range (Steckel *et al.* 1997).

The ASES process has been used for the co-precipitation of pharmaceutical compounds with additives or surfactants in order to improve delivery efficiency. Two drugs, terbutaline sulphate and ipratropium bromide, were co-precipitated with albumin producing particles less than 500 nm in size, non-cohesive and with low agglomeration in air flow, thus leading to effective and consistent lung deposition (Kim and Shing 2008). The precipitation of itraconazole with 2-hydroxypropyl-β-cyclodextrin (HP-β-CD) as an additive produced agglomerates of very fine (100–500 nm) particles with high aqueous solubility and dissolution rate (Lee *et al.* 2008a).

Micro-encapsulation of drugs with polymers by using the ASES technique can produce controlled drug delivery. The dissolution rate of drugs from the composites can be enhanced or retarded depending on the properties of selected polymers. Polyethylene glycol, a hydrophilic polymer, formed a composite with itraconazole (poorly water soluble antifungal drug) which resulted in a fivefold increase in the dissolution rate of the active ingredient in the first 100 min when compared to the unprocessed itraconazole (Barrett *et al.* 2008). On the other hand, the dissolution of acetazolamide, a highly water-soluble drug, from Eudragit® RS composites was slower than the raw material. As Eudragit® RS-types are poorly water permeable, the diffusion of the drug out of the system is slow, leading to lower drug concentrations in solution (Duarte *et al.* 2007). The combination of poly-lactic acid (a hydrophobic polymer) and polyethylene glycol in drug formulations of insulin produced a microencapsulated composite that exhibited a slow but constant insulin release for more than 60 days (Elvassore *et al.* 2001a).

Table 10.3 **Pharmaceutical compounds processed by the ASES/SAS/PCA process.**

Process	Solute	Solvents	Antisolvent	Observation	References
ASES	**Micronization**				
	Ipratropium bromide	DMF	CO_2	2–3 μm Suitable for inhalation therapy	Kim *et al.*, 2008a
	Atenolol	MeOH, EtOH	CO_2	Increased dissolution rate	Kikic *et al.*, 2006
	Budesonide	Methylenechloride	CO_2	1.6–15.8 μm	Steckel *et al.*, 2004
	Lysozyme Albumin Insulin rhDNase	H_2O	CO_2/EtOH	100–500 nm Substantially denatured rhDNase,	Bustami *et al.*, 2000
	rhDNase	H_2O	CO_2/ethanol-triethylamine	100–500 nm Complete recovery of biological integrity of the rhDNase powder	Bustami *et al.*, 2003
	d-Glucosamine hydrochloride	MeOH	CO_2	1.3–33.8 μm	Gimeno *et al.*, 2006
	d-Glucosamine sulphate			1.2–21.9 μm	
	Cu-indomethacin	DMF	CO_2	< 10 μm	Warwick *et al.*, 2002
	Co-precipitation				
	Terbutaline sulphate–albumin Ipratropium bromide–albumin	EtOH/DMF	CO_2	Less cohesive and less agglomerated in air flow, more effective and consistent lung deposition	Kim et al, 2008c
	Micro-encapsulation				
	Itraconazole/PEG	DMF	CO_2	Threefold increase in dissolution rate	Barrett *et al.*, 2008
		DCM		Twofold increase in dissolution rate	
		THF		Threefold increase in dissolution rate	
		Ac		Fivefold increase in dissolution rate	
	Acetazolamide/eudragit RS	Ac	CO_2	8–40 μm, slower released rate	Duarte et al,. 2007
PCA	**Micronization**				
	α-Chymotrypsin	H_2O	CO_2/EtOH	0.2–0.6 μm	Sarkari *et al.*, 2003
	Lysozyme	DMSO	CO_2	< 100 nm	Fusaro *et al.*, 2009
	N-trimethylchitosan chloride	DMSO	CO_2	1–10 μm	Perez de Diego *et al.*, 2006
SAS	**Micronization**				
	Amoxicillin	NMP, DMSO, DMSO+EtOH	CO_2	0.25–0.8 μm	Reverchon *et al.*, 2000 and 2003a; Kalogiannis *et al.*, 2005; Tenorio *et al.*, 2007

Cephalosporins	DMSO	CO_2	0.1–14 µm	Reverchon and De Marco., 2006
Nalmefene HCl	Ethyl alcohol EtOH	CO_2	0.9–4.5 µm	Adami *et al.*, 2007
Ampicillin	NMP	CO_2	0.26 µm	Tenorio *et al.*, 2007
Cilostazol	DCM	CO_2	0.90–4.52 µm	Kim *et al.*, 2007
Rifampicin	DMSO	CO_2	0.85–1.14 µm	Reverchon *et al.*, 2002
Cefonicid	DMSO	CO_2	0.3–1.35 µm	Reverchon and De Marco, 2004
Minocycline hydrochloride	EtOH	CO_2	0.1–1 µm	Cardoso *et al.*, 2008
Atorvastatin calcium	MeOH	CO_2	152–863 nm	Kim *et al.*, 2008b
Camptothecin	DMSO	CO_2	0.25 µm	Zhao *et al.*, 2010
Yttrium acetate	DMSO	CO_2	50–120 nm	Reverchon *et al.*, 2007
Zinc acetate	DMSO		75–150 nm	
Europium acetate	DMSO		55–150 nm	
Gadolinium acetate	DMSO		47–150 nm	
Samarium acetate	DMSO		90–130 nm	
Neodymium acetate,	DMSO		45–67 nm	
Astemizole,	DMSO		95–115 nm	
Nitrotriazole	DMSO		65 nm	
Cellulose Acetate	Ac		125 nm	
Micro-encapsulation				
Paclitaxel/PLA	DCM		465–935 nm Encapsulation efficiencies up To 83.5% and controlled release of paclitaxel for more than 30 days	Lee *et al.*, 2008b
Insulin/dextran Insulin/N-trimethyl chitosan	HCl/DMSO		1–30 µm Dry flowing particles suitable for pulmonary administration	Amidi *et al.*, 2008
Co-precipitation				
Simvastatin/ hydroxypropyl-b-cyclodextrin	MeOH		Solubility of the composites 9 times greater than that of the raw drug	Jun *et al.*, 2007

EtOH: ethanol, MeOH: methanol, Ac: acetone, DMSO: dimethyl sulfoxide, DMF: dimethyl formamide, DCM: dichloromethane, THF: tetrahydrofuran, PEG: polyethylene glycol, PLA: poly-*l*-lactide, NMP: N-methyl pyrrolidone.

10.3.3 Solution-enhanced dispersion by supercritical fluids

The solution enhanced dispersion by supercritical fluids (SEDS) process was developed by researchers at Bradford University, United Kingdom, in order to achieve smaller droplet size and increased transfer rates in particle formation processes. The process employs a highly turbulent flow of solvent and CO_2, leading to a very fast mixing or dispersion, hence mass transfer is not limited by molecular diffusion or convective phenomena (Rehman 2001). The intense mixing of the DG and a solution is obtained via a coaxial nozzle (Hanna and York 2006). Dense gases are used both as antisolvents for their chemical properties and as 'spray enhancers' by mechanical effect. The geometry of the nozzles allows the process to achieve good mixing and small droplet size as a function of the fluid velocity. In addition, the use of a mixing chamber increases the mixing surface area resulting in an increase in the mass transfer rate (Pasquali *et al.* 2008). The high rate of mass transfer leads to a fast nucleation rate and the production of particles with small size and little aggregation (Palakodaty and York 1999). The schematic diagram of SEDS process is presented in Figure 10.5a.

The SEDS technique has been further developed to process pharmaceutical compounds with low solubilities in organic solvents, such as sugars and proteins. For example, lactose has very low solubility in conventional organic solvents, but it dissolves easily in water. Proteins can be prepared in organic solvents, but it can bring undesirable effects such as protein unfolding and denaturation (Jung and Perrut 2001). The modified SEDS process allows the processing of water-based solutions overcoming the limitation of poor miscibility between water and CO_2. The modification includes the use of a nozzle with three coaxial passages (Figure 10.5b) for the simultaneous introduction of three streams into the precipitation chamber: (1) a solution or suspension of active substance in solvent 1 (normally aqueous solvent), (2) an additional solvent or solvent 2, miscible with solvent 1 and soluble in the DG, and (3) a DG. At the point where the active substance solution and solvent 2 are mixed, hydrogen bonding and/or similar interactions that are formed between both solvents promote co-extraction of both solvents by the DG (Jung and Perrut 2001). Table 10.4 shows a list of some pharmaceutical compounds re-engineered by the SEDS process.

The SEDS process has been used for micronisation of various pharmaceutical compounds. Nanosized recombinant human growth hormone particles with a narrow particle size distribution were produced by this process (Pyo *et al.* 2007). The SEDS process has been applied to the precipitation of microsized lysozyme particles, with minimal amounts of residual solvent (DMSO). Within a defined range of conditions, the tertiary structure and full biological activity of the protein were retained (Moshashaee *et al.* 2000).

The SEDS process has been efficient in producing pure polymorphs of various pharmaceutical compounds. Polymorph control is very important

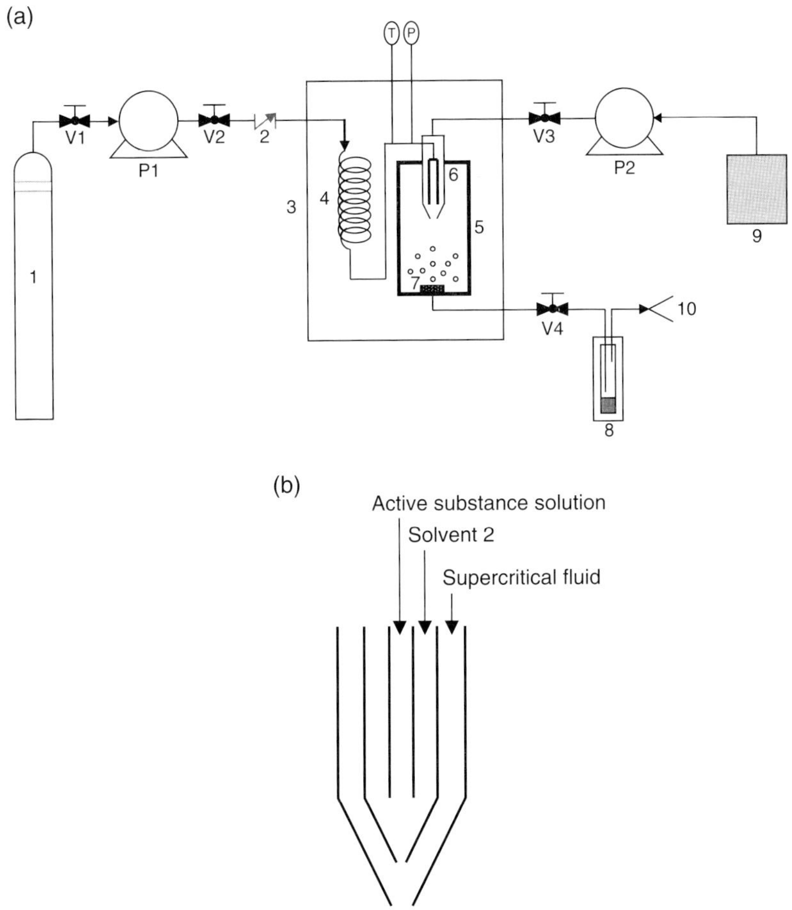

Figure 10.5 (a) Schematic diagram of the SEDS. (b) Three coaxial nozzles.

as polymorphism can affect shelf life, bioavailability, activity and even toxicity of APIs (Knapman 2000). Depending on the experimental conditions, the SEDS process produced pure and specific polymorphs of carbamazepine (Edwards *et al.* 2001), sulfathiazole (Kordikowski *et al.* 2001), flunisolide (Bartolomei 2000), salmeterol xinafoate (Tong *et al.* 2001), turbutaline sulphate (Rehman *et al.* 2004), hydrocortisone (Velaga *et al.* 2002a) and β-carotene (Franceschi *et al.* 2008 and 2009).

Impregnation of many drugs in polymer matrices has been performed by the SEDS process. Hydrocortisone was entrapped in DL-PLG microparticles with an encapsulation efficiency of 22%, the result is quite promising since no attempts were made to optimize the entrapment process (Ghaderi *et al.* 2000; Bodmeier *et al.* 1995). Microspheres of indomethacin in PLA/PLG matrixes were produced by the SEDS process with a mean particle size of about 2 μm and smooth surfaces. The release rate of

Table 10.4 Pharmaceutical compounds re-engineered by the SEDS process.

Solute	Solvents	Antisolvent	Particle size and observation	References
Micronization				
Lysozyme	DMSO	CO_2	1–5 μm, activity depending on the experimental conditions	Moshashaee *et al.*, 2000
Astaxanthin	DCM	CO_2	0.5–6.0 μm,	Hong *et al.*, 2009
Budesonide Flunisolide	Ac, MtOH	CO_2	<4 μm, suitable for inhalation	Velaga *et al.*, 2002b
Recombinant Human growth hormone	H_2O	CO_2/EtOH	118–218 nm	Pyo *et al.*, 2007
Puerarin	EtOH	CO_2	0.5–1 μm	Li *et al.*, 2008
Puerarin	EtOH/DCM	CO_2	0.19 μm Good spherical shape and smooth surface	Chen *et al.*, 2009a
Hydrocortisone	Ac, MtOH		New polymorphs I and II	Velaga *et al.*, 2002a
Micro-encapsulation				
Carotene/PEG	DCM	CO_2	1–10 μm, decrease degradation rate	He *et al.*, 2007
Puerarin/ phospholipids	EtOH	CO_2	2.80–10.43 μm	Li *et al.*, 2008
5-fluorouracil-SiO-PLA	EtOH	CO_2	536 nm, high encapsulation efficiency and slow drug release	Chen *et al.*, 2009b
Paclitaxe/PLA/ PGA	DCM	CO_2	ratio of PLA:PLG determined the drug release rate	Kang *et al.*, 2008b
Indomethacin/ PLA/PGA	DCM	CO_2	Longer sustained inhibition activity on proliferation of the non-small-cell lung cancer	Kang *et al.*, 2008a

EtOH: ethanol, MeOH: methanol, Ac: acetone, DMSO: dimethyl sulfoxide, DCM: dichloromethane, PEG: polyethylene glycol, PLA: poly-*l*-lactide, and PGA: polyglycolide or polyglycolic acid.

indomethacin from the composites was initially controlled by the diffusion of the drug through the polymer matrix and by polymer erosion at subsequent stages; overall the micronized composites produced a constant release of the API (Kang *et al.* 2008a).

A modification of the SEDS process has been developed which is referred to as SEDS with prefilming atomisation (SEDS-PA). In the SEDS-PA process the solution to be atomized flows into the nozzle as a film; once the solution reaches the atomisation edge, the liquid film separates and droplet formation occurs (Sattelmayer and Wittig 1986). Ephedrine, bixin and natural carotene were successfully micronized by the SEDS-PA process (He *et al.* 2004 and 2006).

10.3.4 Atomized rapid injection for solvent extraction

The atomized rapid injection for solvent extraction (ARISE) is a novel process developed by Foster and Sih (2006). The ARISE process employs energised rapid release of organic solutions and antisolvent capability of DG

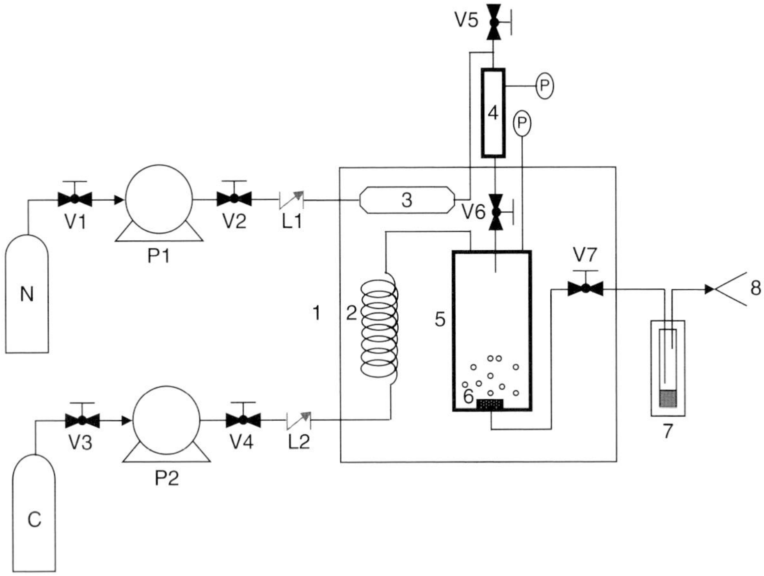

Figure 10.6 Schematic diagram of the ARISE process.

CO_2 to affect precipitation of pharmaceutical compounds from organic solvents. In the ARISE process, a vessel containing organic solution with dissolved pharmaceutical compounds is pressurized with nitrogen and sealed. By rapidly depressurising the vessel through a small tube, the organic solution is effectively atomized into a precipitation chamber containing DG CO_2 under quasi-isothermal and isochoric conditions. The extraction of organic solution and the precipitation of previously dissolved pharmaceutical compounds occur at this stage. The chamber containing the precipitates is next flushed with DG CO_2 to remove residual organic solvent from the system. A schematic diagram of ARISE process is presented in Figure 10.6.

As a recently developed technique, there are a limited number of ARISE-processed pharmaceutical compounds reported in the literature. Particles of insulin (Sih 2008) and tobramycin (unpublished data) produced by the ARISE process consisted of highly aggregated submicron individual particles with 99.96% of particles within the prescribed range for inhalation ($<5\mu m$). These characteristics make the ARISE products suitable for pulmonary administration. The ARISE process exhibits several competitive advantages, including greater throughputs than other DG antisolvent techniques, readily scalable for volume production, tuneable and narrow particle size range, particle morphology suitable for pulmonary or novel drug delivery and low levels of residual solvent in the products.

10.4 SCFs as co-solvents

10.4.1 Depressurisation of an expanded liquid organic solvent

Ventosa *et al.* (2000) obtained a new patent about the depressurisation of an expanded liquid organic solution process (DELOS), where a DG is used for the production of micron-sized and submicron-sized crystalline particles from organic solutions. This process differs from the other high-pressure techniques as the DG acts as a co-solvent that is completely miscible with the organic solution of the solute to be crystallized. The role of the DG is to produce homogeneous subcooling of the solution through its vapourisation, which in turn causes precipitation of solid particles (Ventosa *et al.* 2001a, b, 2003).

In the DELOS process, the solutes are first dissolved in an organic solvent that is then loaded into a high-pressure chamber. A DG is then added to the chamber resulting in a volumetric expansion of the mixture at the desired temperature and pressure. At this stage, the solute concentration must be kept below the saturation limit in the expanded mixture of organic solvent and the DG, otherwise the GAS process will occur. Finally, the system is depressurized into a precipitation chamber at atmospheric pressure through a nonreturn valve. During the depressurisation, the pressure of the solution inside the vessel is maintained constant at the desired pressure by a continuous addition of pressurized nitrogen gas from the top of

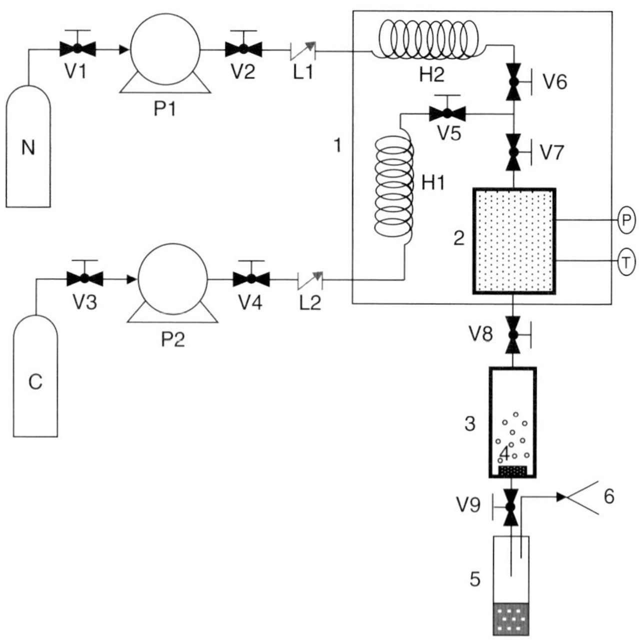

Figure 10.7 Schematic diagram of the DELOS apparatus.

the vessel. The sudden depressurisation of the expanded solution causes a large, fast and homogeneous decrease in the solution temperature, leading to a homogenous increase in the supersaturation ratio of the solutions. Consequently, a rapid nucleation is induced resulting in precipitation of fine crystalline particles with a narrow particle size distribution (Ventosa *et al.* 2003). The schematic diagram of the DELOS process is presented in Figure 10.7.

The DELOS process presents the challenge of quantitatively removing the conventional solvent to produce dry particles free from residual solvent. The application of DELOS to pharmaceutical processing includes the production of submicron particles of acetylsalicyclic acid, acetaminophen and mathenamine (Ventosa *et al.* 2001 a, b).

10.5 Dense gases as aerosolisation aids (spray-drying assistance)

Dense gases are also utilized to facilitate the nebulisation or aerolisation of liquid solutions. The carbon dioxide assisted nebulisation with a bubble dryer (CAN-BD) and the supercritical fluid assisted atomisation (SAA) processes are two methods which employ the aerosolisation principle to aid in micronisation processes.

10.5.1 *Carbon dioxide–assisted nebulisation with a bubble dryer*

The carbon dioxide–assisted nebulisation with a bubble dryer (CAN-BD) process is used to produce dry pharmaceutical microparticles. In this process, a drug solution and a DG are first mixed to form an emulsion and then rapidly expanded through a capillary restrictor to form aerosols. There are two methods of mixing the drug solution with the DG: the dynamic and the static methods. In the static method, the DG is bubbled through the bottom of a vessel filled with the aqueous solution, mixed and allowed to reach equilibrium. In the dynamic method, the drug-containing liquid solution and the DG are pumped simultaneously into a low-dead-volume tee. The emulsion is then expanded through a capillary restrictor resulting in a rapid and immediate depressurisation and aerosolisation of the solution into fine drug particles. The drug particles formed are passed through and dried in a heated bubble drier where warm, dry nitrogen or air flows. Nitrogen is preferred in cases where the drug particles pose fire and explosion hazards (Sievers *et al.* 2000). The dried particles are then collected on a filter. The whole process is conducted at drying temperatures lower than in conventional methods. Water and CO_2 are usually used as solvents in this process giving rise to an environmentally benign process. A schematic diagram of the CAN-BD dynamic process is shown in Figure 10.8.

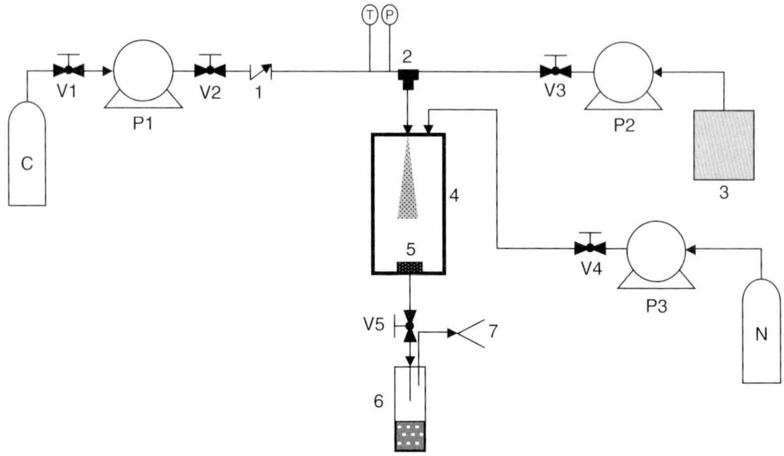

Figure 10.8 Schematic diagram of the dynamic CAN-BD process.

The CAN-BD has been used to produce proteins, vaccines, antibiotics and antivirals. Examples of powders generated by CAN-BD include hepatitis B surface antigen protein vaccine, a live-attenuated measles vaccine, α_1-antitrypsin and trypsinogen (Sievers *et al.* 2007; Cape *et al.* 2008). Additional examples are cromolyn sodium, albuterol sulfate, α-lactose, rifampin, relenza®, amphothericin B, bethamethasone-17,21-dipropionate, budesonide, mannitol, moxifloxacin, naproxen and tobramycin sulfate (Sievers *et al.* 1998, 2000, 2002, 2003, 2007; Villa *et al.* 2005; Cape *et al.* 2008). Table 10.5 lists a number of compounds re-engineered by the CAN-BD method. In most cases, these pharmaceutical compounds were dissolved in water, ethanol or water–ethanol mixtures. pH buffers and appropriate stabilising excipients such as surfactants and disaccharides were added to stabilize and retain the biological activities of proteins (Cape *et al.* 2008). The submicron particles produced under the CAN-BD method tend to be spherical or near-spherical (Sievers *et al.* 2000, 2003; Cape *et al.* 2008). The particle size ranged from 1 μm to 5 μm, which is optimal for pulmonary delivery of pharmaceuticals where particles of less than 5 μm are required to reach the deep alveoli region in the lungs (Sellers *et al.* 2001; Kayrak *et al.* 2003).

Preparations of solid dry pharmaceuticals and neutraceuticals are desirable to ensure longer storage stability, hence labile therapeutics often undergo dehydration processes. Freeze drying is the most commonly used method; however, it tends to induce irreversible damages to the therapeutic compounds. Possible damages include freeze-induced and thermally induced denaturation. Moreover, an additional milling stage after the dehydration process is required to break up the freeze-dried cakes (Sellers *et al.* 2001). On the other hand, the CAN-BD method micronizes and dries

Table 10.5 Pharmaceutical compounds re-engineered by the CAN-BD and SAA process.

Substances	Solvents in dense gas phase	Solvents used in the drug solution	References
CAN-BD			
Albuterol sulfate	CO_2	H_2O	Sievers *et al.*, 2000
Amphothericin B	CO_2	H_2O, EtOH, H_2O+EtOH	Sievers *et al.*, 2003
Bethamethansone-17,21-dipropionate	CO_2	EtOH	Villa *et al.*, 2005
Budesonide	CO_2	EtOH	Sievers *et al.*, 2003
Cromolyn sodium	CO_2	H_2O	Sievers *et al.*, 2000
Hepatitis B surface antigen protein vaccine	CO_2	H_2O	Sievers *et al.*, 2007
α-lactose	CO_2	H_2O	Sievers *et al.*, 2000
Lactose	CO_2	H_2O	Villa *et al.*, 2005
Live measles vaccine	CO_2	H_2O	Sievers *et al.*, 2007; Burger *et al.*, 2008
Mannitol	CO_2	H_2O	Sievers *et al.*, 2002
Relenza®	CO_2	H_2O	Sievers *et al.*, 2007
Rifampin	N_2	Ethyl acetate	Sievers *et al.*, 2007
Tobramycin sulfate	CO_2	H_2O	Sievers *et al.*, 1998
SAA			
Ampicillin	CO_2	H_2O, EtOH, MeOH	Reverchon *et al.*, 2003d
Ampicillin trihydrate	CO_2	H_2O	Reverchon, 2002
Amipicillin trihydrate + chitosan (carrier)	CO_2	1% v/v acetic acid aqueous solution	Reverchon and Antonacci, 2007a
Ampicilin trihydrate + hydroxypropyl methylcellulose (carrier)	CO_2	Phosphate buffer solution	Revercho *et al.*, 2008
Carbamazepine	CO_2	MeOH	Reverchon, 2002
Cefadroxil	CO_2 + Ethanol	H_2O	Li *et al.*, 2009
α-cyclodextrin	CO_2	H_2O	Reverchon and Antonacci, 2006
Dexamethasone	CO_2	Ac	Reverchon, 2002
Erythromycin		EtOH, MeOH, Ac	Reverchon and Spada, 2004
Griseofulvin	CO_2	Ac	Reverchon *et al.*, 2004
Hydroxypropyl-β-cyclodextrin	CO_2	H_2O	Reverchon and Antonacci, 2006
Hydroxypropyl methylcellulose	CO_2	Phosphate buffer solution	Reverchon *et al.*, 2008
Levofloxacin hydrochloride	CO_2	MeOH	Cai *et al.*, 2008
Lysozyme	CO_2, N_2	H_2O	Rodrigues *et al.*, 2009
Poly(methyl methacrylate)	CO_2	Ac	Reverchon and Antonacci, 2007b
Poly-l-lactide	CO_2	DCM	Reverchon and Antonacci, 2007b
Rifampicin	CO_2	MeOH	Reverchon and Della Porta, 2003b
Terbutaline	CO_2	H_2O	Reverchon and Della Porta, 2003b
Tetracyclin	CO_2	H_2O	Reverchon and Della Porta, 2003b
Triclabenzadol	CO_2	MeOH	Reverchon, 2002

EtOH = ethanol, MeOH = methanol, Ac = acetone, DCM = dichloromethane.

therapeutic compounds at temperatures lower than conventional drying methods, consequently micronising and preserving the stability and shelf life of therapeutic peptides and proteins simultaneously (Sievers *et al.* 2007). CAN-BD processed hepatitis B surface antigen protein vaccine, and dry powders of live-attenuated measles vaccine virus were proven to retain viral activities in the presence of appropriate stabilizers (Sievers *et al.* 2007; Cape *et al.* 2008).

The CAN-BD method, however, suffers the drawback of capillary blockage due to a nonnegligible pressure drop in the capillary resulting in solute precipitation (Reverchon 2002; Li *et al.* 2009). In addition, the mixing ratio of the drug solution and the SCF in the low-volume tee can only be approximated (Reverchon 2002; Li *et al.* 2009). Recent modifications of the CAN-BD process include the replacement of the low-dead-volume tee with a cross to mix two liquid streams with a DG to form composite microparticles, and the supercritical fluid assisted atomisation (SAA) work developed by Reverchon (2002). The utilisation of a cross is advantageous in the micronisation of heterogeneous compounds containing one solid soluble in water and the other in an organic solvent. The SAA process is explained in the following section.

10.5.2 Supercritical fluid assisted atomisation

The supercritical fluid assisted atomisation (SAA) process works on the same principle of the CAN-BD where a drug-containing aqueous stream and a DG are mixed together to aid solution atomisation. However, in the SAA process, the DG is first solubilized in the liquid solution and mixed before going through the aerosolisation stage. A thermostated packed tower is added to improve the mixing of the mixture (Reverchon 2002). The thermostated packed tower provides sufficient contacting surface and residence time for the DG CO_2 to saturate the liquid solution allowing for an equilibrated CO_2 solubilisation in the solution (Reverchon 2002). The DG CO_2–liquid solution is then sprayed through a thin wall injector and precipitated into a chamber containing flowing heated nitrogen. This is a two-step aerosolisation process where the primary particles formed are broken into secondary particles from the sudden release of the CO_2 (Reverchon *et al.* 2009). The heated nitrogen helps with the evaporation of the liquid solvent. Although the thin wall injector avoids the capillary blockage problem found in the CAN-BD process, the frit used in the precipitator can also plug up. A schematic diagram of the SAA process is given in Figure 10.9.

The SAA process has been shown to produce micro- and nanosized particles; in most cases, spherical particles are formed (Reverchon and Della Porta 2003b). An improved dissolution rate of SAA-produced griseofulvin compared to a jet-milled product has been reported (Reverchon *et al.* 2004). The morphology and particle size of SAA-treated

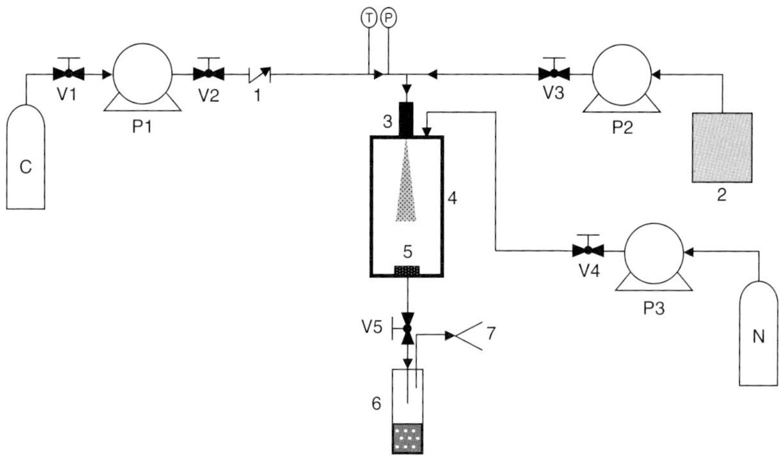

Figure 10.9 Schematic diagram of the SAA process.

compounds are greatly affected by the precipitation temperature, the solvent used and the mass flow ratio of CO_2 to the liquid solution while the operating temperature and pressure of the saturator affect the efficiency of the process (Reverchon and Della Porta 2003b, c; Reverchon and Antonacci 2007a). Changes to the drug concentration in the liquid solution also result in changes to the mean particle size and PSDs (Reverchon and Della Porta 2003 b, c; Reverchon *et al.* 2004). Hence, the morphology of therapeutics can be tailored through manipulation of concentration to produce drugs with specific particle size range that target specific sites for drug delivery. Pharmaceuticals re-engineered by the SAA process include micronized cefadroxil and lysozyme and the co-precipitated amipicilin trihydrate with chitosan (Reverchon and Antonacci 2007a; Li *et al.* 2009; Rodrigues *et al.* 2009). A semicontinuous pilot plant had also been successfully operated for the production of rifampicin microparticles with particle sizes ranging between 0.3 and 0.4 μm and product recovery of up to 91% (Reverchon *et al.* 2006). The list of SAA-treated pharmaceutical compounds is given in Table 10.5.

10.6 Conclusion

The application of DGT to the re-engineering of APIs has been proven a powerful tool for the production of controlled morphologies and composites. The generic advantages of the technology include the combination of product formation and purification in a single process. Other aspects are the reduction or elimination of the use of toxic solvents and the possibility to work under mild conditions, suitable for thermally and chemically labile compounds.

Through the DGT, microparticles with controlled morphologies and crystal structures have been generated. The successful re-engineering of APIs is a key aspect in determining their therapeutic effectiveness and possible administration routes. The aspect is even more critical for newly developed therapeutic entities that are often highly hydrophobic and present low bioavailability. The DGT have produced API powders with high surface areas and improved solubility and dissolution profile; both factors can increase the bioavailability of hydrophobic APIs. On the other side, the re-engineering of established APIs can improve the formulations currently on the market allowing pharmaceutical companies to extend the patent life of the APIs with potentially noticeable economic returns.

Within the applications of the DGT to pharmaceutical processing, the successes in the re-engineering of protein APIs – traditionally labile molecules – and the generation of powders for inhalation delivery – an application requiring strict control over particle morphology – are particularly interesting because of their possible impact on the pharmaceutical market. The extent of the impact of DG processing on the pharmaceutical market will ultimately depend on the scalability of micronisation and encapsulation processes. One aspect hindering the use of DG processes on the commercial scale is the non-ideal nature of the DGs which limit the applicability of theoretical models. Additional challenges to the scale-up of DG micronisation and encapsulation processes derive from the complex fluid dynamics of most particle formation processes based on the DGT. The development of simplified processes, such as the ARISE process, can contribute to the successful operation on the commercial scale, thus capturing the advantages and possibilities of DG processing of pharmaceutics.

11 Industrial Applications Using Supercritical Carbon Dioxide for Food

James T.C. Yuan and John S. Novak

Abstract: The number of publications and patents demonstrates that although supercritical CO_2 has not been commercially applied to a significant extent, the technology is not conceptually new. Nonthermal food processes such as supercritical CO_2 and high hydrostatic pressure result from consumers' demands for natural, fresh-like foods with microbial safety and extended shelf life. Among nonthermal technologies, high hydrostatic pressure has achieved commercial status. Supercritical CO_2 has certain advantages, namely, lower operating pressures combined with greater microbial safety, less change in food sensory quality, and lower capital and operating costs. The aim of the present chapter is to provide industrial pilot-scale research data to support the advantages of applying DPCD technology for the pasteurization of foods. A review of the industrial pilot-scale systems is given, together with microbial and organoleptic results of the processed foods. In addition, a general overview presenting the hurdles that supercritical CO_2 processing may need to overcome in order to become commercially viable is also discussed.

Keywords: commercialization; capital and operating costs.

11.1 Overview

Industrial applications using supercritical carbon dioxide ($scCO_2$) for food disinfection purposes have been quite limited until now. Extensive research has been performed to highlight the benefits of this technology in maintaining the organoleptic properties of foods while reducing detrimental effects of the microorganisms present.

Under standard atmospheric conditions, carbon dioxide (CO_2) exists in a gaseous phase. In this state gaseous CO_2 is found to be more bacteriostatic

Dense Phase Carbon Dioxide: Food and Pharmaceutical Applications, First Edition. Edited by Murat O. Balaban and Giovanna Ferrentino.

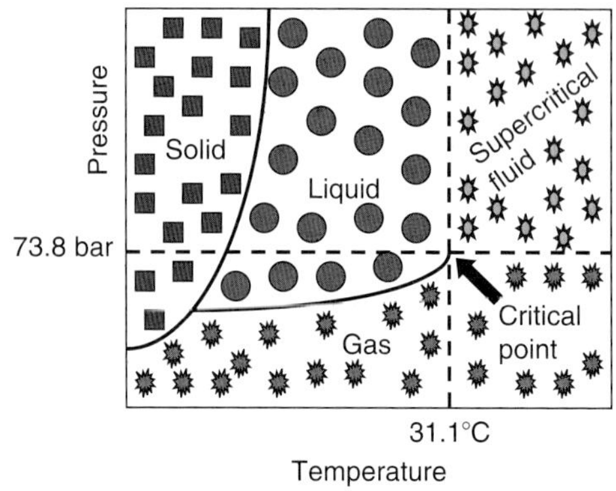

Figure 11.1 The representative phase diagram for CO_2 shows the relationship among different physical states of CO_2 and temperature and pressure. The critical point value of 31.1°C and 7.38 MPa for CO_2 is designated by an arrow.

to most microorganisms present in or on foods. Even though sterilization requires the complete inactivation of all viable cells, mixed results have been attained using a comprehensive collection of microorganisms due to individual differences in susceptibility to CO_2 (Haas *et al.* 1989; Jones and Greenfield 1982; Ishikawa *et al.* 1997; Doyle 1983; Enfors and Molin 1980; Molin 1983). Complete sterility would be very difficult to achieve when the main goal is to maintain food quality.

Low temperatures enable CO_2 phase changes from a gas to a liquid to a solid (Figure 11.1). As pressure and temperature increase above a critical point (temperature at 31.1°C and pressure at 7.38 MPa), CO_2 can exist as a single supercritical phase that has the unique ability to diffuse through solids with lower surface tension as a gas or solubilize compounds as a liquid (Figure 11.1). This critical point is the threshold at which CO_2 becomes $scCO_2$ and enables unique properties above those of gaseous or liquid CO_2 alone. It has been well documented that $scCO_2$ enables the extractability of various organic and inorganic molecules while inactivating a range of microorganisms (Dillow *et al.* 1999; Ishikawa *et al.* 1995b; Spilimbergo *et al.* 2003b; Taran *et al.* 2002; Wilkes *et al.* 2000; Yener 2001).

11.2 Past development

It should be noted that although $scCO_2$ technology has not been commercially applied to a significant extent, the technology certainly is not conceptually new. In the early 1950s, Fraser reported the disruption of bacterial cells by the rapid release of CO_2 from a pressure of 34 atmospheres to ambient atmospheric pressure (Fraser 1951; Foster *et al.* 1962). Nearly 2 decades later, the first commercial US patent for food sterilization using CO_2 was awarded to Swift & Co., Chicago, IL, in 1969 (Kauffman *et al.* 1969). Consumer demands for natural, unadulterated foods combined with concerns for microbial safety have led to fresh-like food products preserved for extended shelf lives using nonthermal food-processing technologies such as high hydrostatic pressure (HHP). Although HHP has the current commercial attention, $scCO_2$ has certain advantages over the more widely applied technology in lower pressures combined with greater microbial safety, greater food sensory qualities, and lower capital/operating costs. This chapter will provide industrial pilot-scale research data to support these advantages.

11.3 Mechanism of microbial inactivation

Even though a number of theories have been presented to describe the mechanisms of CO_2 inactivation of microorganisms, many of these remain hypothetical to this day. The most prevalent concepts include the solubilization of pressurized CO_2 and the rapid decompression of CO_2 within cells (Meyssami *et al.* 1992; Duan and Sun 2003; Fraser 1951; Foster *et al.* 1962), extraction of vital cellular constituents (Lin *et al.* 1993; Kamihira *et al.* 1987), acidification leading to the precipitation of intracellular salts as a result of bicarbonate conversion to carbonate as well as metabolic enzyme inactivations (Kamihira *et al.* 1987), explosive rupture of the cell membrane (Nakamura *et al.* 1994) and changes in membrane fluidity and permeability (Isenschmid *et al.* 1995), as well as interference with cellular electrolyte balance (Lin *et al.* 1993). An excellent discussion of these hypothetical inactivation mechanisms have been covered in a recent review and will not be extensively mentioned here further (Garcia-Gonzalez *et al.* 2007).

11.3.1 Effect of other gases on microbial inactivation

Pressurized gases other than CO_2 have been attempted for food pasteurization applications. Nitrogen (N_2) was found to have little or no effect on microbial survival in or on foods (Fraser 1951; Haas *et al.* 1989; Wei *et al.* 1991; Lin *et al.* 1992, 1993; Nakamura *et al.* 1994; Enomoto *et al.* 1997b; Debs-Louka *et al.* 1999; Hong and Pyun 1999). Similar results were found for tetrafluoroethane (TFE) despite similar supercritical properties such as CO_2 (Dillow *et al.* 1999). Argon (Ar) and a mixture of N_2 and oxygen (O_2) were found to be less effective against microorganisms than CO_2 (Fraser 1951; Enomoto *et al.* 1997b; Debs-Louka *et al.* 1999). Some species-dependent success was obtained using nitrous oxide (N_2O), but even these results did not match $scCO_2$ in most cases (Fraser 1951; Enomoto *et al.* 1997b).

11.4 $scCO_2$ commercialization activities

In terms of $scCO_2$ technology commercialization, the most notable public commercialization efforts have been performed by Praxair Inc., which licensed the technology from the University of Florida (Balaban *et al.* 1995; Balaban 1998). Although a number of mobile demonstration units (capable of processing 1.5 L/min liquid foods) have been constructed, only one commercial scale unit with a processing capacity of 150 L/min has been tested at an actual juice processing plant for a commercial supplier.

Entrepreneur Marc Sims (Porocrit LLC, Berkeley, CA) designed a hollow-fiber membrane contactor for continuous $scCO_2$ pasteurization of pulp-free juices that improved the current technology due to its CO_2 saturation enhancements. Sims licensed the technology to the Chicago Research Center of Air Liquide (Countryside, IL) for further development from 2000 until his untimely death in 2005 (Figure 11.2; Sims 2000).

Air Liquide has since abandoned research interests including commercialization efforts, and without Sims, Porocrit has also ceased operations. The only other $scCO_2$ industrial activity involves pilot-scale $scCO_2$ equipment for continuous treatment of liquid foods and manufactured by Mitsubishi Kakoki Co. (Tokyo, Japan) for Shimadzu Co. (Kyoto, Japan). Maximal flow rates of 3.0 kg/h and 20.0 kg/h were capable of being pumped through a 5.8 L holding treatment vessel, but the commercial development of the technology in Japan currently remains at the laboratory research level only (Garcia-Gonzalez *et al.* 2007).

11.5 Porocrit process

The commercial utility and overall capabilities of the Porocrit LLC $scCO_2$ pilot plant demonstration system were evaluated during the early 2000s at the Chicago Research Center of Air Liquide in collaboration with originator Marc Sims (Figure 11.2). Originally water, orange, apple–lemon–ginger and carrot juices were $scCO_2$-processed by the continuous flow-through system of microporous polypropylene membranes under conditions of 1180 psia (7.3 MPa), 40°C, 6.5 g CO_2/100 ml with a 100 ml/min product flow rate. Processed carrot juice appeared foamier than orange juice, whereas apple–lemon–ginger did not foam most probably due to the low protein and solid content of the juice. Carrot juice contained the highest concentration of indigenous microorganisms (Table 11.1).

Apple–lemon–ginger juice contained the highest number of spore-forming bacteria at 1.75 $\log_{10}$ CFU/ml. The $scCO_2$ was very effective in inactivating indigenous yeast and molds and lactic acid bacteria in orange, carrot, and apple–lemon–ginger juices (Table 11.1). Total aerobic bacteria and spore formers were more resistant, especially in the apple–lemon–ginger juice. The generation of carbonic acid during the $scCO_2$

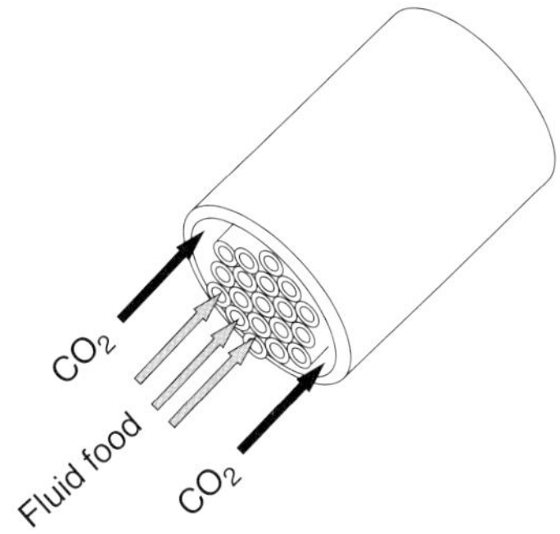

Figure 11.2 The Porocrit LLC $scCO_2$ system (designed by Marc Sims, Berkeley, CA) consists of microporous polypropylene membranes through which liquid product is pumped. CO_2 diffuses across the membranes to inactivate microorganisms in the food product under supercritical conditions of 7.3 MPa and 30–45°C.

Table 11.1 The effect of $scCO_2$ processing[1] on indigenous microorganisms in various juices[2].

Juice	Microorganisms	Log_{10}CFU/ml (unprocessed)	Log_{10}CFU/ml (processed)	Log_{10}CFU/ml Reductions
Orange	Total aerobic bacteria	2.79 ± 1.36	0.88 ± 1.49	1.93 ± 0.37
Orange	Lactic acid bacteria	2.02 ± 0.31	0.39 ± 0.68	1.63 ± 0.86
Orange	Yeast and molds	2.67 ± 1.23	0	2.67 ± 1.23
Orange	Spore formers	0	0	0
Apple–lemon–ginger	Total aerobic bacteria	2.74 ± 0.22	2.83 ± 0.37	–0.10 ± 0.15
Apple–lemon–ginger	Lactic acid bacteria	3.64 ± 0.06	0.64 ± 0.91	3.00 ± 0.84
Apple–lemon–ginger	Yeast and molds	3.79 ± 0.53	0.45 ± 0.64	3.33 ± 0.11
Apple–lemon–ginger	Spore formers	1.75	1.26	0.49
Carrot	Total aerobic bacteria	5.00	2.88 ± 0.20	2.13 ± 0.22
Carrot	Lactic acid bacteria	5.05 ± 0.06	0.69 ± 0.30	4.36 ± 0.23
Carrot	Yeast and molds	3.98 ± 0.31	0	3.98 ± 0.31
Carrot	Spore formers	0.06	0.64	–0.58

[1] 1180 psia (8.1 MPa), 40°C, 100 ml/min product flow rate with 6.5 g CO_2/100 ml consumed.
[2] mean values ± standard deviations.

process led to pH changes in the apple–lemon–ginger juice from 6.6 to 6.0 after depressurization and in carrot juice from 6.3 to 5.8. However, the already acidic orange juice (pH 3.9) was not significantly affected by the process. For all of the juices tested, Brix (% sucrose) remained unchanged as a result of the $scCO_2$ process (apple lemon ginger, 11.6%; orange, 12.7%; and carrot 9.9%, respectively).

11.5.1 *Impact on juice quality*

As a measure of juice flavor quality, solid-phase microextraction gas chromatography (SPME-GC) analysis was used to monitor volatile flavor compounds in orange juice following $scCO_2$ processing using the Porocrit LLC system (Figure 11.3).

These results were consistent with published results on the distribution of flavor compounds present in oranges (Tetra Pak Processing Systems A B, 1998). The SPME-GC analysis detected significant reductions in volatile flavor compounds in the freshly squeezed orange juice treated by the Porocrit process (Figure 11.3). The compounds affected most by the process were limonene, linalool, and α-pinene which were detected at 14–24%, 15–28%, and 18–32% of original values, respectively, dependent upon the time of sampling. Only myrcene was retained at 71–128% throughout $scCO_2$ process sampling, although the flavor compound also exhibited the greatest variation (Figure 11.3). Valencine was retained at 42–56%. Ethanol and ethyl butylate flavors were completely lost. The retention of flavor compounds was not related to the % saturation of CO_2 nor the time of sampling (Figure 11.3). CO_2 was released from the product in a two-stage process first from 1200 psi (8.2 MPa) to 300 psi (2 MPa) and then in the second stage down to atmospheric pressure. It is thought that minimization of CO_2 off-gassing would retain flavor compounds. Unfortunately, there is a fine balance between acceptable dissolved CO_2 in any given

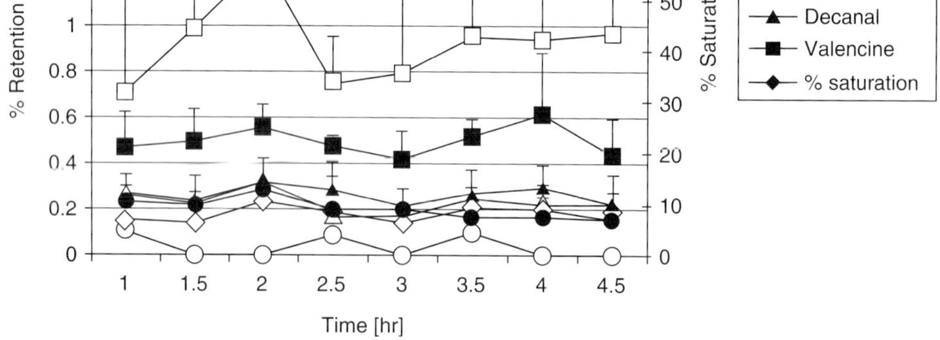

Figure 11.3 Volatile flavor compounds in freshly squeezed orange juice following $scCO_2$ processing detected by solid phase micro-extraction and gas chromatography. Results are the average of four separate trials. The % saturation of the juice with carbon dioxide was measured using a mass flow meter. Bars indicate standard error.

product and loss of natural flavors. Some of these flavor compounds detected by SPME-GC analysis are not clearly distinguishable by human taste test panels or an instrument (electronic tongue, Alpha MOS) used to standardize human sensory responses (unpublished data). SPME-GC differences do not appear to be significant in that the measurements were found to be beyond the sensitivity limits detectable by human taste sensory perceptions.

11.5.2 Impact on nutrient values

With respect to retention of nutrient values and $scCO_2$ processing, vitamin C was chosen as an indicator nutrient due to its risk of breakdown in the presence of photo or natural oxidation. Samples were $scCO_2$-processed, flash-frozen at −80°C, and quantified by HPLC analysis of vitamin C using a third party supplier (National Food Laboratory, Dublin, CA).

Table 11.2 depicts a marginal 3% loss of vitamin C from fresh orange juice following $scCO_2$ processing. When residual CO_2 was removed from the orange juice using inert gases such as Ar or N_2 to sparge, vitamin C increased from 7% to 13% dependent upon type of sparge gas and the duration of sparging (Table 11.2). Overall, the $scCO_2$ processing only marginally affected a very unstable nutrient such as vitamin C, so the $scCO_2$ process was considered relatively inert and insignificant to other more stable nutrients as well.

11.5.3 Impact on microbial inactivation

The $scCO_2$ process proved to be 28–100% effective in inactivating indigenous microorganisms in various liquid foods. It was hypothesized that inoculating water, orange, or carrot juice to a worst-case scenario of 6 to 8 $\log_{10}$ CFU/ml of concerned microorganisms would give a better estimate of the process sanitizing efficiency and applicability. Lactic acid bacteria have been suggested to be the most resistant vegetative cells to $scCO_2$. In tap water, 3.54 $\log_{10}$ CFU/ml of *Lactobacillus plantarum* and 6.30 $\log_{10}$

Table 11.2 The retention and stability of vitamin C in orange juice that has been $scCO_2$ processed.

Treatment	Vitamin C (mg/100 ml)	% total
Fresh–squeezed orange juice (FSOJ)	35.23 ± 6.20	100
$scCO_2$ processed FSOJ	34.30 ± 5.60	97.36
$scCO_2$ processed FSOJ + 15 min Ar	32.88 ± 4.35	93.33
$scCO_2$ processed FSOJ + 30 min Ar	32.30 ± 3.67	91.68
$scCO_2$ processed FSOJ + 30 min N_2	25.97 ± 0.81	87.15

CFU/ml of *E. coli* representative three-strain cocktails were inactivated at 40°C using 5–6 g CO_2 consumed per 100 ml (Table 11.3).

Usually food products increase organic load, thereby decreasing process efficiencies. However, the low pH of the orange juice appeared to increase the antimicrobial efficiency of the CO_2 treatment. In orange juice at 40°C, 5.20 $\log_{10}$ CFU/ml of *L. plantarum* and 7.87 $\log_{10}$ CFU/ml of *E. coli* were inactivated with a CO_2 consumption rate of 6.5 g/100 ml (Table 11.3). These results were only slightly enhanced by raising the treatment temperature to 45°C (6.03 and 7.91 $\log_{10}$ CFU/ml reductions for *L. plantarum* and *E. coli*, respectively).

Carrot juice proved to be a more problematic substrate as the pH was not as acidic as for orange juice. In carrot juice, *L. plantarum* was only reduced 0.91 $\log_{10}$ CFU/ml at 40°C and 6.5 g CO_2/100 ml, whereas *E. coli* was inactivated by 4.67 $\log_{10}$ CFU/ml (Table 11.3). Increasing CO_2 consumption to 8.5 g/100 ml only added an additional 1 to 2 $\log_{10}$ CFU/ml decrease, respectively. Despite the heightened resistance of the microorganisms in carrot juice, microbial inactivations 4.76 $\log_{10}$ CFU/ml of *L. plantarum* and 8.26 $\log_{10}$ CFU/ml of *E. coli* were achieved upon increasing the treatment temperature to 45°C with a CO_2 consumption rate of

Table 11.3 Inactivation of product inoculated to 6–8 $\log_{10}$ CFU/ml with *L. plantarum* and *E. coli* followed by $scCO_2$ processing.

Product	Temperature (°C)	CO_2 (g/100 ml)	*L. plantarum* $\log_{10}$CFU/ml reductions	*E. coli* $\log_{10}$CFU/ml reductions
H_2O	40	2.13	0.13	ND
H_2O	40	2.24	0.61	ND
H_2O	40	2.52	0.64	ND
H_2O	40	3.03	3.23	ND
H_2O	40	3.73	ND	5.34
H_2O	40	4.58	ND	5.49
H_2O	40	5.18	3.54	ND
H_2O	40	5.38	ND	4.52
H_2O	40	5.84	ND	6.01
H_2O	40	6.02	ND	6.30
Orange juice	15	5.00	0.53	0.91
Orange juice	27.5	5.00	0.25	3.15
Orange juice	33	5.70	0.90	4.95
Orange juice	40	6.50	5.20	7.87
Orange juice	45	6.50	6.03	7.91
Carrot juice	40	5.50	0.09	3.47
Carrot juice	40	6.50	0.91	4.67
Carrot juice	40	8.50	1.71	6.60
Carrot juice	45	5.50	2.22	6.70
Carrot juice	45	6.50	1.42	7.59
Carrot juice	45	8.50	0.89	6.34
Carrot juice	45	9.50	4.76	8.26

ND: not determined.

Table 11.4 Inactivation of *B. coagulans* spores inoculated to 5 $\log_{10}$ CFU/ml in water followed by $scCO_2$ processing and elevated temperatures.

Temperature (°C)	CO_2 (g/100 ml)	Product flow rate (ml/min)	$\log_{10}$CFU/ml reductions
50	4.6	100	0.43 ± 0.01
50	6.5	100	0.50 ± 0.04
70	4.0	50	4.87 ± 0.07
70	6.5	100	3.47 ± 0.23
75	3.9	100	2.15 ± 0.31
75	4.0	50	4.43 ± 0.06
75	3.9	75	5.06 ± 0.13
75	2.0	100	2.86 ± 0.02
75	3.9	100	4.41 ± 0.47
80	3.8	50	4.43 ± 0.06

[1] Mean values ± standard deviations.

9.5 g/100 ml (Table 11.3). Additional concerns to carrot juice spoilage include spore formers and process parameters that may need to be modified in order to increase process efficiency. CO_2 saturation appears to be a key factor in microbial inactivation efficiency. For a recent study using the $scCO_2$ system and *L. plantarum* survival with respect to increased CO_2 saturation rates in apple cider, the reader is referred to Yuk and Geveke (2010).

Generally, Gram-negative bacteria were found to be more sensitive to $scCO_2$ treatment as compared with Gram-positive bacteria due to membrane differences (Garcia-Gonzalez *et al.* 2007). Because of the resiliency of bacterial spores, some temperature increases were necessary in order to increase inactivation levels. Although in so doing, the $scCO_2$ process also becomes less of a nonthermal process. Nonetheless, the results achieved for spore inactivation were significantly better than could be achieved at the same temperatures without CO_2 treatment. Spore inactivation levels for *B. coagulans* spores in water were 0.5 $\log_{10}$ CFU/ml at 50°C and 3.47 $\log_{10}$ CFU/ml at 70°C for the same CO_2 levels of 6.5 g/100 ml (Table 11.4).

If the product flow rate was decreased by 50% to 50 ml/min, $\log_{10}$ CFU/ml reductions for spores increased to 4.9 $\log_{10}$ CFU/ml for a CO_2 level of 4.0 g/100 ml (Table 11.4). Increasing the temperature to 75°C with a product flow rate of 75 ml/min at 3.9 g CO_2/100 ml resulted in 5.1 $\log_{10}$ CFU/ml reductions for *B. coagulans* spores (Table 11.4). A temperature of 75°C without CO_2 treatment would be expected to do little more than serve as a stimulant for spore germination.

It could be expected that CO_2 saturation and penetration would become impeded transitioning from liquid to solid food products with an anticipated accompanying decrease in microbial inactivation efficiencies. Whole liquid egg provided a more viscous substrate for flow and $scCO_2$ processing than water or fruit juices. Using whole eggs, *Salmonella enterica* was still found to be reduced 4.64 $\log_{10}$ CFU/ml at 40°C (Table 11.5).

Table 11.5 The reduction of *S. enterica* in whole eggs using $scCO_2$ at 1100 psig (7.6 MPa), 5% CO_2 and 100 ml/min flow rate.

Treatment temp (°C)	Log_{10}CFU/ml reductions
28	1.02 ± 0.07
33	1.76 ± 0.17
35	3.61 ± 0.22
38	4.50 ± 0.42
40	4.64 ± 0.15
43	4.86 ± 0.24

At 43°C, even though microbial inactivation approaches 5 log_{10} CFU/ml, the higher temperature combined with $scCO_2$ processing at elevated pressure resulted in some protein coagulation and denaturation (Table 11.5).

11.5.4 Impact on microbial inactivation for solid foods

With regard to solid products and $scCO_2$ processing, the literature reports that *E. coli* treated at 42.5°C and 4500 psi (30.9 MPa) resulted in 7 log_{10} reductions in phosphate buffer after 10 min and a 1 log_{10} reduction in ground beef after 180 min (Sirisee *et al.* 1998). Therefore, the potential or proof of concept for microbial inactivation even on solid products exists, although $scCO_2$ processing conditions need optimization in order to result in desired pasteurization effects without damage to the food quality and integrity. Table 11.6 depicts the relationship among $scCO_2$ processing time and increased log_{10} CFU/g reductions for *E. coli* on ground beef after 15 min (0.23 log_{10} CFU/g) as compared to 60 min (3.22 log_{10} CFU/g).

However, after 30 min $scCO_2$ processing time, ground beef moisture decreased 7% (Table 11.6). Again process optimization may be the key to commercialization as the holding vessel size proportion ratio to meat

Table 11.6 $ScCO_2$ treatment of *E. coli* on beef inoculated to 7 log_{10} CFU/g.

Treatment	Average log_{10}CFU/g feduction	a_w	Moisture (%)
Untreated	None	0.984 ± 0.007	78.90 ± 1.16
43°C for 5 min	−0.06 ± 0.08	ND	ND
$scCO_2$ processed for 5 min	0.14 ± 0.15	ND	ND
$scCO_2$ processed for 15 min	0.23	ND	ND
$scCO_2$ processed for 30 min	2.19	0.966	70.45
$scCO_2$ processed for 60 min	3.22 ± 0.07	0.943	63.71
$scCO_2$ processed for 60 min (250 g sample–10× others)	1.44	0.980	74.20

ND: not determined.

quantity being processed may decrease product moisture loss at the expense of a 2 $\log_{10}$ CFU/ml reduction (Table 11.6). Ultimately, product integrity may determine the usefulness of scCO_2 technology application and not microbial safety.

11.5.5 scCO_2 processing efficiencies

Several factors have been shown to be important with respect to increasing scCO_2 processing efficiencies. CO_2 solubility is a key to CO_2 saturation and could be influenced by treatment surface area, product agitation, water content, and fat content. Working with solid food products in a limited-sized reaction vessel has shown that a lower sample volume or a lower reaction volume ratio has resulted in a higher scCO_2 processing effect. In addition, higher microbial inactivation is seen for lower microbial contamination loads. In a combination or hurdle approach, lower food pH increases the effectiveness of scCO_2 processing and increased microbial inactivation efficiencies.

11.6 Conclusions

In summary, CO_2 was used under supercritical conditions to inactivate various foodborne pathogens. Microbial inactivation using scCO_2 was product specific and microorganism dependent. Indigenous yeast and mold were successfully inactivated in orange and carrot juices. Lactic acid bacteria have been shown to be more resistant to scCO_2. *E. coli* $\log_{10}$ CFU/ml reductions were 7.9 $\log_{10}$ CFU/ml in orange juice and 8.3 $\log_{10}$ CFU/ml in carrot juice as opposed to *L. plantarum* $\log_{10}$ CFU/ml reductions of 6.0 in orange juice and 4.8 $\log_{10}$ CFU/ml in carrot juice, respectively. *B. coagulans* spores required temperature increases combined with scCO_2 conditions in order to accelerate spore inactivation from 45°C (<1.0 $\log_{10}$ CFU/ml reduction) to 75°C (5 $\log_{10}$ CFU/ml reduction). The flavors of scCO_2-processed orange and carrot juices were milder than unprocessed samples. Apple–lemon–ginger juice was least affected by the process. Changes in pH were more apparent in neutral as compared to acidic juices. The Brix (% sucrose) remained unchanged throughout processing for each juice. Overall, flavor and color effects were less using scCO_2 as compared to conventional heat processing. Solid foods were capable of being scCO_2 processed although with reduced effectiveness due mostly to needed product CO_2 pasteurization optimization adjustments. Although from a safety perspective, microbial inactivation heavily outweighs any product flavor changes. The tailoring of exact scCO_2 process conditions to specific consumer products needs emphasis on individual customer preferences.

Ultimately, commercial applications of a new process for large-scale food preparation require market acceptability, product quality, economic production feasibility, and finally product microbial safety. Usually a thermal process already in use is difficult to transform to a nonthermal process largely due to the cost associated with commercial process overhaul and production restart or downtime. Consumer preferences are acknowledged only if they result in increased product sales. If consumers accept lower quality products as a result of cost, commercial changes will not be justified. These are the real hurdles that scCO_2 processing may need to overcome in order to become commercially viable.

12 Outlook and Unresolved Issues

Luc Van Ginneken, Linsey Garcia-Gonzalez, Kathy Elst, and Frank Devlieghere

Abstract: Published results show that dense phase CO_2 processing technology has great potential for guaranteeing food safety and maintaining food quality. Weighing difficulties against opportunities indicates that, in the near future, the technology could become a key food sanitation application. However, to meet this high expectation, more extensive and accurate data are required to convince consumers and stakeholders about the improvements this emergent technology offers. Some technological and regulatory challenges, such as adequate mathematical modeling for process optimization, proper scale-up, simulating and calculating process economics, the acquisition of more sensory and shelf life data, quality assurance, compliance with existing regulations, and hygienic standards, all need to be addressed. The aim of the final chapter is to provide the reader with an outlook for the future of dense phase CO_2 processing, and major issues still to be resolved before it can be adopted by the industry.

Keywords: potentials; unresolved issues; research needs.

12.1 Introduction

Traditionally foods have been preserved by using heat, such as applied in commercial sterilization, pasteurization, and blanching processes. Foods are heated for a number of reasons, the main one being to inactivate pathogenic and/or spoilage microorganisms (Raso and Barbosa-Cánovas 2003). Another reason includes the inactivation of undesirable enzymes, because foods may also change and become unacceptable due to enzyme-catalyzed reactions (e.g. the loss of cloud by pectinesterase in orange juice or the discoloration of fruit and vegetables by polyphenol oxidases). Although traditional thermal processing of foods has the ability to inactivate microorganisms and enzymes to the required levels,

Dense Phase Carbon Dioxide: Food and Pharmaceutical Applications, First Edition. Edited by Murat O. Balaban and Giovanna Ferrentino.

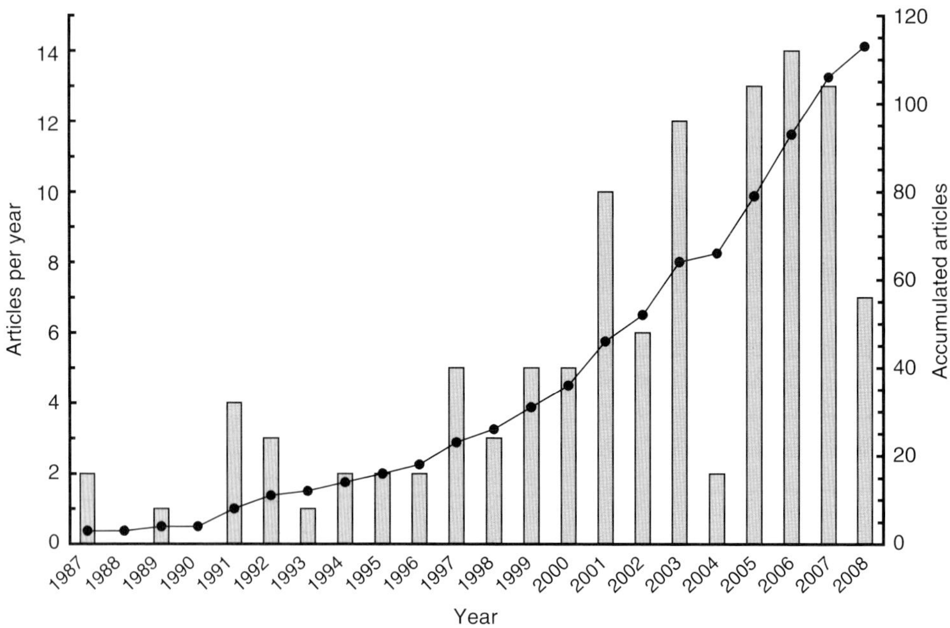

Figure 12.1 Number of journal articles (bars) and cumulative number of articles (line) published per year in the field of dense phase carbon dioxide inactivation of microorganisms. Articles published in scientific literature from 1987 to 2008 were taken into account, and were retrieved from different sources (such as Web of Science™, Medline/PubMed, and different search engines on the Internet). Reproduced from Garcia-Gonzalez *et al.*, copyright 2007, with permission of Elsevier.

heat produces deleterious changes in the flavor, color, texture and nutritional qualities of sensitive foods (Raso and Barbosa-Cánovas 2003). In the last decades, therefore, intensive research efforts have focused on the development of alternative, nonthermal preservation processes in the food industry.

Alternative processing technologies for inactivating pathogenic and spoilage microorganisms in foods without relying on heat are not new concepts, but their development for use as food preservation treatments has received considerable attention only recently, in response to consumer demands for more "fresh-like" and "natural" food products (Ross *et al.* 2003). In the last 2 decades, a large number of *in vitro* studies on DPCD preservation have been performed by food scientists (predominantly on reconstituted food samples) and valuable results have been achieved, as demonstrated in the preceding chapters of this book. From 2001 on, the number of published journal articles related to food preservation using DPCD has significantly increased, reaching a total number of approximately 115 journal articles published by the end of December 2008 (Figure 12.1). Despite these intensified research efforts and significant advances during

Table 12.1 Important opportunities and drawbacks of the DPCD preservation technique.

Opportunities	Threats
Natural image	Inactivation mechanism not entirely clear
High fresh-like organoleptic quality.	Extraction of and interaction with food ingredients
Spores can be inactivated when combined with, for example, heat, pressure pulsing, and acid environment.	Occurrence of HPCD-resistant mutants still needs to be investigated
Applicable to acid foods	Inactivation conditions dependent on type of food and type of microorganism
Upscaling of equipment has been developed for other applications (e.g. extraction).	Large investment costs
Continuous processing for liquid foods is possible.	Processing-related problems for solid foods (discontinuous in nature, limited diffusion of CO_2, packaging after treatment, etc.)

Reproduced from Garcia-Gonzalez *et al.*, copyright 2007, with permission of Elsevier.

the last eight years, however, the DPCD preservation technique thus far has not been implemented on a commercial scale by food producers and processors.

In the preceding chapters of this book, underlying the principles behind the production of safe and high quality DPCD-treated foods, a wealth of information on DPCD preservation of foods has been given. The most important opportunities and threats of the DPCD preservation technology that could be derived from all this information are summarized in Table 12.1. Striking a balance by weighing threats against opportunities indicates that in some years DPCD treatment could become a key food sanitation technology. However, to meet this high expectation, more extensive and accurate data will be required to convince consumers and stakeholders about the improvements this emergent technology offers. In addition, reliable and unbiased information, including the potentially negative aspects of the technology and its limitations, will have to be released. Although the present book shows that DPCD processing technology has great potential for guaranteeing food safety and improving food quality, some technological and regulatory challenges (such as adequate mathematical modeling for process optimization, proper scale-up, simulation and calculation of process economics, acquisition of more sensory and shelf life data, quality assurance, compliance with existing regulation and hygienic standards, etc.) still remain (Garcia-Gonzalez *et al.* 2007). The aim of this chapter is to supply the reader the future outlook of DPCD processing and major issues still to be resolved before the supply chain can receive the benefits of DPCD-treated foods.

12.2 Unresolved issues

12.2.1 *Inactivation mechanism of DPCD*

12.2.1.1 General

Although published reports (Damar and Balaban 2006; Zhang *et al.* 2006b; Garcia-Gonzalez *et al.* 2007) and the present book demonstrate the huge potential of DPCD for food preservation purposes, its exact inactivation mechanism is not clear. However, based on the observed trends in literature, a hypothetical inactivation mechanism (for simple food systems) has been proposed in recent years (Spilimbergo and Bertucco 2003; Damar and Balaban 2006; Garcia-Gonzalez *et al.* 2007). In this hypothetical model (Figure 12.2), different levels of interaction between CO_2 and the bacterial cells can be identified and summarized as follows:

1. When dense-phase CO_2 is applied to a food (or broth) system, it may partly dissolve in the water phase of the food (or broth), resulting after equilibrium in a certain concentration of dissolved CO_2 in the water phase of the product, and hence in a lowered extracellular pH (pH_{ex}).

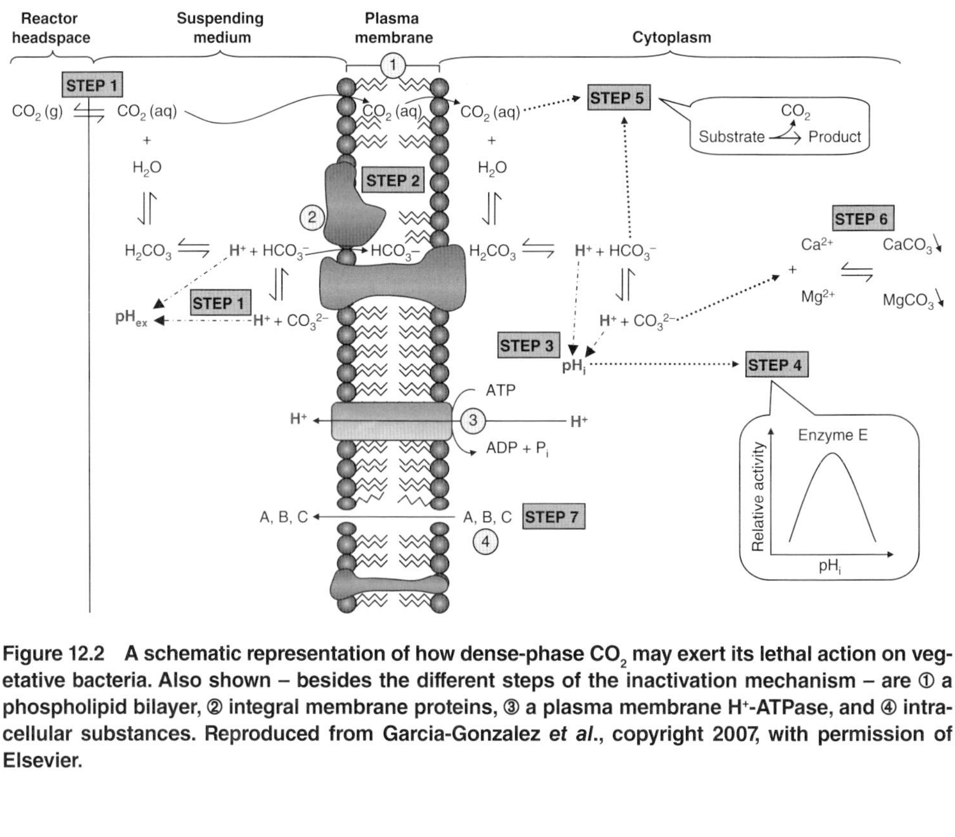

Figure 12.2 A schematic representation of how dense-phase CO_2 may exert its lethal action on vegetative bacteria. Also shown – besides the different steps of the inactivation mechanism – are ① a phospholipid bilayer, ② integral membrane proteins, ③ a plasma membrane H⁺-ATPase, and ④ intracellular substances. Reproduced from Garcia-Gonzalez *et al.*, copyright 2007, with permission of Elsevier.

2. Dissolved CO_2 may diffuse into the cellular membrane of bacteria, where it (chemically) modifies the membrane fluidity and hence increases membrane CO_2 permeability.
3. As a result of the increased membrane permeability, dense-phase CO_2 easily penetrates through the bacterial cell membrane and accumulates in the cytoplasm, where it may be able to produce a rapid lowering of the internal pH of the cell.
4. This lowered intracellular pH (pH_i) may inactivate certain enzymes, through which various essential metabolic pathways are disrupted.
5. Carbon dioxide may also exert a more direct inhibitory effect on certain intracellular enzyme systems by product inhibition or by an equilibrium-based mass action effect.
6. In addition, the carbonate ion may precipitate inorganic electrolytes (such as Ca^{2+} and Mg^{2+}) and calcium-binding proteins, which are essential for the regulation of many important intracellular activities, as carbonates.
7. Accumulated CO_2 may also extract vital constituents (such as phospholipids and hydrophobic compounds) from the cells or cell membranes upon pressure release, which lethally disturbs or alters the balance of biological systems.

Further research and more experimental evidence about the prevailing mechanism of biological inactivation by DPCD needs to be generated in order to potentially overcome some of the limitations to practical use and commercial implementation of the DPCD preservation process. Once the inactivation mechanism is more elucidated, mathematical modeling of the different (most important) steps in the DPCD inactivation process (see Figure 12.2) could be attempted, which should allow further process optimization. In this way, the DPCD preservation method could be adapted to achieve food pasteurization or sterilization at milder process conditions, which is important from both an economic and a food quality point of view. In addition, it would allow better identifying the target foods (on the basis of composition of their matrix and their native bacterial flora), for which DPCD preservation might prove especially suitable and advantageous. It would also allow further optimizing the most promising design and processing implementation of how contact between the DPCD and the foods must be ideally achieved to circumvent potential diffusion limitations and mass transfer restrictions in real-life food systems.

12.2.1.2 Research needs

12.2.1.2.1 General

In order to unravel the inactivation mechanism of DPCD, the different levels of interaction as outlined above (and as graphically represented in Figure 12.2) need to be studied in more detail. First, it seems of the utmost importance to unravel how the concentration (or, more precisely, the chemical activity) of dissolved CO_2 in the aqueous phase of foods drives the DPCD process, and how the CO_2 solubility rate in the aqueous phase, the rate of pH_{ex} decrease (see Step 1 in Figure 12.2), the rate of pH_i decrease (see Step 3 in Figure 12.2), and the microbial inactivation rate all are related. The relationship between pH_{ex} and the microbial inactivation rate is not very straightforward, as demonstrated by (among others) Spilimbergo (2002). She showed that, while pH_{ex} in water is strongly affected by CO_2, the CO_2 effect in orange juice is totally buffered, probably because of the presence of other substances in such a complex substrate (implying that CO_2 strongly affects pH in "pure" water but has no effect in the orange juice considered). It must be noted, however, that a high microbial inactivation efficiency could be obtained in orange juice as well as in physiological saline after DPCD treatment (Spilimbergo 2002). Thus, it can be concluded that apparently changes in pH_{ex} cannot be responsible alone for the microbial reduction, and other mechanisms are involved to explain this phenomenon. It is the internal (cytoplasmic) pH (i.e. the pH_i), rather than the external one, that is ultimately responsible of cellular viability.

The fact that there exists no straightforward relationship between pH_{ex} and cellular viability, is apparent because the synergistic effect of dense-phase CO_2 and (predominantly) temperature on the cell membrane should also be taken into account (see Step 2 in Figure 12.2). The generally observed acceleration of cellular inactivation at higher temperatures seems to be manifest at two levels (Oulé *et al.* 2006; Garcia-Gonzalez *et al.* 2007). Firstly, temperature acts on the physical properties of CO_2, decreasing its viscosity, to facilitate its diffusion in the cells. Secondly, temperature may stimulate the diffusion of CO_2 into the cells, acting on the integrity of cellular membrane and increasing its fluidity. So apart from the "chemical" component (i.e. the CO_2 speciation in the external medium, and hence the resulting pH_{ex}), the physiological component (i.e. the effect pressure and temperature have on the biological membrane) is also extremely important.

From the above discussion, it should be clear that the interrelation between dissolved CO_2, extracellular pH, intracellular pH, and cellular viability is complex, and not very clear at the moment. In addition, more research is needed to establish whether the observed selective enzyme inactivation (see Step 4 in Figure 12.2) is the primary cause of cell death or if it is a result of the decreased cell viability after DPCD treatment. Research effort also needs to be strengthened to test the assumed mechanism of CO_2 extraction of cell contents (see Step 7 in Figure 12.2). The anticipated and specific research needs to unravel the DPCD inactivation mechanism(s) will be briefly discussed in the next sections.

12.2.1.2.2 Modeling of extracellular pH

In the past, improved thermodynamic models (using an equation-of-state approach) have been developed to accurately predict CO_2 solubility in aqueous solutions at elevated pressure. For instance, see Edwards *et al.*

(1978), Duan and Sun (2003), Spilimbergo *et al.* (2005), and Duan *et al.* (2006) for H_2O–CO_2; Rumpf *et al.* (1994), Duan and Sun (2003), and Duan *et al.* (2006) for H_2O–CO_2–salt systems; and Van Halsema *et al.* (1998) for systems of H_2O–CO_2 with organic acids. In addition, mathematical models have been presented to predict the pH in CO_2-pressurized (complex) aqueous systems, such as pure water (Meyssami *et al.* 1992; Spilimbergo *et al.* 2005; Li and Duan 2007) and milk (Hofland *et al.* 1999). These models can be used as a first estimation to predict pH in other complex aqueous systems. However, complex aqueous systems (such as liquid foods and/or broths) often contain buffering components, which influence the equilibrium pH in the solution. This buffering capacity must also be taken into account, implying that the existing models need to be adapted to cope with the buffering capacity of the food or broth solution in order to accurately calculate and predict the pH_{ex} in these complex environments. Very recently, we developed a simple model, linking conventional titration curves with a thermodynamic description of the phase behavior of the water–CO_2 system, that allowed one to predict pH_{ex} (as a function of pressure, temperature and buffer capacity) in complex aqueous food systems (Garcia-Gonzalez *et al.* 2010a). Our mathematical model, however, needs further refinement to properly take into account non-ideality of the liquid phase.

In addition, the CO_2-to-food product volume ratio in the DPCD treatment system will have an influence on the kinetics (mass transfer) of CO_2, dissolving in the aqueous phase of the food product. Indeed, the concentration of dissolved CO_2 (and hence pH_{ex}) will also be dependent on the mass transfer kinetics of CO_2 in the food/broth system, and hence will be dependent on, for example, the contact time, stirring rate, and CO_2-to-food product volume ratio. The theoretical time needed by CO_2 to reach equilibrium in an aqueous (food) solution under the operating DPCD conditions can be estimated by modeling the CO_2 dissolution in the liquid phase as a function of time (see e.g. Elvassore *et al.* 2000; Hofland *et al.* 2003). Further development and improvement of both equilibrium and dynamic models is of paramount importance to predict the pH_{ex} at a certain contact time at the prevailing DPCD conditions in complex, buffered food and broth solutions.

12.2.1.2.3 Measurement of extracellular pH

It is important that, besides modeling and calculation of the pH_{ex} using different thermodynamic and dynamic models, these calculations are validated with *in situ* measurements of the pH_{ex} in both simple and complex (buffered) solutions. Some reports found in literature measured the pH_{ex} after depressurization under atmospheric pressure (Erkmen 2000e). However, this does not give the pH during the DPCD treatment, because CO_2 will vaporize and pH_{ex} will increase relatively rapidly after depressurization to restore close to its initial value. A limited number of studies determined the pH_{ex} decrease during DPCD treatment in simple

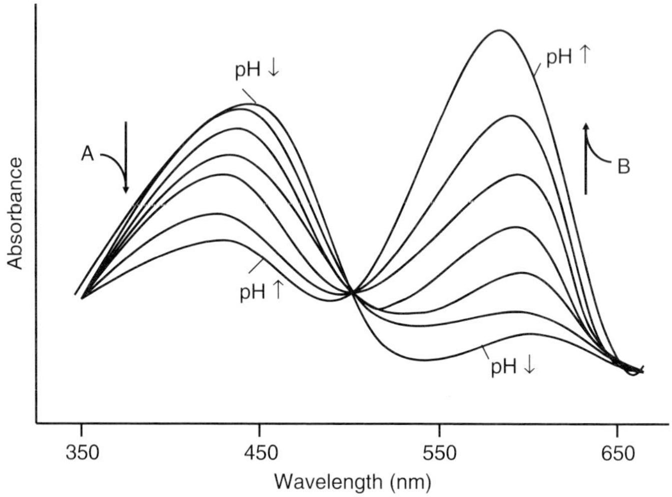

Figure 12.3 Schematic absorption spectra of bromophenol blue indicator in buffer solutions. Arrow A indicates decrease in absorbance of the acid form of the indicator as buffer pH increases. Arrow B indicates increase in absorbance of the base form of the indicator as buffer pH increases.

solutions under different pressures and temperatures by using a pressure-resistant pH probe (Haas *et al.* 1989; Balaban *et al.* 1991b; Meyssami *et al.* 1992; Hofland *et al.* 1999, 2000; Korshid *et al.* 2007). Pressure balancing needs and liquid junctions make it, however, practically inconvenient to use hydrogen and/or glass electrodes for routine measurements in high-pressure and/or high-temperature systems (Raghuraman *et al.* 2007).

A more suitable approach might be the spectroscopic measurement of pH using pH-sensitive dyes (Raghuraman *et al.* 2007) that can exist in an acid or base form. The optical absorbance spectra of pH-sensitive dyes change as they convert from their acid to base form. The fraction of the dye present in the acid and base forms depends on the pH of the solution (Raghuraman *et al.* 2007). This spectroscopic approach was used by Toews *et al.* (1995) to measure the pH of pure water in contact with supercritical CO_2 by observing the spectra of bromophenol blue with a UV-vis spectrophotometer. For this purpose, a high-pressure view cell, equipped with quartz side windows was used. Citric acid buffer solutions ranging from pH 2.6 to 3.6 were spiked with bromophenol blue and calibrated against a pH meter. The citric acid buffers were individually placed in a UV-vis cell, and the UV-vis spectra were then collected. Bromophenol blue shows a color transformation over the pH range 3.0–4.6; the acidic form of the indicator has an electronic transition at 430 nm, while the basic form has an electronic transition at 590 nm (Toews *et al.* 1995; Parton *et al.* 2002). With increasing pH, the absorbance at 590 nm increases, while the absorbance at 430 nm decreases (Figure 12.3). Because the absorbance ratio (A_{430}/A_{590}) is proportional to the hydrogen ion concentration, the pH of an aqueous solution pressurized with CO_2 can be determined by comparison with the corresponding calibration curve (i.e. a plot of absorbance ratio A_{430}/A_{590} versus H^+ concentration).

The approach of Toews *et al.* (1995) was also successfully applied by Spilimbergo (2002) and Parton *et al.* (2002) for a simple aqueous solution. More research, however, is needed to implement this very promising spectroscopic technique into DPCD research for more relevant (but hence more complex) food systems (such as fruit juices). This might require modification of the spectroscopic procedure because the food matrix itself might not allow an accurate UV-determination (due to interference with the food's solids content) or might exhibit a strong absorbance at one of the wavelengths used for the recording of the UV-vis spectra (Spilimbergo 2002). Very recently, we designed a spectrofluorometric method using calcein as fluorescent pH indicator for the *in situ* measurement of the pH of complex aqueous systems in contact with DPCD as function of pressure, temperature, initial pH, working volume ratio, and broth composition (Garcia-Gonzalez *et al.* 2010a). In general, there was a good agreement between the measured and predicted (i.e. modeled) pH values. We therefore believe the use of fluorescent pH-indicators might prove to be very promising to extend the range of usable pH-sensitive dyes (and hence the usable and sensitive pH range).

12.2.1.2.4 *Measurement of intracellular pH*

In the last decade, a number of experimental techniques using fluorescent probes have been proposed to determine the intracellular pH of different bacteria at atmospheric conditions (Breeuwer *et al.* 1996; Riondet *et al.* 1997; Chitarra *et al.* 2000; Corvini *et al.* 2000). One of the most promising methods (Breeuwer *et al.* 1996) is based on the intracellular conjugation of the fluorescent probe 5-(and 6-) carboxyfluorescin succinimidyl ester (cFSE) which is taken up by bacteria in the form of its diacetate ester, 5-(and 6-) carboxyfluorescin diacetate succinimidyl ester (cFDASE), which is subsequently hydrolyzed by esterases to cFSE in the cytoplasm. The free, unbound cFSE probe is eliminated by a short incubation of the cells in the presence of a fermentable sugar, and the intracellular pH in the bacteria can then be determined from the ratio of the fluorescence signal of the pH-sensitive wavelength (490 nm) and the fluorescence signal at the pH-insensitive wavelength (440 nm). This method was successfully used to determine the intracellular pH in Gram-positive (Breeuwer *et al.* 1996) and Gram-negative (Riondet *et al.* 1997) bacteria under normal and stress conditions at atmospheric circumstances. Moreover, the application of flow cytometry (FCM) with fluorescent cFSE-treated cells allowed the determination of the pH_i of individual cells within a population (Chitarra *et al.* 2000). The major advantage of the FCM technique with respect to fluorescence spectrofluorometry is its high sensitivity compared to spectrofluorometry, the capability of detecting low numbers of cells and the possibility to distinguish treated and untreated cells based on the absence or the presence of a pH gradient, respectively (Chitarra *et al.* 2000; Spilimbergo *et al.* 2005).

Spilimbergo *et al.* (2005) were the first to extend the use of fluorescent probes to DPCD research. These authors determined that, using flow cytometry in combination with the fluorescent cFSE probe, the pH_i of a *Bacillus subtilis* suspension drastically decreased to a value of 3.3 upon DPCD exposure (P = 8 MPa, T = 30°C, t = 5 min), while the pH_{ex} (calculated theoretically by means of the statistical associating fluid theory (SAFT) equation-of-state) reached a value of 3.2. Under these treatment conditions, a 5-decimal reduction of *B. subtilis* cells could be demonstrated. Since the values of pH_i and pH_{ex} were very similar, these findings seem to support the hypothesis that under certain DPCD exposure conditions microbial cells might be unable to maintain a favorable cytoplasmic pH homeostasis. In addition, in the framework of a collaboration between Professor Spilimbergo from the University of Trento, Italy, and our own laboratory, we evaluated the pH_i of *Listeria monocytogenes* cells during a DPCD treatment by cell fluorescent staining with cFDASE and 5-(and 6-) carboxy-2′7′-dichlorofluorescin diacetate succinimidyl ester (cDCFDASE) and by spectrofluorometric analysis (Spilimbergo *et al.* 2010). Different sets of experiments were carried out in batch mode at 25°C and 6.8 MPa at different treatment times. The data analysis indicated a progressive intracellular pH decrease inside the cell caused by the action of DPCD, followed by a further pH_i decrease (down to the pH_{ex}) after cellular death. This result supports the hypothesis that the cellular death is due to the CO_2-induced permeabilization of the phospholipid bilayer of the cellular membrane.

In the work of Spilimbergo *et al.* (2005, 2010), although very interesting and innovative in nature, the intracellular pH in the bacteria was not measured online (i.e. during DPCD treatment), but after the DPCD treatment and after the bacterial suspension was already depressurized to ambient pressure. Molina-Gutierrez *et al.* (2002) determined in situ and in real time the intracellular pH of *Lactococcus lactis* and *Lactobacillus plantarum* during a hydrostatic pressure treatment. The cFSE fluorescence of the sample during pressure treatment was determined in real time using an optical fiber, connected to the sapphire window of the pressure chamber and the excitation and emission monochromators of the fluorimeter. To our knowledge, however, not much research effort has been devoted to use this online fluorescence concept for DPCD research. Although quite challenging, work in this exciting area is in progress in our own laboratory, using a pressure chamber equipped with cylindrical sapphire windows and an optical fiber to measure intracellular pH online and in situ with the aid of different fluorescent probes (Figure 12.4).

12.2.1.2.5 *Measurement of cell membrane injury or modification*

Several techniques in DPCD research have been used to quantify the level of killing, and to characterize changes in integrity of the cell membrane,

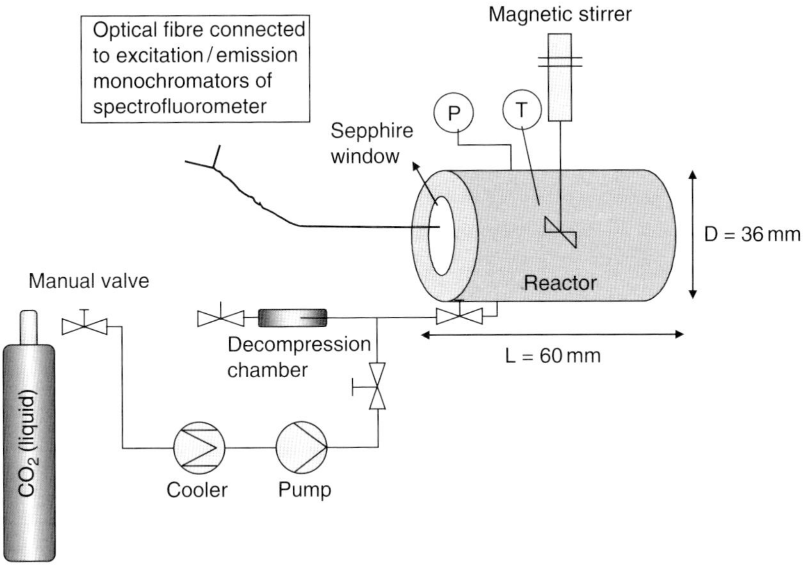

Figure 12.4 Schematic diagram of a high-pressure device used for real-time fluorescence analysis under dense-phase CO_2 conditions.

the permeability of the cell wall, and morphological changes (as elegantly reviewed by Zhang *et al.* 2006b) in order to unravel the inactivation mechanism of CO_2. A very interesting paper to mention in this respect is the work of Oulé and coworkers (2006), who evaluated the effect of different DPCD treatments on the cells by plating and by transmission and scanning electron microscopy observation.

Gänzle and Vogel (2001) established an online assay for determination of pressure mediated outer membrane damage based on 1-*N*-phenylnaphtylamine (NPN) fluorescence to distinguish between reversible and irreversible outer membrane damage in *E. coli*. The fluorescence of the sample during hydrostatic pressure treatment was determined in real time using an optical fiber, connected to the sapphire window of the pressure chamber and monochromators of the fluorimeter. Recently, CO_2 diffusion and cell permeabilization caused by DPCD was also experimentally evaluated by real-time measurement of propidium iodide uptake by spectrometric analysis (Spilimbergo *et al.* 2009). The reported data evidenced a correlation between cellular death and CO_2 permeabilization inside the cells. In a very recent publication, we also provided experimental evidence that the cellular membrane modification and damage is an important mechanism in the inactivation of microbial cells by DPCD (Garcia-Gonzalez *et al.* 2010b). Supporting electron microscopic examination in our study revealed that a DPCD treatment influenced membrane permeabilization by locally introducing pores which resulted in a loss of cytoplasmic content. In addition, we also observed aggregation and uneven distribution of the cytoplasmic content of DPCD-treated cells. More research in this field is needed, however, because knowledge of cellular membrane permeabilization by DPCD is useful to derive new strategies to manipulate membrane fluidization by DPCD with the goal of optimizing CO_2 inactivation technologies (Spilimbergo *et al.* 2009).

12.2.2 *Food quality and storage*

Published reports focus on either the microbial inactivation or (in only a minority of reported cases) the effects on the organoleptic and nutritional properties of liquid foods *immediately after* DPCD treatment (Damar and Balaban 2006; Garcia-Gonzalez *et al.* 2007). The quality of a food product that is ultimately being evaluated by the consumer will, however, depend upon the culmination of reactions that have taken place not only during processing but also during raw material storage, packaging operations, and subsequent storage (Lewis and Heppell 2000). In terms of the effect on product quality, storage is by far the most variable and therefore the most indeterminate of these processes, because there is less scope for the food producer to be able to control it (Lewis and Heppell 2000). It is therefore of the utmost importance to establish what a DPCD-processed food product will be like at the end of its shelf life (preferably under the most unfavorable storage conditions).

12.2.2.1 Storage and microbial stability

Conventional pasteurization is a relatively mild heat treatment resulting in products with a limited shelf life at refrigerated conditions. To replace thermal preservation techniques, nonthermal processing techniques (such as DPCD) must not only maintain food quality but also promote an equivalent or enhanced shelf life (Raso and Barbosa-Cánovas 2003). This implies that the main aim of processing is to control the activity of (psychrotrophic) spoilage and pathogenic microorganisms that are capable of growth at low temperatures during storage.

The most obvious way to improve the shelf life would be to increase the severity of the DPCD treatment, in order to inactivate or control those organisms that survive conventional processing. However, care should be taken not to unduly change the sensory characteristics of the processed foods.

Despite its importance, the keeping quality of DPCD-treated foods is hardly investigated during its refrigerated storage. In this respect, only Park *et al.* (2002), Kincal *et al.* (2005), and Del Pozo-Insfran *et al.* (2006) seemed to have performed a shelf life study to evaluate the effect of DPCD treatment on bacterial growth during refrigerated storage. Park *et al.* (2002) performed a shelf-life study to evaluate the effect of a consecutive DPCD (0.1–4.90 MPa, 5 min, 5°C) and high hydrostatic pressure (HHP) (400 MPa, 5 min, 25°C) treatment on microbial quality of carrot juice. During storage at 4°C, they observed no reactivation of aerobes in the pressure-treated carrot juice, while the control showed 8.4 log cfu/ml after 4 weeks. Kincal *et al.* (2005), performing a shelf life study on DPCD-treated orange juice, observed an increase in the bacterial number during storage at 1.7°C of the orange juice treated with continuous DPCD (107 MPa, CO_2–juice ratio of 1.0, residence time of 10 min), while at the beginning of storage no culturable organisms seemed to be present. By 6 weeks of storage, the microbial counts in the DPCD-treated juice reached 10^6 cfu/ml, considered unacceptable for juice quality. Two possible explanations for this observation were given by the authors: (1) injury and repair with the microorganisms could have taken place, or (2) postcontamination could have occurred, although every attempt was made to maintain aseptic conditions (Kincal *et al.* 2005). Dagan and Balaban (2006), however, reported that yeast populations in DPCD-processed beer remained unchanged after 1 month of storage at 1.67°C, suggesting that there may not be an injury or repair mechanism due to the nature of DPCD pasteurization.

Del Pozo-Insfran *et al.* (2006) noted only comparable microbial counts between DPCD-treated (34.5 MPa, 8 and 16% CO_2, 6 min) and thermally pasteurized (75°C, 15 s) muscadine grape juice during the first five weeks of storage at 4°C. Throughout subsequent storage, yeast and mold counts for DPCD-treated juices continuously increased whereas no changes were observed for heat-pasteurized juices. The number of total aerobic microorganisms, however, did not change during storage for both thermally pasteurized and DPCD-pasteurized muscadine grape juice.

12.2.2.2 Storage and nutritional damage, and changes in sensory characteristics

A very limited number of storage studies was also performed to evaluate the effect of DPCD treatment on physicochemical and sensory attributes during refrigerated storage. Folkes (2004), using a continuous system for DPCD pasteurization of beer, compared its physical and sensory attributes with those of fresh and heat-pasteurized beer. Although heat-treated beer was significantly different in taste and aroma from fresh beer after 1 month of storage at 1.67°C, the aroma and flavor of DPCD-treated beer were found not to be significantly different. A similar conclusion was reached by Damar and Balaban (2005), who pasteurized a coconut water–based beverage with DPCD using a continuous-flow system (34.5 MPa, 25°C, 13% CO_2 in the juice, 6 min) and conducted a shelf life study during 9 weeks of refrigerated (4°C) storage. Taste panels were used to evaluate overall likeability of fresh control, DPCD-pasteurized, and heat-pasteurized samples. Overall likeability of DPCD-treated coconut water was not significantly different from that of fresh control, whereas heat-pasteurized samples were rated significantly lower. Results from informal evaluations also suggested that color intensity, aroma, and flavor of DPCD-processed muscadine grape juice were comparable to control juices throughout the first 6 weeks of storage, after which the presence of a yeasty aroma negatively impacted their sensory attributes (Del Pozo-Insfran *et al.* 2006).

Apart from microbial stability and sensory attributes, Del Pozo-Insfran *et al.* (2006) also investigated the phytochemical stability of DPCD-processed muscadine grape juice during storage as compared to thermally pasteurized refrigerated juice. The DPCD-processed juices retained higher total anthocyanins, total soluble phenolics, and antioxidant capacity than thermally pasteurized juices after 10 weeks of storage at 4°C.

Dagan and Balaban (2006) investigated the effects of DPCD treatment (10% CO_2, 5 min, 21°C) on beer quality attributes, such as haze, foam capacity, foam stability, aroma, and flavor. After 30 days of storage at 1.67°C, haze was slightly reduced by DPCD pasteurization at 20.7 MPa from 432 nephelometric turbidity units (NTU) for fresh beer samples to 399 NTU for the DPCD-treated samples, but was not affected by DPCD pasteurization at 27.6 MPa (turbidity of 431 NTU). DPCD pasteurization also had a minimal effect on foam capacity during storage: heated and fresh beer samples exhibited the highest foam capacity with 332% and 327%, respectively, while CO_2-processed samples had a foam capacity of 370%. This change in foam capacity, however, would most likely be unnoticed by beer consumers. Foam stability throughout storage was not affected by DPCD treatment: fresh and CO_2-processed samples were not significantly different with fresh beer foam stability at 44%, and 27.6 MPa and 20.7 MPa at 43% and 41%, respectively. Aroma and flavor of DPCD-treated beer samples throughout storage were also not considered significantly different from fresh beer samples. As such, DPCD pasteurization is a valid alternative to heat pasteurization because of its ability to maintain fresh beer characteristics.

From all of the above, it is clear that further research is essential to demonstrate and explain the effect of DPCD treatment on the microbial, sensory and nutritional quality of both liquid and solid foods throughout storage.

12.2.3 Target foods

The microbial inactivation rate that can be achieved by a DPCD treatment is strongly affected by the constituents of the suspending media or the nature of the food matrix, as strong protective effects on microbial cells were observed in complex physicochemical environments of food systems when compared to simple solutions. Several investigators attributed the increased resistance to DPCD inactivation of vegetative cells suspended in complex media to the lipid and fat components of the media, suggesting that the presence of fat in growth and suspending media leads to a decreased CO_2 penetration into cells by changing the structure of cell walls and membranes (Garcia-Gonzalez *et al.* 2007). Furthermore, equal numbers of bacteria

Table 12.2 Influence of water activity, a_w (induced by different concentrations of NaCl, sucrose, or glycerol in the suspending medium), on the DPCD-induced inactivation of *P. fluorescens*. The experiments were performed at 10.5 MPa and 35°C during 20 min of treatment. Mean values ± standard deviations (N ≥3) are shown.

Type of solute	% (w/v)	a_w	Log_{10} (N_0/N_i)
Glycerol	0	0.997	4.1 ± 0.70
	2.9	0.989	4.7 ± 0.10
	20	0.950	4.9 ± 0.61
	32.8	0.924	2.6 ± 0.15
NaCl	0	0.997	4.1 ± 0.70
	8	0.950	4.8 ± 0.67
	16	0.888	6.9 ± 0.30
	23	0.834	7.7 ± 0.18
Sucrose	0	0.997	4.1 ± 0.70
	7.5	0.993	4.4 ± 0.70
	26	0.975	4.0 ± 0.20
	52.3	0.950	1.5 ± 0.46

suspended in simple solutions at the same initial pH were not inactivated with the same ease under exactly the same treatment conditions. Apart from a difference in buffering capacity, this finding can also be attributed to the different constituents of the suspending media (Garcia-Gonzalez *et al.* 2007).

Microbial inactivation also strongly depends on the water content (or the water activity, a_w) of the medium in which the cells are suspended during DPCD treatment. In general, experiments performed with different water contents demonstrated that vegetative cells with low water content are poorly inactivated, and that their resistance to inactivation increases with decreasing water content. The reason why wet microbial cells are more prone to HPCD inactivation is probably the direct result of an increased CO_2 solubility (and hence an increased formation and dissociation of carbonic acid), which liberates more H^+ ions that subsequently reduce the pH_{ex} of the suspending medium to lower values (Garcia-Gonzalez *et al.* 2007). In addition, the synergistic effect of water is also attributed to swollen cell walls and membranes due to the presence of water, by which these biological barriers are expanded to become more penetrable by CO_2 (Dillow *et al.* 1999).

Recently, we investigated the influence of water activity on the DPCD reduction of *Pseudomonas fluorescens* at 10.5 MPa and 35°C during 20 min of treatment in suspensions from which the a_W was adjusted with sucrose, NaCl, or glycerol (Garcia-Gonzalez *et al.* 2009). Our results (Table 12.2), however, indicated that a_W as such is not the determining factor influencing the inactivation, but that the kind of soluble solute used to lower water activity (i.e. NaCl, sucrose, or glycerol) and its concentration in the suspending media greatly influenced the inactivation of *P. fluorescens* cells. High concentrations of NaCl increased the susceptibility of cells towards

DPCD processing, while high sucrose and glycerol concentrations significantly inhibited DPCD inactivation of *P. fluorescens*.

Industrial application of the DPCD technique requires, among others, more in-depth information on the effect that the composition and the properties of a food matrix have on the inactivation efficacy of different relevant food pathogens and spoilage organisms. The influence of different food components (such as NaCl, oil, starch, proteins, and emulsifiers) and food properties (such as pH, fluid viscosity, and water activity) on the inactivation efficacy of DPCD, therefore, needs to be determined more thoroughly. This information will allow to gain more insight in (1) the interaction of DPCD with food components and (2) the relative effect that different food properties have on the DPCD inactivation efficacy. Ultimately, this will allow proper selection of target foods and niche food markets for which DPCD preservation especially might generate added value.

12.2.4 Process equipment and intellectual property

12.2.4.1 Modes of DPCD treatment

Several batch, semibatch as well as continuous systems of the DPCD pasteurization and sterilization process have been tested for various foods (Damar and Balaban 2006; Garcia-Gonzalez *et al.* 2007). In a batch system, CO_2 and the food to be treated are stationary in an autoclave during treatment. In a semicontinuous system, a continuous flow of CO_2 through the autoclave (which contains the food to be treated) is effected, while a continuous system allows flow of both CO_2 and the (liquid) food through the system. A semicontinuous system is much more efficient than a batch one, because it takes at least one order of magnitude less time to reach the same CO_2 concentration in the water phase of the food, as compared to a batch operation device (Elvassore *et al.* 2000). Continuous implementations, however, hold the most promise with regard to industrial implementation.

12.2.4.2 Patents and intellectual property

Regarding intellectual property, many patents were filed and are pending for processes that can be considered to use the DPCD pasteurization/sterilization concept for microbial inactivation (Garcia-Gonzalez *et al.* 2007). The existing, rather complex situation, however, may refrain companies to use these very promising techniques, as patent infringement might be feared by potential users.

Different batch, semi-continuous and continuous implementations of the DPCD pasteurization/sterilization process have been claimed (Table 12.3). Batch and semi-continuous (autoclave) reactors are typically used only for processing plants having lower capacities. Tubular (flow) reactor processes,

Table 12.3 Relevant patents in the field of HPCD pasteurization and sterilization.

Assignee	Patent number	Mode of DPCD treatment	Publication date	Inventors
Franz Juchem GmbH (DE)	DE 3734025 A1	Batch	1989	Lehmann and Juchem
University of Florida (US)	WO 90/02799, US 5,393,547	Batch	1990, 1995	Balaban *et al.*
Franz Juchem GmbH (DE)	EP 0 488 012 B1	Batch	1995	Juchem
Nippon Tansan Company, Ltd. (JP)	US 5,520,943	“Static” and “dynamic” batch using microbubbles	1996	Osajima *et al.*
Nippon Tansan Company, Ltd. (JP)	US 5,667,835	“Static” and “dynamic” batch using microbubbles	1997	Osajima *et al.*
Massachusetts Institute of Technology (US)	WO 99/066960, US 6,149,864,	Batch	1999, 2000	Dillow *et al.*
Safe-Fresh Technologies, LLC (US)	WO 03/101210 A1	Batch	2003	Garwood
LBG Invest & Consulting nv (BE)	WO 2004/039180 A1	Batch	2004	Van Ginneken *et al.*
NovaSterilis Inc. (US)	US 2005/0025667	Batch	2005	Christensen *et al.*
Air Liquide	US 2008/0171116 A1	Batch	2008	Rasanayagam and Yuan
Shimadzu Corporation and Nippon Tansan Company, Ltd. (JP)	US 5,704,276	Continuous using microbubbles	1998	Osajima *et al.*
Shimadzu Corporation and Nippon Tansan Company, Ltd. (JP)	US 5,869,123	Continuous using microbubbles	1999	Osajima *et al.*
Shimadzu Corporation and Nippon Tansan Company, Ltd. (JP)	EP 0 786 513 B1	Continuous using microbubbles	1999	Osajima *et al.*
Porocrit, LLC (US)	WO 00/41805, US 6,331,272 B1,	Continuous using a membrane contactor	2000, 2001	Sims
Praxair Technology, Inc. (US)	EP 0 979 657 A1, US 2002/0122860, WO 02/03816 A1	Continuous	2000, 2002	Wildasin *et al.*
Shimadzu Corporation and Nippon Tansan Company, Ltd. (JP)	US 6,616,849 B1	Continuous using microbubbles	2003	Osajima *et al.*
Kyushu TLO Company, Ltd. (JP)	WO 2004/000434 A1	Continuous using microbubbles	2003	Shimoda
Shimadzu Corporation and Nippon Tansan Company, Ltd. (JP)	US 6,821,481 B1	Continuous	2004	Osajima *et al.*
University of Florida Research Foundation (US)	US 6,723,365 B2, US 2004/0131739 A1	Continuous	2004	Balaban
Kyushu TLO Company, Ltd. (JP)	WO 2004/045316 A1	Continuous	2004	Shimoda
University of Florida (US)	US 2004/0234661 A1, WO 2005/034655 A1	Continuous	2004, 2005	Balaban
—	US 2005/0084581 A1	Continuous	2005	Sato

on the other hand, exhibit higher capacities for processing and manufacturing foods than do autoclave processes. Typical patents are as follows:

- In 1988, Balaban *et al.* (1995) were the first to file a method for inactivating enzymes in food products by exposing the food to pressurized CO_2 to produce a carbonic acid solution in the food product having a pH sufficiently low to irreversibly inactivate the enzymes. The method of the invention could be applied in either a batch mode or in a continuous mode. In this patent, the bactericidal effect of the CO_2 treatment was also demonstrated.
- Dillow *et al.* (2000) patented a batch method for sterilizing materials for medical applications, particularly polymers for drug delivery and implantation. In their method, the material is contacted with supercritical CO_2 in a high-pressure extraction vessel while agitating the contents of the vessel. By partially depressurizing and repressurizing (i.e. pressure cycling) the extraction vessel several times throughout the sterilization process, killing of microorganisms is enhanced by providing a driving force for CO_2 mass transport across the cell wall of the microorganisms. After maintaining the temperature and pressure (for a certain residence time), the extraction vessel is depressurized to atmospheric pressure and the sterilized material collected. Addition of water is also claimed to enhance the sterilization method.
- Van Ginneken *et al.* (2004) filed a patent for a batch method to inactivate biological contaminants in liquid high fat and/or protein containing food or feed products, such as liquid eggs. Their method comprises the subsequent steps of (1) introducing pressurized CO_2 (e.g. up to 6 MPa) in a reactor vessel at a predetermined temperature, (2) introducing the liquid food in this reactor while stirring, (3) introducing an additional amount of CO_2 to attain a final CO_2 pressure (e.g. 10 MPa) at a predetermined temperature (e.g. 40°C), (4) holding the final CO_2 temperature and pressure inside the vessel constant for a certain amount of time while stirring the mixture, and (5) releasing CO_2 pressure and collecting the treated liquid food. In the semicontinuous implementation of the process, the treated liquid food is led out of the reactor at the prevailing CO_2 pressure (e.g. 10 MPa) into a separator in order to separate the dissolved CO_2 from the liquid food. In this embodiment of the process, a large amount of CO_2 remains in the reactor with minimal pressure loss, and hence can be re-used to process the next batch of liquid.
- In a series of patents, Osajima and coworkers patented different methods and apparatus for inactivating enzymes and microorganisms in liquid foodstuffs. Inactivation in all these methods is obtained by contacting the liquid foodstuff with micron-sized CO_2 particles (mostly) in a supercritical state (so-called "micro bubbles"), which are generated by supplying pressurized CO_2 through a filter with a mesh size of not more than 100 μm in average diameter. In the "static" batch

implementation (Osajima *et al.* 1996, 1997), micro-bubbles of supercritical CO_2 are continuously fed to an autoclave until the pressure reaches the desired level, after which the CO_2 flow is stopped and treatment is continued. In the "dynamic" batch implementation (Osajima *et al.* 1996, 1997), a continuous flow of CO_2 micro-bubbles through the liquid foodstuff is maintained throughout the entire treatment period. In the continuous implementation of the process (Osajima *et al.* 1998, 1999a, 1999b), microbubbles of supercritical CO_2 and the liquid foodstuff are simultaneously pumped through a treatment vessel in respective flow rates (parallel-flow contact). After having enough contact with each other in the parallel flow, the liquid material and the supercritical CO_2 are separated into different phases at the top of the processing tank. In a more recent implementation of the continuous process (Osajima *et al.* 2003, 2004), microbubbles of liquefied (rather than supercritical) CO_2 are first dissolved in the liquid foodstuff, after which the liquid (taken out from an outlet port) is introduced into a warming pipe to transform the liquid CO_2 in its supercritical state. This considerably seems to shorten the processing time and makes pre-heating of the liquid material unnecessary.

- Sims (2000) patented a method and apparatus for preserving a liquid by contacting this liquid with CO_2 in a membrane contactor, which consists of several tubular hollow fiber porous membranes. The liquid, pressurized by a liquid feed pump, is directed along a first surface of a porous membrane, while pressurized CO_2 flows along the opposite surface of the membrane so that the liquid and CO_2 are contacted at pores in the membrane in a nondispersive manner. The CO_2 is continuously recirculated through the membrane contactor while maintaining the dense CO_2 in the pressurized state throughout the recirculation. The pressurized treated liquid, delivered to a first receiver tank held at high pressure, is continuously released into a second receiver tank to separate the CO_2 from the treated liquid at atmospheric pressure. A claimed advantage of the system is that, since the process can be run continuously and isobarically, it is possible to save on energy and equipment costs.
- In another implementation, Balaban (2004a) developed (in cooperation with Praxair) a continuous process, in which the liquid food product is mixed with liquid CO_2 under pressure (e.g. in a static mixer). The pressure and temperature in the flow regions are maintained at a level which is sufficient to keep the CO_2 in a continuous liquid state, without freezing the liquid product. After passing along a tubular reaction zone for a sufficient time, the liquid is depressurized in one or more expansion stages, causing the liquid to separate from the vaporized CO_2, killing microorganisms to sufficient levels. In addition, Balaban (2004b) also filed a patent for reducing microorganisms in a liquid beer or wine product with his continuous method.

12.2.4.3 Equipment design needs

12.2.4.3.1 Recovery of stripped volatiles

An important issue to be addressed with regard to DPCD pasteurizer design is the simultaneous effect of CO_2 as an extracting agent, as it could interfere with the quality of the final product. This means that volatiles, stripped from the foods during a DPCD pasteurization treatment as a result of the high solvating power of DPCD, should be recovered and returned to the food product in order not to deleteriously affect the food quality. A claimed advantage of the continuous process of Osajima *et al.* (1998, 1999a, b) is that their DPCD pasteurizer is provided with a system whereby flavor components extracted from the liquid foodstuff during the inactivation process can be retrieved, and returned to the liquid. Recently, a modification of this continuous implementation was also filed (Shimoda 2004). In this modified implementation, an inert gas is passed additionally through the liquid food after CO_2 depressurization, thereby removing dissolved CO_2 to minimize the loss of scent of components. PoroCrit also claimed that, in their system, aromatic constituents of the liquid which are soluble in CO_2 largely stay in the liquid after the recirculating CO_2 flow becomes saturated with them, retaining the original flavor of the treated liquid (Sims 2000). Praxair states in its patents that methods such as condensation or absorption may be utilized to recover aromas or flavors from the CO_2 gas, resulting after depressurization. So although in some patents this issue seems to be addressed for liquid foods, more research is needed to assess the importance of stripping of volatiles for food quality and how to deal with it in equipment design.

12.2.4.3.2 Processing of solid foods

Although the DPCD inactivation method is claimed to be applicable to any food product having an aqueous phase, or being surrounded by a sufficient amount of an aqueous phase to dissolve sufficient CO_2 to inactivate the target microorganisms and/or enzyme(s), the majority of research was conducted on liquid foods, mainly beverages such as fruit juices (Damar and Balaban 2006; Garcia-Gonzalez *et al.* 2007). This probably has to do with the fact that, from an industrial-scale point of view, continuous DPCD processes, in which solid materials have to be fed into and out of the pressure reactor, are much more difficult to handle as compared to liquid foods. An interesting design in this respect might be the supercritical CO_2 extrusion process developed by the Agrotechnology and Food Sciences Group of Wageningen University and Research Center (Langelaan *et al.* 1999; Bartels *et al.* 2002) to isolate valuable or undesired components from solids in a continuous manner. The process is claimed to be flexible and easily controllable, and has low energy costs. The high pressure in the extruder is maintained by creating two material plugs in the beginning and end of the extruder (Figure 12.5). Between the material plugs, the DPCD is exercising

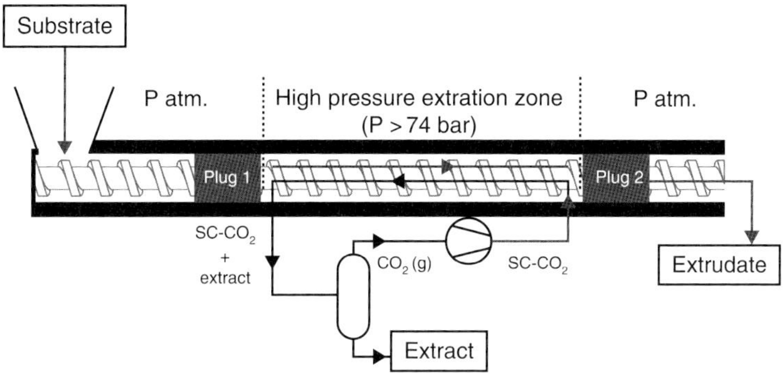

Figure 12.5 Schematic diagram of a supercritical CO_2 extrusion process developed by the Agrotechnology and Food Sciences Group of Wageningen University and Research Center (Langelaan *et al.* 1999; Bartels *et al.* 2002). Patent number (WO 99/26707). Reproduced with permission from Remco Hamoen and Jeroen Willemsen of Agrotechnology and Food Sciences Group of Wageningen University and Research Center.

its effect on the solid food. Although initially designed for extraction purposes, this extrusion system might be adaptable to be used as a DPCD pasteurizer for the continuous processing of solid foods. If the system can be redesigned for pasteurization purposes, however, the recovery of extracted volatiles also will have to be dealt with (as previously indicated).

12.2.4.3.3 Connection of packaging equipment to the DPCD processing plant

The aim of the DPCD process is to assure a required decimal reduction. From the moment that the product is depressurized below a lethal pressure in the process, it is vulnerable to recontamination by microorganisms, and the aim of all subsequent downstream operations should be to assure that the product is aseptically packaged without recontamination, which may occur from process equipment surfaces in the depressurization section or filling equipment, or by ingress of nonsterile air, water, or soil (Lewis and Heppell 2000). Careful consideration of hygienic design of the depressurization section and the food contact surfaces of packaging materials is of the utmost importance to maintain product integrity. To date, not much research efforts have been devoted to these issues in the DPCD processing of foods.

12.2.5 Fouling, cleaning, and disinfecting

When foods are treated with DPCD for preservation purposes, different reactions can take place that might give rise to the formation of deposits on the surface of several equipment parts of the DPCD pasteurizer. This phenomenon is termed 'fouling' (or 'scaling'). Such fouling can adversely affect the efficiency of certain equipment parts (e.g. heat exchangers). In addition, if these deposits become dislodged, they may plug narrow parts of the equipment or may end up in the final product (endangering product quality). Therefore where fouling occurs to a significant extent, these deposits must be subsequently removed by the process of cleaning. Following cleaning, surfaces must then be either disinfected or sterilized, prior to processing or shutdown.

12.2.5.1 Fouling

In general, fouling is affected by a number of factors, which can be categorized as process, product or equipment dependent (Lewis and Heppell 2000). Process-dependent variables include DPCD operational parameters, such as flow rate, residence time, turbulence, shear rate, temperature, and pressure. Although it would certainly be beneficial to consider these process-related parameters at the stage of equipment design, there may not be much scope for changing them in order to reduce fouling, because they largely will be predetermined by the required degree of inactivation and food quality issues. Product-dependent variables are related to the composition and viscosity of the product, and the effect CO_2 has on certain chemical, biochemical, and physical reactions. For instance, dissolved CO_2 is known to be able to precipitate proteins and calcium phosphate from milk (Hofland *et al.* 2003), which can deposit on equipment surfaces and initiate fouling. Equipment-dependent variables include the material of plant construction and surface finish, as surface is believed to play an important role in the fouling process. A rough surface is believed to increase fouling by providing more surface to which material may adsorb and by allowing better "keying" of the deposit on the surface (Lewis and Heppell 2000). With regard to fouling, however, not much research has been devoted to this area in relation to DPCD processing of foodstuffs.

12.2.5.2 Cleaning

Cleaning is defined as the removal of deposits from fouled equipment surfaces with the aid of cleaning agents. The aim is to achieve surfaces that are clean, so that the entire processing equipment is brought back to an acceptable starting condition for subsequent processing or shut-down. The cleaning efficiency depends on the type and concentration of the cleaning agents used and on the fluid temperature and time in contact with the equipment. The type of cleaning product(s) to be used to eliminate the fouled deposits depends on the characteristics of the treated food product and the demands of sanitization. The proper combination of pre-rinse, wash, and post rinse steps is also important. It should be noted that cleaning procedures, although of the utmost importance, consume valuable resources (such as energy, water for rinsing, and detergents for cleaning) and also produce effluents, and thus heavily impinge upon the environment.

Ideally, food-processing equipment is designed with cleanability in mind (avoiding crevices and dead spaces in pipelines, valves, and other fittings), as this is fundamental to the concept of producing safe products (Connery *et al.* 2005). A cleaning process is most efficient when the equipment can be cleaned-in-place (CIP) (i.e. when the production equipment can be cleaned without being dismantled). A CIP system must be built using components that are considered to be sanitary. For DPCD systems, working at relatively high pressure, it is not that straightforward to comply with the requirements of frequent cleaning and sanitization. For instance, finding the adequate sanitary components was one of the main hurdles that Praxair's engineers had to face at the beginning of their DPCD pasteurizer development, as most of the available components (at that time) were only rated for use at low pressure (Connery *et al.* 2005). In addition, assembling CIP components in a DPCD system also heavily contributes to equipment cost. So complying with the requirements of CIP is still an important issue that DPCD pasteurizers and sterilizers face.

12.2.5.3 Disinfecting and sterilizing

Following cleaning, water is generally used to remove the cleaning agents from the processing plant, and the surfaces may then be either disinfected (i.e. reducing the microbial load on the equipment surfaces to predetermined low levels) or sterilized (i.e. complete removal of the microbial load), prior to shutdown, startup, and/or processing.

Using heat (in the form of steam or superheated water) for disinfection and sterilization purposes is the simplest and most effective procedure. The main advantages of heat over chemical disinfectors are that there is no residue to wash out and no risk of cross-contamination exists (Lewis and Heppell 2000). For DPCD pasteurizers and sterilizers, the use of dense-phase CO_2 as (the main) disinfecting and sterilizing agent after cleaning seems evident, due to its (1) bacteriostatic nature, (2) availability in huge amounts in the DPCD processing facility, (3) vaporizable nature, and (4) noncorrosive nature.

12.2.6 Occurrence of DPCD-resistant mutants

One of the major concerns about new preservation technologies (such as HHP and DPCD) for the food industry is the occurrence of vegetative bacteria, which become resistant to inactivation after multiple rounds of treatment (Devlieghere *et al.* 2004). Development of resistance towards DPCD could form a potential risk to the safety and stability of DPCD-processed food, predominantly in relation with (insufficient) cleaning of DPCD processing equipment and repeated use of this equipment to process and preserve foods with DPCD.

Hauben *et al.* (1997) were able to select pressure-resistant *E. coli* mutants after numerous repeated cycles of selective HHP treatments and outgrowth of surviving population. With regard to DPCD pasteurization and sterilization, there is hardly any information available on the ability of microorganisms to become resistant to DPCD inactivation after successive treatments of survivors. Only Hong *et al.* (1999), who reported tailing in

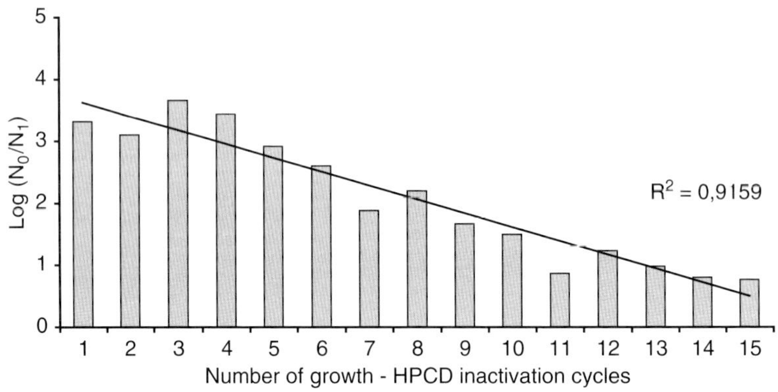

Figure 12.6 Development of resistance of *E. coli* cells to dense phase CO_2 after repetitive exposure to DPCD treatment (at 10.5 MPa and 35°C for 10 min) and regrowth of the surviving cells.

survivor curves when *L. plantarum* cells were treated at 13.8 MPa in MRS broth at temperatures higher than 30°C, looked into this matter. Tailing is often an indication of the existence of a subfraction in the bacterial population with higher resistance to inactivation or with a very small inactivation rate (Barbosa-Cánovas *et al.* 1998). When isolates derived from this subfraction were cultured and subjected again to a DPCD treatment at the same conditions, little difference in DPCD sensitivity (as compared to the original culture) was observed (Hong *et al.* 1999).

Recently, we investigated if resistance to DPCD inactivation could be induced in bacterial cells after successive outgrowth and treatment of survivors (Garcia-Gonzalez *et al.* 2010c). For this purpose, cells of the Gram-negative bacterium, *E. coli*, (and also cells of the Gram-positive bacterium, *L. monocytogenes*) were repeatedly treated at 10.5 MPa and 35°C during 10 min, followed by regrowth of the surviving cell fraction. The obtained results, graphically represented in Figure 12.6, clearly indicate that *E. coli* cells showed a gradual increase in resistance to inactivation after multiple rounds of DPCD treatment. This implies that occurrence of DPCD-resistant mutants might indeed be an important issue with regard to the safety of DPCD-processed foods. Therefore, further research on repeated exposure of survivor populations to DPCD is absolutely necessary to determine the extent of possible adaptation of microorganisms to DPCD inactivation, and the implications this might have for industrial-scale DPCD preservation of foods.

12.2.7 Industrial implementation and process economics

Despite the huge research and development efforts performed in the last 20 years, at present there is no commercial food product pasteurized by DPCD that is marketed, even though several companies have actively worked in this field (e.g. Praxair, PoroCrit LLC, Shimadzu, and l'Air Liquide) (Damar and Balaban 2006; Garcia-Gonzalez *et al.* 2007). To our knowledge, most of the commercialization efforts have been performed by Praxair Inc. (Burr Ridge, IL, US). Based on technology, licensed from the University of Florida (Balaban *et al.* 1995; Balaban 2004a), Praxair developed a continuous DPCD process and system (Wildasin *et al.* 2002). This system, commercialized under the Trade Mark "Better Than Fresh (BTF)", was advertised as an adequate nonthermal alternative to thermal pasteurization to produce fresh-tasting juice (Connery *et al.* 2005). Praxair constructed different mobile BTF units for demonstration purposes (Garcia-Gonzalez *et al.* 2007), and in 2003 the company announced that Sun Orchard, Inc. (Tempe, AZ, US) would install the first BTF nonthermal processing system at its Florida facility to process juice (*Chemical Engineering*, April 2003, p. 17). Once installed at Sun Orchard's Florida facility at the second quarter of 2003, it was foreseen to process juice at a rate of 40 gallons/min (about 150 l/min). Sun Orchard, however, never implemented Praxair's BTF technology on a commercial scale and the juice-producing company did not pursue it any further, using its own proprietary ultralight heat pasteurization process instead. The provider of industrial gasses itself also seemed to have completely abandoned the BTF program, as it is not a current product line at Praxair anymore.

Although it has often been stated that "commercialization could be a matter of time" or that "it is going to be a reality within a short time," the DPCD pasteurization process is still looking for its first commercial success. The following factors seem to determine successful commercialization of the technology (Damar and Balaban 2006): (1) significant differentiation of the DPCD product from those existing in the market regarding taste, quality, and shelf life is needed; (2) niche areas where no other conventional processing is possible need to be found, such as tropical juices that are very sensitive to heat; and (3) competing nonthermal technologies may be more appropriate for certain products, for example semisolid or solid foods.

Also regulatory hurdles need to be considered, as previously discussed in depth by Garcia-Gonzalez *et al.* (2007). But above all, process economics seems to be the biggest hurdle the technology is facing with regard to commercial implementation, as the per-gallon (or per-liter) cost of the DPCD process is higher than the one for thermal processing. Although it was expected a few years ago that cost engineering, process changes and the use of different (equipment) components could drive down the difference, at present this still does not seem to be the case.

12.3 Future outlook and conclusions

In the future, DPCD treatment could become one of the most available emergent preservation technologies. However, to meet this high expectation, consumers and stakeholders must be convinced about the improvements this new technology represents over competing technologies.

The DPCD technology clearly seems to represent some distinctive opportunities, as it does not change organoleptic and nutritional attributes to a detrimental level and does not leave residues in the treated food products. In addition, DPCD can be used as a single technology, as a "hurdle" in combined methods, or as a complementary step with mild thermal processes. DPCD preservation can also be very attractive to the food industry, because it might provide the opportunity to introduce new products in the food market.

On the "threats" side, process economics and compliance with CIP seems to be the biggest hurdles. There's a capitalization component and a processing component, and cost depends on system scale, utilization rate and other factors. So although it is difficult to fix the total cost for DPCD processing, the per-gallon (or per-liter) cost is certainly higher than traditional heat pasteurization. Occurrence of DPCD-resistant mutants is also a point of attention. In addition, more research is needed to understand the mechanism of bacteriostatic action.

It is clear that some technological and some regulatory hurdles still need to be overcome before the supply chain can receive the benefits of DPCD preservation. The major problem that novel, nonthermal preservation methods (such as DPCD) face in gaining widespread acceptance is that thermal processes are so firmly established and are capable of producing foods that are safe and of a high quality and nutritional value in large volumes at very low processing costs. We are, however, confident that the present book will represent a step further towards the production of foods industrially processed and preserved by this new technology.

12.4 Acknowledgements

The authors would like to thank Remco Hamoen and Jeroen Willemsen of the Agrotechnology and Food Sciences Group of Wageningen University and Research Center for giving us the permission to reproduce Figure 12.5.

References

Abraham, G., Debray, E., Candau, Y., and Piar, G. 1990. Mathematical model of thermal destruction of *Bacillus stearothermophilus* spores. *Applied and Environmental Microbiology* 56:3073–3080.

Adachi, Y., and Sugie, H. 1986. A new mixing rule-modified conventional mixing rule. *Fluid Phase Equilibria* 23:103–118.

Adami, R., Reverchon, E., Järvenpää, E., and Huopalahti, R. 2007. Supercritical antisolvent micronization of nalmefene HCl on laboratory and pilot scale. *Powder Technology* 179:163–170.

Aitken, A. 1990. *Identification of Protein Consensus Sequences*. Chichester: Ellis Horwood Press.

Alessi, P., Cortesi, A., Kikic, I., and Foster, N.R. 1997. Determination and correlation of solubility of pharmaceutical products in supercritical carbon dioxide. *Chemical and Biochemical Journal* 11:19–23.

Alessi, P., Cortesi, A., Kikic, I., Foster, N.R., and McNaughton, S.J. 1996. Particle production of steroid drugs using supercritical fluid processing. *Industrial & Engineering Chemistry Research* 35(12):4718–4726.

Amidi, M., Pellikaan, H.C., de Boer A.H., Crommelin, D.J.A., Hennink, W.E., and Jiskoot W. 2008. Preparation and physicochemical characterization of supercritically dried insulin-loaded microparticles for pulmonary delivery. *European Journal of Pharmaceutics and Biopharmaceutics* 68:191–200.

Amigo, L., Olano, A., and Calvo, M.M. 1995. Preservation of raw milk with carbon dioxide. *Zeitschrift für Lebensmitteluntersuchung und -Forschung A* 200(4):293–296.

Anderson, G.K. 2002. Solubility of carbon dioxide in water under incipient clathrate formation conditions. *Journal of Chemical & Engineering Data* 47(2):219–222.

Andrews, A.T., Anderson, M., and Goodenough, P.W. 1987. A study of the heat stabilities of a number of indigenous milk enzymes. *Journal of Dairy Research* 54:237–246.

Angus, S., Armstrong, B., and De Reuck, K.M. 1976. *International Thermodynamic Tables of the Fluid State – International Union of Pure and Applied Chemistry*, 3rd ed. New York: Pergamon Press.

Arreola, A.G. 1990. Effects of supercritical CO_2 on some quality attributes of single strength orange juice. MSc thesis, University of Florida. Gainesville, FL.

Arreola, A.G., Balaban, M.O., Marshall, M.R., Peplow, A.J., Wei, C.I., and Cornell, J.A. 1991. Supercritical carbon dioxide effects on some quality attributes of single strength orange juice. *Journal of Food Science* 56(4):1030–1033.

Dense Phase Carbon Dioxide: Food and Pharmaceutical Applications, First Edition. Edited by Murat O. Balaban and Giovanna Ferrentino.

Bae, Y.Y., Lee, H.J., Kim, S.A., and Rhee, M.S. 2009. Inactivation of *Alicyclobacillus acidoterrestris* spores in apple juice by supercritical carbon dioxide. *International Journal of Food Microbiology* 136:95–100.

Baggerman, W.I. 1984. Heat resistance of yeast cells and fungal spores. In: R.A. Samsom, E.S. Hoekstra, and C.A.N. Van Oorschot (eds.), *Introduction of Food-borne Fungi*, pp. 227–231. Baarn: Centraalbureau voor Schimmelcultures.

Bakhbakhi, Y., Charpentier, P.A., and Rohani, S. 2006. Experimental study of the GAS process for producing microparticles of beclomethasone-17, 21-dipropionate suitable for pulmonary delivery. *International Journal of Pharmaceutics* 309:71–80.

Balaban, M.O. 2004a. Method and apparatus for continuous flow reduction of microbial and/or enzymatic activity in a liquid product using carbon dioxide. US Patent 6,723,365 B2 (April 20, 2004) and US Patent Application 2004/0131739 A1 (July 8, 2004).

Balaban, M.O. 2004b. Method and apparatus for continuous flow reduction of microbial and/or enzymatic activity in a liquid beer product using carbon dioxide. US Patent Application 2004/0234661 A1.

Balaban, M.O., Kincal, D., Hill, S., Marshall, M.R., and Wildasin, R. 2001. The synergistic use of carbon dioxide and pressure in nonthermal processing of juices. Paper presented at the Institute of Food Technologists (IFT) annual meeting, New Orleans, LA, 23–27 June.

Balaban, M.O., Marshall, M.R., and Wicker, L. 1995. Inactivation of enzymes in foods with pressurized CO_2. US Patent 5,393,547.

Balaban, M.O., Pekyardimci, S., Chen, J.S., Arreola, A., Zemel, G., and Marshall, M.R. 1993. Enzyme inactivation by pressurized carbon dioxide. In: *Science for the Food Industry of the 21st Century*, pp. 235–252. Mount Prospect, IL: Yalpani Press.

Balaban, M.O., Pekyardimci, S., Marshall, M.R., Arreola, A., Chen, J.S., and Wei, C.I. 1991a. Effects of high pressure CO_2 treatment on enzyme activity in model systems and orange juice. In: *Proceedings of the 2nd International Symposium on Supercritical Fluids*, pp. 114–117, Boston, MA, 20–22 May.

Balaban, M.O., Arreola, A., Marshall, M.R., Peplow, A., Wei, C.I., and Cornell, J. 1991b. Inactivation of pectinesterase in orange juice by supercritical carbon dioxide. *Journal of Food Science* 56(3):743–750.

Ballestra, P., and Cuq, J-L. 1998. Influence of pressurized CO_2 on the thermal inactivation of bacterial and fungal spores. *Lebensmittel-Wissenschaft und-Technologie* 31:84–88.

Ballestra, P., Dasilva, A.A., and Cuq, J.L. 1996. Inactivation of *Escherichia coli* by carbon dioxide under pressure. *Journal of Food Science* 61:829–831.

Bamberger, A., Sieder, G., and Maurer, G. 2000. High-pressure (vapor + liquid) equilibrium in binary mixtures of (carbon dioxide+water or acetic acid) at temperatures from 313 to 353 K. *Journal of Supercritical Fluids* 17(10):–110.

Bando, S., Takemura, F., Nishio, M., Hihara, E., and Akai, M. 2003. Solubility of CO_2 in aqueous solutions of NaCl at (30 to 60) °C and (10 to 20) MPa. *Journal of Chemical & Engineering Data* 48(3):576–579.

Barbosa-Cánovas, G.V., Pothakamury, U.R., Palou, E., and Swanson, B.G. 1998. *Nonthermal Preservation of Foods*. New York: Marcel Dekker.

Barrett, A.M., Dehghani, F., and Foster, N.R. 2008. Increasing the dissolution rate of itraconazole processed by gas antisolvent techniques using polyethylene glycol as a carrier. *Pharmaceutical Research* 25:1274–1289.

Bartels, P.V., Hulleman, S.H.D., and Langelaan, H.C. 2002. Method for the extraction of a substance from a starting material. US Patent 6,491,892 B1.

Bartholomé, E., and Friz, H. 1956. Solubility of carbon dioxide in water at high pressures. *Chemical Engineering & Technology* 28:706–708.

Bartolomei, M. 2000. Solid-state studies on the hemihydrate and the anhydrous forms of flunisolide. *Journal of Pharmaceutical and Biomedical Analysis* 24:81–93.

Battino, R., and Clever, L.H. 1965. The solubility of gases in liquids. *Chemical Reviews* 66(4): 395–463.

Bauer, C., Steinberger, D.J., Schlauer, G., Gamse, T., and Marr, R. 2000. Activation and denaturation of hydrolases in dry and humid supercritical carbon dioxide. *Journal of Supercritical Fluids* 19:79–86.

Bayne, H.G., and Michener, H.D. 1979. Heat resistance of *Byssochlamys* ascospores. *Applied and Environmental Microbiology* 37:449–453.

Beret, S., and Prausnitz, J.M. 1975. Perturbed hard-chain theory: an equation of state for fluids containing small and large molecules. *AIChE Journal* 21:1123–1132.

Bertoloni, G., Bertucco, A., De Cian, V., and Parton, T. 2006. A study on the inactivation of microorganisms and enzymes by high pressure CO_2. *Biotechnology and Bioengineering* 95:155–160.

Bodmeier, R., Wang, H., Dixon, D.J., Mawson, S., and Johnston, K.P. 1995. Polymeric microspheres prepared by spraying into compressed carbon dioxide. *Pharmaceutical Research* 12:1211–1217.

Bolz, U., and Stephan, K. 1991. A new synthetic method for high pressure phase equilibria of ternary mixtures. In: *Proceedings of the 2nd International Symposium on Supercritical Fluids*, Boston, MA, 20–22 May.

Bouzidi, A., Majewski, W., Hober, D., and Perrut, M. 1998. Supercritical fluids improve tolerance and viral safety of fresh frozen plasma. In: *Proceedings of the 5th Meeting on Supercritical Fluids*, pp. 717–722, 23–35, Nice, France, March.

Breeuwer, P., Drocourt, J-L., Rombouts, F.M., and Abee, T. 1996. A novel method for continuous determination of the intracellular pH in bacteria with the internally conjugated fluorescent probe 5 (and 6-)-carboxyfluorescein succinimidyl ester. *Applied and Environmental Microbiology* 62:178–183.

Breland, E., and Englezos, P. 1996. Equilibrium hydrate formation data for carbon dioxide in aqueous glycerol solutions. *Journal of Chemical & Engineering Data* 41(1):11–13.

Briones, J.A., Mullins, J.C., Thies, M.C., and Kim, B.U. 1987. Ternary phase equilibria for acetic acid–water mixtures with supercritical carbon dioxide. *Fluid Phase Equilibria* 36:235–246.

Brunner, G., and Teich, J. 1994. Phase equilibria in systems containing hydrogen, carbon dioxide, water and hydrocarbons. *Fluid Phase Equilibria* 100:253–268.

Brunner, G., Steffen, A., and Dohrn, R. 1993. High pressure liquid–liquid equilibria in ternary systems containing water, benzene, toluene, n–hexane and n-hexadecane. *Fluid Phase Equilibria* 82:165–172.

Buhler, E., Dobrynin, A.V., DeSimone, J.M., and Rubinstein, M. 1998. Light-scattering study of diblock copolymers in supercritical carbon dioxide: CO_2 density-induced micellization transition. *Journal of Macromolecular Science* 31:7347–7355.

Burger, J.L., Cape, S.P., Braun, C.S., McAdams, D.H., Best, J.A., Bhagwat, P., Pathak, P., Rebits, L.G, and Sievers, R.E. 2008. Stabilizing formulations for inhalable powders of live-attenuated measles virus vaccine. *Journal of Aerosol Medicine and Pulmonary Drug Delivery* 21(1):25–34.

Bustami, R.T., Chan, H.K., Sweeney, T., Dehghani, F., and Foster, N.R. 2003. Generation of fine powders of recombinant human deoxyribonuclease using the aerosol solvent extraction system. *Pharmaceutical Research* 20: 2028–2035.

Bustami, R.T., Chan, H.K., Dehghani, F., and Foster N.R. 2001. Recent applications of supercritical fluid technology to pharmaceutical powder systems. *Kona* 19:57–70.

Bustami, R.T., Chan, H.K., Dehghani, F., and Foster, N.R. 2000. Generation of micro–particles of proteins for aerosol delivery using high pressure modified carbon dioxide. *Pharmaceutical Research* 17:1360–1366.

Button, J.K., and Gubbins, K.E. 1999. SAFT prediction of vapour–liquid equilibria of mixtures containing carbon dioxide and aqueous monoethanolamine or diethanolamine. *Fluid Phase Equilibria* 158–160:175–181.

Byrappa, K., Ohara, S., and Adschiri, T. 2008. Nanoparticles synthesis using supercritical fluid technology – towards biomedical applications. *Advanced Drug Delivery Reviews* 60:299–327.

Cai, M-Q., Guan, Y-X., Yao, S-J., and Zhu, Z-Q. 2008. Supercritical fluid assisted atomization introduced by hydrodynamic cavitation mixer (SAA-HCM) for micronization of levofloxacin hydrochloride. *Journal of Supercritical Fluids* 43(3):524–534.

Calderone, M., and S. Tallon 2008. Particle formation by rapid expansion from solution using near-critical dimethyl-ether. *Journal of Supercritical Fluids* 45(2):245–252.

Calix, T.F. 2008. Design and testing of an apparatus to measure carbon dioxide solubility in liquid foods. MSc dissertation, University of Florida, Gainesville.

Calix, T.F., Ferrentino, G., and Balaban, M.O. 2008. Measurement of high–pressure carbon dioxide solubility in orange juice, apple juice and model liquid solutions. *Journal of Food Science* 73(9):E439–E445.

Calvo, L., Muguerza, B., and Cienfuegos-Jovellanos, E. 2007. Microbial inactivation and butter extraction in a cocoa derivative using high pressure CO_2. *Journal of Supercritical Fluids* 42:80–87.

Calvo, M.M., and Balcones, E. 2001. Inactivation of microorganisms and changes of proteins during treatment of milk with subcritical carbon dioxide. *Milchwissenschaft Milk Science International* 56(7):366–369.

Calvo, M.M., Montilla, A., and Cobos, A. 1999. Lactic acid production and rheological properties of yogurt made from milk acidified with carbon dioxide. *Journal of the Science of Food and Agriculture* 79(9):1208–1212.

Calvo, M.M., and De Rafael, D. 1995. Deposition formation in a heat exchanger during pasteurization of CO_2 acidified milk. *Journal of Dairy Research* 62:641–644.

Cape, S., Villa, J., Huang, E., Yang, T-H., Carpenter, J., and Sievers, R.2008. Preparation of active proteins, vaccines and pharmaceuticals as fine powders using supercritical or near-critical Fluids. *Pharmaceutical Research* 25(9):1967–1990.

Capilla, V., Manez, M., Moreno, M.J., and Jimenez, R. 2003. Disinfection and disinfection effect of CO_2 under pressure on food matrix. In: *Proceedings of the 5th International Smposium on Supercritical Fluids*, pp. 2095–2100, Versailles.

Cardoso, M.A.T., Monteiro, G.A., Cardoso, J.P., Prazeres, T.J.V., Figueiredo, J.M.F., Martinho, J.M.G., Cabral, J.M.S., and Palavra, A.M.F. 2008. Supercritical antisolvent micronization of minocycline hydrochloride. *Journal of Supercritical Fluids* 44:238–244.

Cardoso, M., and O'Connell, J.P. 1987. Activity coefficients in mixed solvent electrolyte solution, *Fluid Phase Equilibria* 33:315–326.

Carnahan N.F., and Starling K.E. 1972. Intermolecular repulsion and the equation of state for fluids. *AIChE Journal* 18:1184–1189.

Carnahan N.F., and Starling K.E. 1969. Equation of state for non-attracting rigid spheres. *Journal of Chemical Physics* 51:635–636.

Carroll, J.J., Slupsky, J.D., and Mather, A.E. 1991. The solubility of carbon dioxide in water at low pressure. *Journal of Physical and Chemical Reference Data* 20:1201–1209.

Castier, M., and Sandler, S.I. 1997a. Critical points with the Wong–Sandler mixing rule – I. Calculations with the van der Waals equation of state. *Chemical Engineering Science* 52:3393–3399.

Castier, M., and Sandler, S.I. 1997b. Critical points with the Wong–Sandler mixing rule – II. Calculations with a modified Peng–Robinson equation of state. *Chemical Engineering Science* 52:3579–3588.

Castor, T.P., and Hong, G.T. 1995. Supercritical fluid disruption of and extraction from microbial cells. Aphios Corp., US Patent 5,380,826.

Caussette, M., Gaunand, A., Planche, H., Colombie, S., Monsan, P., and Lindet, B. 1999. Lysozyme inactivation by inert gas bubbling: kinetics in a bubble column reactor. *Enzyme Microbial Technology* 24:412–418.

Chang, C.J. 1992. The solubility of carbon dioxide in organic solvents at elevated pressures. *Fluid Phase Equilibria* 74:235–242.

Chang, C.J., and A.D. Randolph 1989. Precipitation of microsize organic particles from supercritical fluids. *AIChE Journal* 35(11):1876–1882.

Chapman, W.G., Gubbins, K.E., Jackson, G., and Radosz, M. 1990. New reference equation of state for associating liquids. *Industrial & Engineering Chemistry Research* 29:1709–1721.

Chapman, W.G., Jackson, G., and Gubbins, K.E. 1988. Phase equilibria of associating fluids – chain molecules with multiple bonding sites. *Molecular Physics* 65:1057–1079.

Chapoy, A., Mohammadi, A.H., Chareton, A., Tohidi, B., and Richon, D. 2004. Measurement and modeling of gas solubility and literature review of the properties for the carbon dioxide–water system. *Industrial & Engineering Chemistry Research* 43(7):1794–1802.

Charoenchaitrakool, M., Polchiangdee, C., and Srinophakun, P. 2009. Production of theophylline and polyethylene glycol 4000 composites using gas anti-solvent (GAS) process. *Materials Letters* 63:136–138.

Charoenchaitrakool, M., Dehghani, F., Foster, N.R., and Chan, H.K. 2000. Micronization by rapid expansion of supercritical solutions to enhance the dissolution rates of poorly water-soluble pharmaceuticals. *Industrial & Engineering Chemistry Research* 39(12):4794–4802.

Charpentier, P.A., Jia, M., and Lucky, R.A. 2008. Study of the RESS process for producing beclomethansone-17,21-dipropionate particle suitable for pulmonary delivery. *AAPS PharmSciTech* 9(1):39–46.

Chen, A.Z., Lia, Y., Chau, F.T., Lau, T.Y., Hua, J.Y., Zhao, Z., and Mok, D.K. 2009a. Application of organic nonsolvent in the process of solution-enhanced dispersion by supercritical CO_2 to prepare puerarin fine particles. *Journal of Supercritical Fluids* 49:394–402.

Chen, A.Z., Li, Y., Chen, D., and Hu, J.Y. 2009b. Development of core–shell microcapsules by a novel supercritical CO_2 process. *Journal of Materials Science: Materials in Medicine* 20:751–758.

Chen, C.C., and Song, Y. 2005. Extension of non-random two-liquid segment activity coefficient model for electrolytes. *Industrial & Engineering Chemistry Research* 44:8909–8921.

Chen, C.C., Bokis, C.P., and Mathias, P. 2001. Segment-based excess Gibbs energy model for aqueous organic electrolytes. *AIChE Journal* 47:2593–2602.

Chen, C.C., and Evans, L.B. 1986. A local composition model for the excess Gibbs energy of aqueous electrolyte systems. *AIChE Journal* 32:444–454.

Chen, J.H., and Hotchkiss, J.H. 1993. Growth of *Listeria monocytogenes* and *Clostridium sporogenes* in cottage cheese in modified atmosphere packaging. *Journal of Dairy Science* 76(4):972–977.

Chen, J.H., Hotchkiss, J.H., and Lawless, H. T. 1992b. Sensory and microbiological quality of cottage cheese packaged in high-barrier film with added carbon dioxide (abstr.). *Journal Dairy Science* 75(Suppl. 1):95.

Chen, J.H., and Hotckiss, J.H. 1991a. Effect of dissolved carbon dioxide on the growth of psychrotrophic organisms in cottage cheese. *Journal of Dairy Science* 74(9):2941–2945.

Chen, J.H., and Hotckiss, J.H. 1991b. Long shelf-life cottage cheese through dissolved carbon dioxide and high-barrier packaging (abstr.). *Journal of Dairy Science* 74(Suppl. 1):125.

Chen, J-L., Zhang, J., Song, L., Jiang, Y., Wu, J., and Hu, X.S. 2010. Changes in microorganism, enzyme, aroma of hami melon (*Cucumis melo* L.) juice treated with dense phase carbon dioxide and stored at 4°C. *Innovative Food Science and Emerging Technologies* 11(4):623–629. doi: 10.1016/j.ifset.2009.06.003

Chen, J.S., Balaban M.O., Wei, C.I., Gleeson, R.A., and Marshall, M.R. 1993. Effect of carbon dioxide on the inactivation of Florida spiny lobster polyphenol oxidase. *Journal of the Science of Food and Agriculture* 61:253–259.

Chen, J.S., Balaban M.O., Wei, C.I., Marshall, M.R., and Hsu, W.Y. 1992a. Inactivation of polyphenol oxidase by high pressure carbon dioxide. *Journal of Agricultural and Food Chemistry* 40:2345–2349.

Chikere, A.C., Galunsky, B., Overmeyer, A., Brunner, G., and Kasche, V. 2000. Activity and stability of immobilised soybean lipoxygenase-1 in aqueous and supercritical carbon dioxide media. *Biotechnology Letters* 22:1815–1821.

Chingunpitak, J., Puttipipatkhachorn, S., Tozuka, Y., Moribe, K., and Yamamoto, K. 2008. Micronization of dihydroartemisinin by rapid expansion of supercritical solutions. *Drug Development and Industrial Pharmacy* 34(6):609–617.

Chiou, A.H-J., Yeh, M-K., Chen, C-Y., and Wang, D-P. 2007. Micronization of meloxicam using a supercritical fluids process. *Journal of Supercritical Fluids* 42(1):120–128.

Chitarra, L.G., Breeuwer, P., van den Bulk, R.W., and Abee, T. 2000. Rapid fluorescence assessment of intracellular pH as a viability indicator of *Clavibacter michiganensis* subsp. *michigansis*. *Journal of Applied Microbiology* 88:809–816.

Chu, J., Li, G., Row, K.H., Kim, H., and Lee, Y. W. 2009. Preparation of cefpodoxime proxetil fine particles using supercritical fluids. *International Journal of Pharmaceutics* 369:85–91.

Chun, M-K., Yoon, J-H., and Lee H. 1996. Clathrate phase equilibria for the water + deuterium oxide + carbon dioxide and water + deuterium oxide + chlorodifluoromethane (R22) systems. *Journal of Chemical & Engineering Data* 41(5):1114–1116.

Coan, C.R., and King, A.D. 1971. Solubility of water in compressed carbon dioxide, nitrous oxide, and ethane: evidence for hydration of carbon dioxide and nitrous oxide in the gas phase. *Journal of the American Chemical Society* 93(8):1857–1862.

Coghill, D., and Juffs, H.S. 1979. Incidence of psychrotrophic spore-forming bacteria in pasteurized milk and cream products and effects of temperature on their growth. *Australian Journal of Dairy Technology* 34(4):150–153.

Connery, K.A., Shah, P., Coleman, L., and Hunek, B. 2005. Commercialization of Better Than Fresh™ dense phase carbon dioxide processing for liquid food. In: *Proceedings of the International Symposium on Supercritical Fluids* [CD-ROM], Orlando, FL, May.

Coquelet, C., Chapoy, C., and Richon, D. 2004. Development of new alpha function for the Peng–Robinson equation of state: comparative study of alpha function models for pure gases (natural gas components) and water–gas system. *International Journal of Thermophysics* 25:133–158.

Corradini, M.G., and Peleg, M. 2009. A dynamic heat inactivation kinetic model of adapting bacteria. *Applied and Environmental Microbiology* 75:2590–2597.

Corradini, M.G., and Peleg, M. 2007. A Weibullian model of microbial injury and mortality. *International Journal of Food Microbiology* 119:319–328.

Corradini, M.G., and Peleg, M. 2004. Demonstration of the Weibull-Log logistic survival model's applicability to non isothermal inactivation of *E. coli* K12 MG1655. *Journal of Food Protection* 67:2617–2621.

Corradini, M.G., and Peleg, M. 2003. A model of microbial survival curves in water treated with a volatile disinfectant. *Journal of Applied Microbiology* 95:1268–1276.

Corrigan, O. I., and Crean, A. M. 2002. Comparative physicochemical properties of hydrocortisobe-PVP composites prepared using supercritical carbon dioxide by the gas anti-solvent recrystallization process by coprecipitation and by spray drying. *International Journal of Pharmaceutics* 245:75–82.

Corvini, P.F.X., Cautier, H., Rondags, E., Vivier, H., Goergen, J.L., and Germain, P. 2000. Intracellular pH determination of pristinamycin-producing *Streptomyces pristinaespiralis* by image analysis. *Microbiology* 146:2671–2678.

Corwin, H., and Shellhammer, T.H. 2002. Combined carbon dioxide and high pressure inactivation of pectin methylesterase, polyphenol oxidase, *Lactobacillus plantarum* and *Escherichia coli*. *Journal of Food Science* 67(2): 697–701.

Crovetto, R. 1991. Evaluation of solubility data of the system CO_2–H_2O from 273 K to the critical point of water. *Journal of Physical and Chemical Reference Data* 20(3):575–589.

Da Porto, C., Decorti, D., and Tubaro, F. 2010. Effects of continuous dense-phase CO_2 system on antioxidant capacity and volatile compounds of apple juice. *International Journal of Food Science and Technology*. doi:10.1111/j.1365–2621.2010.02339.x

Dagan, G.F., and Balaban, M.O. 2006. Pasteurization of beer by a continuous dense-phase CO_2 system. *Journal of Food Science* 71(3):E164–E169.

Dahl, S., Dunalewicz, A., Fredenslund, A., and Rasmussen, P. 1992. The MHV2 model: prediction of phase equilibria at sub- and supercritical conditions. *Journal of Supercritical Fluids* 5:42–47.

Dahl, S., and Michelsen, M.L. 1990. High-pressure vapor–liquid equilibrium with a UNIFAC-based equation of state. *AIChE Journal* 36:1829–1836.

Damar, S. 2006. Processing of coconut water with high pressure carbon dioxide technology. PhD issertation, University of Florida, Gainesville.

Damar, S., Balaban, M.O., and Sims, C.A. 2009. Dense phase CO_2 processing of coconut water. *International Journal of Food Science & Technology* 44(4):666–673.

Damar, S., and Balaban, M.O. 2006. Review of dense phase CO_2 technology: microbial and enzyme inactivation, and effects on food quality. *Journal of Food Science* 71(1):R1–R11.

Damar, S., and Balaban, M.O. 2005. Cold pasteurization of coconut water with a continuous dense phase CO_2 system. In: *IFT Annual Meeting Book of Abstracts: 2005*, abstract 54F-2, New Orleans, LA, 15–20 July. Chicago: Institute of Food Technologists.

Danesh, A., and Todd, A. C. 1990. A novel sampling method for compositional analysis of high pressure fluids. *Fluid Phase Equilibira* 57(1–2):161–171.

Daniels, J.A., Krishnamurthi, R., and Rizvi, S.S.H. 1985. A review of the effects of carbon dioxide on microbial growth and food quality. *Journal Food Protection* 48:532–537.

Debs-Louka, E., Louka, N., Abraham, G., Chabot, V., and Allaf, K. 1999. Effect of compressed carbon dioxide on microbial cell viability. *Applied and Environmental Microbiology* 65:626–631.

Dehghani, F., and Foster, N. R. 2003. Dense gas anti-solvent process for pharmaceutical formulation. *Current Opinion in Solid State and Materials Science* 7:363–369.

Deiters, U.K., and Schneider, G.M. 1986. High pressure phase equilibria: experimental methods. *Fluid Phase Equilibria* 29:145–160.

Del Pozo-Insfran, D., Balaban, M.O., and Talcott, S.T. 2007. Inactivation of polyphenol oxidase in muscadine grape juice by dense phase CO_2 processing. *Food Research International* 40:894–899.

Del Pozo-Insfran, D., Balaban, M.O., and Talcott, S.T. 2006. Microbial stability, phytochemical retention, and organoleptic attributes of dense phase CO_2 processed muscadine grape juice. *Journal of Agricultural and Food Chemistry* 54:5468–5473.

Dellaglio, F., Dallacasa, V., Venturi, A., Bertucco, A., and Pallado, P. 1999. Effect of supercritical carbon dioxide on some bacteria and yeast of interest in the food industry. In: H. Ludwig (ed.), *Advances in High Pressure Bioscience and Biotechnolog*, pp. 121–124. Heidelberg, Germany: Springer Verlag.

Descoins, C., Mathlouthi, M., Le Moual, M., and Hennequin, J. 2006. Carbonation monitoring of beverage in a laboratory scale unit with on-line measurement of dissolved CO_2. *Food Chemistry* 95(4):541–553.

Devlieghere, F., Vermeiren, L., and Debevere, J. 2004. New preservation technologies: possibilities and limitations. *International Dairy Journal* 14:273–285.

Devlieghere, F., and Debevere, J. 2000. Influence of carbon dioxide on the growth of spoilage bacteria. *Lebensmittel-Wissenschaft und-Technologie* 33(8):531–537.

Devlieghere, F., Debevere, J., and Van Impe, J. 1998. Concentration of carbon dioxide in the water phase as a parameter to model the effect of modified atmosphere on microorganisms. *International Journal of Food Microbiology* 43(1–2):105–113.

Diamond, L.W., and Akinfiev, N.N. 2003. Solubility of CO_2 in water from −1.5 to 100°C and from 0.1 to 100 MPa: evaluation of literature data and thermodynamic modelling. *Fluid Phase Equilibria* 208:265–290.

Di Giacomo, G., Bradani, V., Del Re, G., and Mucciante, V. 1989. Solubility of essential oil components in compressed supercritical carbon dioxide. *Fluid Phase Equilibria* 52:405–411.

Dillow, A.K., Langer, R.S., Foster, N., and Hrkach, J.S. 2000. Supercritical fluid sterilization method. US Patent 6,149,864.

Dillow, A.K., Dehghani, F., Hrkach, J.S., Foster, N.R., and Langer, R. 1999. Bacterial inactivation by using near- and supercritical carbon dioxide. *Proceedings of the National Academy of Sciences of the United States of America* 96:10344–10348.

Dodds, W.S., Stutzman, L.F., and Sollami, B.J. 1956. Carbon dioxide solubility in water. *Ind. Eng. Chem. Eng. Data Series*, 1(1):92–95.

Dohrn, R., and Brunner, G. 1995. High-pressure fluid-phase equilibria: experimental methods and systems investigated. *Fluid Phase Equilibria* 106:213–282.

Dohrn, R., Bunz, A.P., Devlieghere, F., and Thelen, D. 1993. Experimental measurements of phase equilibria for ternary and quaternary systems of glucose, water, CO_2 and ethanol with a novel apparatus. *Fluid Phase Equilibria* 83:149–158.

Domingo, C., Wubbolts, F.E., and van Rosmalen, G.M. 1999. Solid crystallization by rapid expansion of supercritical ternary mixtures. *Journal of Crystal Growth* 198–199(Part 1):760–766.

Domingo, C., Berends, E., and van Rosmalen, G.M. 1997. Precipitation of ultrafine organic crystals from the rapid expansion of supercritical solutions over a capillary and a frit nozzle. *Journal of Supercritical Fluids* 10(1):39–55.

Domingo, C., Berends, E.M., and van Rosmalen, G.M. 1996. Precipitation of ultrafine benzoic acid by expansion of a supercritical carbon dioxide solution through a porous plate nozzle. *Journal of Crystal Growth* 166(1–4):989–995.

Donohue, M.D., and Prausnitz, J.M. 1978. Perturbed hard chain theory for fluid mixtures: thermodynamic properties for mixtures in natural gas and petroleum technology. *AIChE Journal* 24:849–860.

Doyle, M.P. 1983. Effect of carbon dioxide on toxin production by *Clostridium Botulinum*. Eur. *Applied Microbiology and Biotechnology* 17(1):53–56.

Driks, A. 1999. *Bacillus subtilis* coat. *Microbiology and Molecular Biology Reviews* 63:1–20.

D'Souza, R., and Teja, A. 1988. High pressure phase equilibria in the system glucose + fructose + water + ethanol + carbon dioxide. *Fluid Phase Equilibria* 39:211–224.

Duan, Z., Sun, R., Zhu, C., and Chou, I-M. 2006. An improved model for the calculation of CO_2 solubility in aqueous solutions containing Na^+, K^+, Ca^{2+}, Mg^{2+}, Cl^-, and SO_4^{2-}. *Marine Chemistry* 98:131–139.

Duan, Z., and Sun, R. 2003. An improved model calculating CO_2 solubility in pure water and aqueous NaCl solutions from 273 to 533 K and from 0 to 2000 bar. *Chemical Geology* 193:257–271.

Duarte, A.R.C., Roy, C., Gonzalez, A.V., Duarte, C.M.M., and Paternault, P.S. 2007. Preparation of acetazolamide composite microparticles by supercritical anti-solvent techniques. *International Journal of Pharmaceutics* 332:132–139.

Dunford, N.T., and Temelli, F. 1996. Effect of supercritical CO_2 on myrosinase activity and glucosinolate degradation in canola. *Journal of Agricultural and Food Chemistry* 44(8):2372–2376.

Duthie, C.M., Shipe, W.F., and Hotchkiss, J.H. 1985. Effect of low level carbonation on the keeping quality of processed milk (abstr.). *Journal of Dairy Science* 68(Suppl. 1):69.

Economou, I.G., and Tsonopoulos, V. 1997. Associating models and mixing rules in equations of state for water/hydrocarbon mixtures. *Chemical Engineering Science* 52:511–525.

Edwards, A.D., Shekunov, B.Y., Kordikowski, A., Forbes, R.T., and York, P. 2001. Crystallization of pure anhydrous polymorphs of carbamazepine by Solution Enhanced Dispersion with Supercritical Fluids (SEDS). *Journal of Pharmaceutical Sciences* 90:1115–1124.

Edwards, T.J., Maurer, G., Newman, J., and Prausnitz, J.M. 1978. Vapor–liquid equilibria in multicomponent aqueous solutions of volatile weak electrolytes. *AIChE Journal* 24:966–976.

Elvassore, N., Bertucco, A., and Caliceti, P. 2001a. Production of insulin-loaded poly(ethylene glycol)/poly(l-lactide) (PEG/PLA) nanoparticles by gas antisolvent techniques. *Journal of Pharmaceutical Sciences* 90:1628–1636.

Elvassore, N., Bertucco. A., and Caliceti, P. 2001b. Production of protein loaded polymeric micro-capsules by supercritical CO_2 in mixed solvent. *Industrial & Engineering Chemistry Research*. 40:795–800.

Elvassore, N., Sartorello, S., Spilimbergo, S., and Bertucco, A. 2000. Micro-organisms inactivation by supercritical CO_2 in a semi-continuous process. *Proceedings of the 7th Meeting on Supercritical Fluids*, pp. 773–778, Antibes, France, December.

Enfors, S.O., and Molin, G. 1980. Effect of high concentrations of carbon dioxide on growth rate of *Pseudomonas fragi*, *Bacillus cereus*, and *Streptococcus cremoris*. *Journal of Applied Bacteriology* 48(3):409–416.

Enomoto, A., Nakamura, K., Hakoda, M., and Amaya N. 1997a. Lethal effect of high-pressure CO_2 on a bacterial spore. *Journal of Fermentation and Bioengineering* 83(3):305–307.

Enomoto, A., Nakamura, K., Nagai, K., Hashimoto, T., and Hakoda, M. 1997b. Inactivation of food microorganisms by high-pressure carbon dioxide treatment with or without explosive decompression. *Bioscience, Biotechnology and Biochemistry* 61:1133–1137.

Erkmen, O., and Bozoglu, T.F. 2008a. *Food Microbiology I: Microorganisms in Foods, Microbial Growth, Foodborne Diseases and Detection of Microorganisms and Their Toxins*. Ankara: İlke Publishing Company.

Erkmen, O., and Bozoglu, T.F. 2008b. *Food Microbiology III: Food Presevation*. Ankara: İlke Publishing Company.

Erkmen, O. 2007. *Basic Methods for the Microbiological Analysis of Foods*. Ankara: Nobel Publishing Company.

Erkmen, O. 2002. Effect of carbon dioxide pressure on *Saccharomyces cerevisiae*. *Food Science and Technology International* 8(6):361–364.

Erkmen, O. 2001a. Antimicrobial effect of pressurized carbon dioxide on *Yersinia enterocolitica* in broth and foods. *Food Science and Technology International* 7(3):245–250.

Erkmen, O. 2001b. Mathematical modeling of *Escherichia coli* inactivation under high-pressure carbon dioxide. *Journal of Bioscience and Bioengineering* 92:39–43.

Erkmen, O. 2000a. Antimicrobial effects of carbon dioxide on *Brochothrix thermosphacta* in broth and foods. *Journal of the Science of Food and Agriculture* 80(9):1365–1370.

Erkmen, O. 2000b. Antimicrobial effect of pressurized carbon dioxide on *Enterococcus faecalis* in physiological saline and foods. *Journal of the Science of Food and Agriculture* 80(4):465–470.

Erkmen, O. 2000c. Effect of carbon dioxide pressure on *Listeria monocytogenes* in physiological saline and foods. *Food Microbiology* 17(6):589–596.

Erkmen, O. 2000d. Effects of high pressure carbon dioxide on *Escherichia coli* in nutrient broth and milk. *International Journal of Food Microbiology* 65(1–2):133–137.

Erkmen, O. 2000e. Inactivation of *Salmonella typhimurium* by high pressure carbon dioxide. *Food Microbiology* 17(2):225–232.

Erkmen, O. 1997. Antimicrobial effect of pressurized carbon dioxide on *Staphylococcus aureus* in broth and milk. *Lebensmittel-Wissenchaft und Technologie* 30:826–829.

Evelein, K.A., Moore, R.G., and Heidemann, R.A. 1976. Correlation of the phase behavior in the systems hydrogen sulfide–water and carbon dioxide–water. *Industrial and Engineering Chemistry Process Design and Development* 15:423–428.

Fadıloğlu, S., Erkmen, O., and Sekeroglu, G. 2004. Thermal and carbon dioxide inactivation of alkaline phosphatase in buffer and milk. *Food Technology and Biotechnology* 42(1):27–32.

Fadıloğlu, S., and Erkmen, O. 2002. Inactivation of lipase by carbon dioxide under atmospheric pressure. *Journal of Food Engineering* 52:331–335.

Fages, J., Jean, E., Frayssinett, P., Poirier, B., Mathon, D., Autefage, A., and Larzul, D. 1998. Bone allografts and supercritical processing: effects on osteointegration and viral safety. *Journal of Supercritical Fluids* 13:351–356.

Fan, S-S., and Guo, T-M. 1999. Hydrate formation of CO_2-rich binary and quaternary gas mixtures in aqueous sodium chloride solutions. *Journal of Chemical & Engineering Data* 44(4):829–832.

Farber, J.M. 1991. Microbiological aspects of modified atmosphere packaging technology review. *Journal of Food Protection* 54:58–70.

Fedio, W.M., Macleod, A., and Ozimek, L. 1994. The effect of modified atmosphere packaging on the growth of microorganisms in cottage cheese. *Milchwissenschaft* 49(11):622–629.

Feeherry, F.E., Munsey, D.T., and Rowley, D.B. 1987. Thermal inactivation and injury of *Bacillus stearothermophilus* spores. *Applied and Environmental Microbiology* 53:365–370.

Fermeglia, M., and Kikic, I. 1993. The perturbed hard chain theory for the prediction of supercritical fluid extraction: binary mixtures. *Chemical Engineering Science* 48:3889–3896.

Ferrentino, G. 2009. Microbial stabilization of liquid food with carbon dioxide under pressure. PhD dissertation, University of Salerno, Fisciano, Italy.

Ferrentino, G., Barletta, D., Balaban, M.O., Ferrari, G., and Poletto, M. 2010a. Measurement and prediction of CO_2 solubility in sodium phosphate monobasic solutions for food treatment with high pressure carbon dioxide. *Journal of Supercritical Fluids* 52(1):142–150.

Ferrentino, G., Balaban, M.O., Ferrari, G., and Poletto, M. 2010b. Food treatment with high pressure carbon dioxide: *Saccharomyces cerevisiae* inactivation kinetics expressed as a function of CO_2 solubility. *Journal of Supercritical Fluids* 52(1):151–160.

Ferrentino, G., Barletta, D., Donsì, F., Ferrari, G., and Poletto, M. 2010c. Experimental measurements and thermodynamic modeling of CO_2 solubility at high pressure in model apple juices. *Industrial and Engineering Chemistry Research* 49(6):2992–3000.

Ferrentino, G., Bruno, M., Ferrari, G., Poletto, M., and Balaban, M. 2009a. Microbial inactivation and shelf life of apple juice treated with high pressure carbon dioxide. *Journal of Biological Engineering* 3(Feb. 4):3.

Ferrentino, G., Plaza, M.L., Ramirez-Rodrigues, M., Ferrari, G., and Balaban, M.O. 2009b. Effects of dense phase carbon dioxide pasteurization on the physical and quality attributes of a red grapefruit juice. *Journal of Food Science* 74(6):E333–E341.

Fogh, F., and Rasmussen, P. 1989. Detection of high-pressure dew and bubble points using a microwave technique. *Industrial & Engineering Chemistry Research* 28(3):371–375.

Folkes, G. 2004. Pasteurization of beer by a continuous dense-phase CO_2 system. PhD dissertation, University of Florida, Gainesville.

Fontes, N., Nogueiro, E., Elvas, A.M., de Sampaio, T.C., and Barreiros, S. 1998. Effect of pressure on the catalytic activity of subtilisin Carlsberg suspended in compressed gases. *Biochimica Biophysica Acta* 1383:165–174.

Foster, J.W., Cowan, R.M., and Maag, T.A. 1962. Rupture of bacteria by explosive decomposition. *Journal of Bacteriology* 83:330–334.

Foster, N.R., and Sih, R. 2006. Particle formation. Australian Provisional Patent 2006/905571, 6 October.

Foster, N., Dehghani, F., Charoenchaitrakool, K., and Warwick, B. 2003a. Application of dense gas techniques for the production of fine particles. *AAPS Journal* 5(2):32–38.

Foster, N.R., Ng, A.S., Dehghani, F., and Regtop, H.L. 2003b. Micronisation of tobramycin using dense gas anti-solvent techniques. In: *Proceedings of the 6th International Symposium of Supercritical Fluids*, pp. 1647–1652, Versailles, France.

Foster, N.R., Mammucari, R., Dehghani, F., Barrett, A., Bezanehtak, K., and Coen, E. 2003c. Processing pharmaceutical compounds using dense-gas technology. *Industrial & Engineering Chemistry Research* 42:6476–6493.

Foster, N.R., Dehghani, F., Bustami, R.T., and Chan, H.K. 2003d. Generation of lysozyme–lactose powders using the ASES process. In: *Proceedings of the 6th International Symposium of Supercritical Fluids*, pp. 1831–1836, Versailles, France.

Fox, P.F. 1992. *Advanced Dairy Chemistry: Volume 3. Lactose, Water, Salts and Vitamins*. London: Chapman & Hall Press.

Franceschi, E., Cesaro, A.M.D., Ferreira, S.R.S., and Oliveira, J.V. 2009. Precipitation of β-carotene microparticles from SEDS technique using supercritical CO_2. *Journal of Food Engineering* 95:656–663.

Franceschi, E., Cesaro, A.M.D., Feiten, M., Ferreira, S.R.S., Dariva, C., Kunita, M.H., Rubira, A.F., Muniz, E.C., Corazza, M.L., and Oliveira, J.V. 2008. Precipitation of β-carotene and PHBV and co-precipitation from SEDS technique using supercritical CO_2. *Journal of Supercritical Fluids* 47:259–269.

Frank, S.G., and Ye, C. 2000. Small particle formation and dissolution rate enhancement of relatively insoluble drugs using rapid expansion of supercritical solutions (RESS) processing. In: C. Eckert and A. Teja (eds.), *Proceedings (CD-ROM) of the Fifth International Symposium on Supercritical Fluids*.

Fraser, D. 1951. Bursting bacteria by release of gas pressure. *Nature* 167:33–34.

Frayer, J.M. 1941. The dissolved gases in milk and dye reduction. *Vermont Agricultural Experiment Station Bulletin* 461.

Fredenslund, A., Gmehling, J., and Rasmussen, P. 1977. *Vapour–Liquid Equilibria Using UNIFAC*. Amsterdam: Elsevier.

Fricks, A.T., Souza, D.P.B., Oestreicher, E.G., Antunes, O.A.C., Girardi, J.S., Oliveira, D., and Dariva, C. 2006. Evaluation of radish (*Raphanus sativus* L.) peroxidase activity after high-pressure treatment with carbon dioxide. *Journal of Supercritical Fluids* 38:347–353.

Fromm, H.I., and Boor, K.J. 2004. Characterization of pasteurized fluid milk shelf–life attributes. *Journal of Food Science* 69(8):M207–M214.

Furst, W., and Renon, H. 1993. Representation of excess properties of electrolyte solutions using a new equation of state. *AIChE Journal* 39:335–343.

Furukawa, S., Watanabe, T., Koyama, T., Hirata, J., Narisawa, N., Ogihara, H., and Yamasaki, M. 2006. Effect of high pressure carbon dioxide on the clumping of the bacterial spores. *International Journal of Food Microbiology* 106:95–98.

Furukawa, S., Watanabe, T., Tai, T., Hirata, J., Narisawa, N., Kawarai, T., Ogihara, H., and Yamasaki, M. 2003. Effect of high pressure gaseous carbon dioxide on the germination of bacterial spores. *International Journal of Food Microbiology* 91:209–213.

Fusaro, F., Kluge, J., Mazzotti, M., and Muhrer, G. 2009. Compressed CO_2 antisolvent precipitation of lysozyme. *Journal of Supercritical Fluids* 49:79–92.

Fusaro, F., Hänchen, M., Mazzotti, M., Muhrer, G., and Subramaniam, B. 2005. Dense gas antisolvent precipitation: a comparative investigation of the GAS and PCA techniques, *Industrial & Engineering Chemistry Research* 44: 1502–1509.

Gadermann, M., Kular, S., Al-Marzouqi, A.H., and Signorell, R. 2009. Formation of naproxen-polylactic acid nanoparticles from supercritical solutions and their characterization in the aerosol phase. *Physical Chemistry Chemical Physics* 11(36):7861–7868.

Galindo, A., Davies, L.A., Gil-Villegas, A., and Jackson, G. 1998. The thermodynamics of mixtures and the corresponding mixing rules in the SAFT-VR approach for potentials of variable range. *Molecular Physics* 93:241–252.

Gallagher, P.M., Coffey, M.P., Krukonis, V.J., and Klasutis, N. 1989. Gas antisolvent recrystallization: new process to recrystallize compounds insoluble in supercritical fluids. In: K.P. Johnston and J.M.L. Penninger (eds.), *Supercritical Fluid Science and Technology*, pp. 334–354. Washington, DC: American Chemical Society.

Gangola, P., and Rosen, B.P. 1987. Maintenance of intracellular calcium in *Escherichia coli*. *Journal of Biological Chemistry* 262:12570–12574.

Gänzle, M.G., Margosch, D., Buckow, R., Ehrmann, M.A., Heinz, V., and Vogel, R.F. 2007. Pressure and heat resistance of *Clostridium botulinum* and other spores. In: Christopher J. Doona and Florence E. Feeherry (eds.), *High Pressure Processing of Foods*, pp. 95–114. Ames, IA: Institute of Food Technologists Press – Blackwell Publishing Professional.

Gänzle, M.G., and Vogel, R.F. 2001. On-line fluorescence determination of pressure mediated outer membrane damage in *Escherichia coli*. *Systematic and Applied Microbiology* 24:477–485.

Gao, J., Zheng, D-Q., and Guo, T-M. 1997. Solubilities of methane, nitrogen, carbon dioxide, and a natural gas mixture in aqueous sodium bicarbonate solutions under high pressure and elevated temperature. *Journal of Chemical & Engineering Data* 42(1):69–73.

Garcia-Gonzalez, L., Teichert, H., Geeraerd, A.H., Elst, K., Van Ginneken, L., Van Impe, J.F., Vogel, R.F., and Devlieghere, F. 2010a. Mathematical modelling and *in situ* determination of pH in complex aqueous solutions during high-pressure carbon dioxide treatment. *Journal of Supercritical Fluids*. doi:10.1016/j.supflu.2010.05.024

Garcia-Gonzalez, L., Geeraerd, A.H., Mast, J., Briers, Y., Elst, K., Van Ginneken, L., Van Impe, J.F., and Devlieghere, F. 2010b. Membrane permeabilization and cellular death of *Escherichia coli*, *Listeria monocytogenes* and *Saccharomyces cerevisiae* as induced by high pressure carbon dioxide treatment. *Food Microbiology* 27:541–549.

Garcia-Gonzalez, L., Rajkovic, A., Geeraerd, A.H., Elst, K., Van Ginneken, L., Van Impe, J.F., and Devlieghere, F. 2010c. The development of *Escherichia coli* and *Listeria monocytogenes* variants resistant to high-pressure carbon dioxide inactivation. *Letters in Applied Microbiology* 50:653–656.

Garcia-Gonzalez, L., Geeraerd, A.H., Elst, K., Van Ginneken, L., Van Impe, J.F., and Devlieghere, F. 2009. Influence of type of microorganism, food ingredients and food properties on high-pressure carbon dioxide inactivation of microorganisms. *International Journal of Food Microbiology* 129:253–263.

Garcia-Gonzalez, L., Geeraerd, A.H., Spilimbergo, S., Elst, K., Van Ginneken, L.V., Debevere, J., Van Impe, J.F., and Devlieghere, F. 2007. High pressure carbon dioxide inactivation of microorganisms in foods: the past, the present and the future. *International Journal of Food Microbiology* 117:1–28.

Garnier, S., Neau, E., Alessi, P., Cortesi, A., and Kikic, I. 1999. Modelling solubility of solids in supercritical fluids using fusion properties. *Fluid Phase Equilibria* 158–160:491–500.

Gasperi, F., Aprea, E., Biasioli, F., Carlin, S., Endrizzi, I., Pirretti, G., and Spilimbergo, S. 2009. Effects of supercritical CO_2 and N_2O pasteurisation on the quality of fresh apple juice. *Food Chemistry* 115:19–136.

Gevaudan, S., Lagaude, A., Tarodo de la Fuente, B., and Cuq, J.L. 1996. Effect of treatment by gaseous carbon dioxide on the colloidal phase of skim milk. *Journal of Dairy Science* 79(10):1713–1721.

Ghaderi, R., Artursson, P., and Carlfors, J. 2000. A new method for preparing biodegradable microparticles and entrapment of hydrocortisone in DL-PLG microparticles using supercritical fluids. *European Journal of Pharmaceutical Sciences* 10:1–9.

Giessauf, A., and Gamse, T. 2000. A simple process for increasing the specific activity of porcine pancreatic lipase by supercritical carbon dioxide treatment. *Journal of Molecular Catalysis B: Enzymatic* 9:57–64.

Giessauf, A., Magor, W., Steinberger, D.J., and Marr, R. 1999. A study of hydrolases stability in supercritical carbon dioxide. *Enzyme and Microbial Technology* 24:577–583.

Gillepsie, P.C., and Wilson, G.M. 1982. *Vapor–Liquid and Liquid–Liquid Equilibria: Water–Methane, Water–Carbon Dioxide, Water–Hydrogen Sulfide, Water–Npentane, Water–Methane–Npentane*. Research report RR-48. Tulsa, OK: Gas Processors Association.

Gil-Villegas, A., Galindo, A., Whitehead, P.J., Mills, S.J., Jackson, G., and Burgess, A.N. 1997. Statistical associating fluid theory for chain molecules with attractive potentials of variable range. *Journal of Chemical Physics* 106:4168–4186.

Gimeno, M., Ventosa, N., Boumghar, Y., Fournier, J., Boucher, I., and Veciana, J. 2006. Micronization of the chitosan derivatives d-glucosamine hydrochloride and d-glucosamine sulphate salts by dense gas anti-solvent precipitation techniques. *Journal of Supercritical Fluids* 38:94–102.

Glass, K.A., Kaufman, K.M., Smith, A.L., Johnson, E.A., Chen, J.H., and Hotchkiss, H. 1999. Toxin production by *Clostridium botulinum* in pasteurized milk treated with carbon dioxide. *Journal of Food Protection* 62:872–876.

Gosselin, P.M., Thibert, R., Preda, M., and McMullen, J.N. 2003. Polymorphic properties of micronized carbamazepine produced by RESS. *International Journal of Pharmaceutics* 252(1–2):225–233.

Gould, G.W. 1983. Mechanisms of resistance and dormancy. In: A. Hurst and G.W. Gould (eds.), *The Bacterial Spore*, vol. 2, pp. 173–209. New York: Academic Press.

Gregorowicz, J., Fermeglia, M., Soave, G., and Kikic, I. 1991. The perturbed hard chain theory for the prediction of supercritical fluid extraction: pure component properties. *Chemical Engineering Science* 46:1427–1436.

Gu, F.Y. 1998. Solubility of carbon dioxide in aqueous sodium chloride solution under high pressure. *Chinese Journal of Chemical Engineering* 12(2):118–123.

Gueimonde, M., and de los Reyes-Gavilan, C.G. 2004. Reduction of incubation time in carbonated *Streptococcus thermophilus/ Lactobacillus acidophilus* fermented milks as affected by the growth and acidification capacity of the starter strains. *Milchwissenschaft* 59(5–6):280–283.

Gueimonde, M., Alonso, L., Delgado, T., Bada-Gancedo, J.C., and de los Reyes-Gavilan, C.G. 2002. Quality of plain yoghurt made from refrigerated and CO_2-treated milk. *Food Research International* 36(1):43–48.

Gui, F., Wu, J., Chen, F., Liao, X., Hu, X., Zhang, Z., and Wang, Z. 2007. Inactivation of polyphenol oxidases in cloudy apple juice exposed to supercritical carbon dioxide. *Food Chemistry* 100:1678–1685.

Gui, F., Chen, F., Wu, J., Wang, Z., Liao, X., and Hu, X. 2006a. Inactivation and structural change of horseradish peroxidase treated with supercritical carbon dioxide. *Food Chemistry* 97:480–489.

Gui, F.Q., Wu, J.H., Chen, F., Liao, X.J., Hu, X.S., Zhang, Z.H., and Wang, Z.F. 2006b. Change of polyphenol oxidase activity, color, and browning degree during storage of cloudy apple juice treated by supercritical carbon dioxide. *European Food Research and Technology* 223:427–432.

Gunes, G., Blum, L.K., and Hotchkiss, J.H., 2005. Inactivation of yeasts in grape juice using a continuous dense phase carbon dioxide processing system. *Journal of the Science of Food and Agriculture* 85:2362–2368.

Haas, G.J., Prescott, H.E., Dudley, E., Dik, R., Hintlian, C., and Keane, L. 1989. Inactivation of microorganisms by carbon dioxide under pressure. *Journal of Food Safety* 9:253–265.

Hamman, H., and Sivik, B. 1993. Phase behavior of some pure lipids in supercritical carbon dioxide. *Journal of Supercritical Fluids* 6(4):223–227.

Han, B., Peng, D-Y., Fu, C-T., and Vilcsak, G. 1992. An apparatus for phase equilibrium studies of carbon dioxide + heavy hydrocarbon systems. *Canadian Journal of Chemical Engineering* 70(6):1164–1171.

Hanlin, J.H., Evancho, G.M., and Slade, P.J. 1995. Microbiological concerns associated with MAP and sous vide products. In: J.M. Faber and K.L. Dodds (eds.), *Principles of Modified Atmosphere and Sous Vide Product Packaging*, pp. 69–104. Lancaster, PA: Technomic Press.

Hanna, M., and York P. 2006. Methods and apparatus for the formation of particles. US Patent Application US 2006/0073087 A1.

Hauben, K.J.A., Bartlett, D.H., Soontjens, C.C.F., Cornelis, K., Wuytack, E.Y., and Michiels, C.W. 1997. *Escherichia coli* mutants resistant to inactivation by high hydrostatic pressure. *Applied and Environmental Microbiology* 63:945–950.

He, W., Suo, Q., Hong, H., Shan, A., Li, C., Huang, Y., Li, Y., and Zhu, M. 2007. Production of natural carotene-dispersed polymer microparticles by SEDS-PA co-precipitation. *Journal of Materials Science* 42:3495–3501.

He, W.Z., Suo, Q.L., Hong, H.L., Li, G.M., Zhao, X.H., Li, C.P., and Shan, A. 2006. Supercritical antisolvent micronization of natural carotene by the SEDS process through prefilming atomization. *Industrial & Engineering Chemistry Research* 45:2108–2115.

He, W.Z., Suo, Q.L., Jiang, Z.H., and Hong, H.L. 2004. Precipitation of ephedrine by SEDS process using a specially designed prefilming atomizer *Journal of Supercritical Fluids* 31:101–110.

Helfgen, B., Hils, P., Holzknecht, C., Türk, M., and Schaber, K. 2001. Simulation of particle formation during the rapid expansion of supercritical solutions. *Journal of Aerosol Science* 32(3):295–319.

Hendricks, M.T., and Hotchkiss, J.H. 1997. Effect of carbon dioxide on *Pseudomonas fluorescens* and *Listeria monocytogenes* growth in aerobic atmospheres. *Journal of Food Protection* 60(12):1548–1552.

Hermsdorf, D., Jauer, S., and Signorell, R. 2007. Formation and stabilization of ibuprofen nanoparticles by pulsed rapid expansion of supercritical solutions. *Molecular Physics*. 105:951–959.

Hezave, A.Z., and Esmaeilzadeh, F. 2010. Micronization of drug particles via RESS process. *Journal of Supercritical Fluids* 52: 84–98.

Ho, K-L.G. 2004. Treatment of milk and dairy based beverages by dense phase carbon dioxide processing (abstr.). Paper presented at the Institute of Food Technologists annual meeting, Las Vegas, NV, 12–16 July.

Ho, K-L.G. 2003. Dense phase carbon dioxide processing for juice. Paper presented at the Institute of Food Technologists annual meeting, Chicago.

Hochgeschurtz, T., Hutchenson, K.W., Roebers, J.R., Liu, G.Z., Mullins, J.C., and Thies, M.C. 1993. Production of mesophase pitch by supercritical fluid extraction. In: E. Kiran and J. Brennecke (eds.), *Supercritical Fluid Engineering Science*, pp. 347–362. Washington DC: American Chemical Society Press.

Hoffman, W. 1906. Ueber den Einfluss hohen hohlensauredrucks auf Bakterien im Wasser und in der milch. *Archivs für Hygiene* 57:379–383.

Hofland, G.W., Berkhoff, M., Witkamp, G-J., and van der Wielen, L.A.M. 2003. Dynamics of precipitation of casein with carbon dioxide. *International Dairy Journal* 13:685–697.

Hofland, G.W., de Rijke, A., Thiering, R., van der Wielen, L.A.M., and Witkamp, G-J. 2000. Isoelectric precipitation of soybean protein using carbon dioxide as a volatile acid. *Journal of Chromatography B* 743:357–368.

Hofland, G.W., van Es, M., van der Wielen, L.A.M., and Witkamp, G-J. 1999. Isoelectric precipitation of casein using high-pressure CO_2. *Industrial and Engineering Chemistry Research* 38:4919–4927.

Holdsworth, S.D. 1997. *Thermal Processing of Packaged Food*. London: Blackie Academic.

Hong, H.L., Suo, Q.L., Han, L.M., and Li, C.P. 2009. Study on precipitation of astaxanthin in supercritical fluid. *Powder Technology* 191:294–298.

Hong, S.I., and Pyun, Y.R. 2001. Membrane damage and enzyme inactivation of *Lactobacillus plantarum* by high pressure CO_2 treatment. *International Journal of Food Microbiology* 63:19–28.

Hong, S-I., and Pyun, Y.R. 1999. Inactivation kinetics of *Lactobacillus plantarum* by high pressure carbon dioxide. *Journal of Food Science* 64:728–733.

Hong, S-I., Park, W-S., and Pyun, Y-R. 1999. Non-thermal inactivation of *Lactobacillus plantarum* as influenced by pressure and temperature of pressurized carbon dioxide. *International Journal of Food Science and Technology* 34:125–130.

Hong, S.I., Park, W.S., and Pyun, Y.R. 1997. Inactivation of *Lactobacillus* sp. from kimchi by high pressure carbon dioxide. *Food Science and Technology* 30:681–685.

Hooper, H.H., Michel, S., and Prausnitz, J.M. 1988. High–temperature mutual solubilities for some binary and ternary aqueous mixtures containing aromatic and chlorinated hydrocarbons. *Journal of Chemical & Engineering Data* 33(4):502–505.

Hotchkiss, J.H., Werner, B.G., and Lee, E.Y.C. 2006. Addition of carbon dioxide to dairy products to improve quality: a comprehensive review. *Comprehensive Reviews in Food Science and Food Safety* 5:158–168.

Hotchkiss, J.H., Chen, J.H., and Lawless, H.T. 1998. Effect of direct addition of carbon dioxide on microbial and sensory changes in pasteurized milk. *Journal of Dairy Science* 82:690–695.

Howdle, S.M., Watson, M.S., Whitaker, M.J., Davies, M.C., Shakesheff, K.M., Popov, V.K., Mandel, F.S., and Don Wang, J. 2001. Supercritical fluid mixing: preparation of thermally sensitive polymer composites containing bioactive materials. *Chemical Communications* 109–110.

Hsieh, M.J., Huang, A.S., and Balaban, M.O. 2007. The quality improvement of kava beverage by a non-thermal dense phase CO_2 pasteurization. Paper presented at the Institute of Food Technologists annual meeting, Chicago.

Hu, W., Zhang, Y., Wang, Y., Zhou, L., Leng, X., Liao, X., and Hu, X. 2010. Aggregation and homogenization, surface charge and structural change, and inactivation of mushroom tyrosinase in an aqueous system by subcritical/supercritical carbon dioxide. *Langmuir* 27(3):909–916.

Huang, S.H., and Radosz, M. 1991. Equation of state for small, large, polydisperse, and associating molecules: extension to fluid mixtures. *Industrial & Engineering Chemistry Research* 30:1994–2005.

Huang, S.H., Lin, H.M., Tsai, F.N., and Chao, K.C. 1988. Solubility of synthesis gases in heavy n–paraffins and Fischer–Tropsch wax. *Industrial & Engineering Chemistry Research* 27(1):162–169.

Huang, Z., Sun, G-B., Chiew, Y.C., and Kawi, S. 2005. Formation of ultrafine aspirin particles through rapid expansion of supercritical solutions (RESS). *Powder Technology* 160(2):127–134.

Huron, M.J., and Vidal, J. 1979. New mixing rules in simple equations of state for representing vapor–liquid equilibria of strongly nonideal mixtures. *Fluid Phase Equilibria* 3:255–271.

Hutchenson, K.W., Roebers, J.R., and Thies, M.C. 1990. Vapor–liquid equilibrium for phenanthrene–toluene mixtures at elevated temperatures and pressures. *Fluid Phase Equilibria* 60(3): 309–317.

Hutkins, R.W., and Nannen, N.L. 1993. pH homeostatis in lactic acid bateria. *Journal of Dairy Science* 76:2354–2365.

Isenschmid, A., Marison, I.W., and von Stockar, U. 1995. The influence of pressure and temperature of compressed CO_2 on the survival of yeast cells. *Journal of Biotechnology* 39(3):229–237.

Ishihara, Y., Saito, H., and Takano, J. 1999. Differences in the surface membranes and water content between the vegetative cells and spores of *Bacillus subtilis. Cell Biochemistry and Function* 17(1):9–13.

Ishikawa, H., Yoshimura, T., Kajahira, M., Shimoda, M., Matsumoto, K., and Osajima, Y. 2000a. pH-dependent inactiveation of enzymes by microbubbling of supercritical carbon dioxide. *Food Science and Technology Research* 6(3): 212–215.

Ishikawa, H., Shimoda, M., Yonekura, A., Mishima, K., Matsumoto, K., and Osajima, Y. 2000b. Irreversible unfolding of myoglobin in an aqueous solution by supercritical carbon dioxide. *Journal of Agricultural and Food Chemistry* 48:4535–4539.

Ishikawa, H., Shimoda, M., Tamaya, K., Yonekura, A., Kawano, T., and Osajima, Y. 1997. Inactivation of *Bacillus* spores by the supercritical carbon dioxide micro-bubble method. *Bioscience, Biotechnology, and Biochemistry* 61(6):1022–1023.

Ishikawa, H., Shimoda, M., Yonekura, A., and Osajima, Y. 1996. Inactivation of enzymes and decomposition of alpha-helix structure by supercritical carbon dioxide microbubble method. *Journal of Agricultural and Food Chemistry* 44:2646–2649.

Ishikawa, H., Shimoda, M., Kawano, T., and Osajima, Y. 1995a. Inactivation of enzymes in an aqueous solution by micro-bubbles of supercritical carbon dioxide. *Bioscience, Biotechnology, and Biochemistry* 59:628–631.

Ishikawa, H., Shimoda, M., Kawano, T., Sakamoto, K., and Osajima, Y. 1995b. Inactivation of enzymes in Namazake using micro-bubble supercritical carbon dioxide. *Bioscience, Biotechnology, and Biochemistry* 59:1027–1031.

Ishikawa, H., Shimoda, H., Shiratsuchi, H., and Osajima, Y. 1995c. Sterilization of microorganisms by the supercritical carbon dioxide micro-bubble method. *Bioscience, Biotechnology and Biochemistry* 59:1949–1950.

Iwatsuki, K., Mizota, Y., Sumi, M., Sotoyama K., and Tomita, M. 2000. Studies on the sensory characteristics and physicochemical properties of UHT processed milk. Part III. Effect of pasteurization temperature on the sensory characteristics of UHT processed milk. *Journal of the Japanese Society for Food Science and Technology* 47:538–543.

Jablonka, M.S., Munro, P.A., and Duffy, G.G. 1988. Use of light scattering techniques to study the kinetics of precipitation of mineral acid casein from skim milk. *Journal of Dairy Research* 55:179–188.

Jablonka, M.S., and Munro, P.A. 1987. Effect of precipitation temperature and pH on the continuous pilot-scale precipitation of acid casein curd. *New Zealand Journal of Dairy Science and Technology* 21:111–123.

Jackson, K., Bowman, L.E., and Fulton, J.L. 1995. Water solubility measurements in supercritical fluids and high-pressure liquids using near-infrared spectroscopy. *Analytical Chemistry* 67(14):2368–2372.

Jacobs, M.A. 2005. Measurement and modeling of thermodynamic properties for the processing of polymers in supercritical fluids, PhD thesis, Eindhoven University of Techynology, Eindhoven.

Jangkamolkulchai, A., and Luks, K.D. 1989. Partial miscibility behavior of methane + ethane + n–docosane and the methane + ethane + n–tetradecylbenzene ternary mixtures. *Journal of Chemical & Engineering Data* 34(1):92–99.

Japas, M.L., Chai-Kao, C.P., and Paulaitis, M.E. 1992. Experimental determination of H_2 solubilities in liquid fluorocarbons. *Journal of Chemical & Engineering Data* 37(4):423–426.

Jay, J.M. 1992. *Modern Food Microbiology*, 4th ed. New York: Van Nostrand Reinhold.

Ji X., Tan S.P., Adidharma H., and Radosz M. 2005. SAFT1-RPM approximation extended to phase equilibria and densities of CO_2–H_2O and CO_2–H_2O–NaCl systems. *Industrial & Engineering Chemistry Research* 44:8419–8427.

Jiang, S.Y., Chen, M., and Zhao, Y.P. 2003. A study on micronization of phytosterol by the RESS technique with supercritical CO_2. In: G. Brunner, I. Kikic, and M. Perrut (eds.), *Proceedings of the Sixth International Symposium on Supercritical Fluids*.

Jin, G., and Donohue, M.D. 1988a. An equation of state for electrolyte solutions. 1. Aqueous systems containing strong electrolytes. *Industrial & Engineering Chemistry Research* 27:1073–1084.

Jin, G., and Donohue, M.D. 1988b. An equation of state for electrolyte solutions. 2. Single volatile weak electrolytes in water. *Industrial & Engineering Chemistry Research* 27:1737–1743.

Johnston, K.P. 1994. Safer solutions for chemists. *Nature* 368:187–188.

Jones, R.P., and Greenfield, P.E. 1982. Effect of carbon dioxide on yeast and fermentation. *Enzyme and Microbial Technology* 4:210–223.

Jordan, P.J., Lay, K., Ngan, N., and Rodley, G.F. 1987. Casein precipitation using high-pressure carbon-dioxide. *New Zealand Journal of Dairy Science and Technology* 22(3):247–256.

Jun, S.W., Kim, M.S., Kim, J.S., Park, H.J., Lee, S., Woo, J.S., and Hwang, S.J. 2007. Preparation and characterization of simvastatin/hydroxypropyl-b-cyclodextrin inclusion complex using supercritical antisolvent (SAS) process. *European Journal of Pharmaceutics and Biopharmaceutics* 66:413–421.

Jung, J., and Perrut, M. 2001. Particle design using supercritical fluids: literature and patent survey. *Journal of Supercritical Fluids* 20:179–219.

Kaiser, T., Vomerbaumer, C., and Schweiger, G. 1992. A new approach to the determination of fluid phase equilibria: concentration measurements by Raman spectroscopy. *Berichte der Bunsengesellschaft für Physikalische Chemie* 96:976–980.

Kalogiannis, C.G., Pavlidou, E., and Panayiotou, C.G. 2005. Production of amoxicillin microparticles by supercritical antisolvent precipitation, *Industrial & Engineering Chemistry Research* 44:9339–9346.

Kamat, S., Critchley, G., Beckman, E.J., and Russell, A.J. 1995. Biocatalytic synthesis of acrylates in organic solvents and supercritical fluids: 3. Does carbon dioxide covalently modify enzymes? *Biotechnology and Bioengineering* 46(1):610–620.

Kamihira, M., Taniguchi, M., and Kobayashi, T. 1987. Sterilization of microorganisms with supercritical CO_2. *Agricultural and Biological Chemistry* 51:407–412.

Kaminoga, S., Mizobuchi, H., and Yamauchi, K. 1972. Comparison of Bovine Milk Protease with Plasmin. *Agricultural and Biological Chemistry* 36(12):2163–2167.

Kang, Y., Wua, J., Yin, G., Huang, Z., Yao, Y., Liao, X., Chen, A., Pu, X., and Liao, L. 2008a. Preparation, characterization and in vitro cytotoxicity of indomethacin-loaded PLLA/PLGA microparticles using supercritical CO_2 technique. *European Journal of Pharmaceutics and Biopharmaceutics* 70:85–97.

Kang, Y., Yin, G., Ouyang, P., Huang, Z., Yao, Y., Liao, X., Chen, A., and Pu, X. 2008b. Preparation of PLLA/PLGA microparticles using solution enhanced dispersion by supercritical fluids (SEDS). *Journal of Colloid and Interface Science* 322:87–94.

Karagul-Yuceer, Y., Wilson, J.C., and White, C.H. 2001. Formulations and processing of yogurt affect the microbial quality of carbonated yogurt. *Journal of Dairy Science* 84(3):543–550.

Karagul-Yuceer, Y., Coggins, P.C., Wilson, J.C., and White, C.H. 1999. Carbonated yogurt: sensory properties and consumer acceptance. *Journal of Dairy Science* 82(7):1394–1398.

Karukstis, K.K., and Van Hecke, G.R. 2000. *Chemistry Connections: The Chemical Basis of Everyday Phenomena*. San Diego, CA: Academic Press.

Kauffman, F.L., Shank, J.L., and Urbain, W.M. 1969. Irradiation with CO_2 under pressure. US Patent 3,483,005.

Kayrak, D., Akman, U., and Hortaçsu, Ö. 2003. Micronization of ibuprofen by RESS. *Journal of Supercritical Fluids* 26(1): 17–31.

Kerc, J., Srcic, S., Knez, Z., and Sencar-Bozic, P. 1999. Micronization of drugs using supercritical carbon dioxide. *International Journal of Pharmaceutics* 182:33–39.

Kiepe, J., Horstmann, S., Fischer, K., and Gmehling, J. 2004. Application of the PSRK model for systems containing strong electrolytes. *Industrial & Engineering Chemistry Research* 43:6607–6615.

Kiepe, J., Horstmann, S., Fischer, K., and Gmehling, J. 2002. Experimental determination and prediction of gas solubility data for $CO_2 + H_2O$ mixtures containing NaCl or KCI at temperatures between 313 and 393 K and pressures up to 10 MPa. *Industrial & Engineering Chemistry Research* 41:4393–4398.

Kikic, I., Aless, P., Eva, F., and Moneghini, M., B. 2006. Supercritical antisolvent precipitation of atenolol: the influence of the organic solvent and of the processing approach. *Journal of Supercritical Fluids* 38:434–441.

Kikic, I., Lora, M., and Bertucco, A. 1997. A thermodynamic analysis of three-phase equilibria in binary and ternary systems for applications in rapid expansion of a supercritical solution (RESS), particles from gas-saturated solutions (PGSS), and supercritical antisolvent (SAS). *Industrial & Engineering Chemistry Research* 36:5507–5515.

Kikic, I., Fermeglia, M., and Rasmussen, P. 1991. UNIFAC predictions of vapour–liquid equilibria in mixed solvent–salt systems. *Chemical Engineering Science* 46:2775–2780.

Kilimann, K.V., Hartmann, C., Delgado, A., Vogel, R.F., and Gänzle, M.G. 2006. Combined high pressure and temperature induced lethal and sublethal injury of *Lactococcus lactis* – application of multivariate statistical analysis. *International Journal of Food Microbiology* 109:25–33.

Kim, C.H., Clark, A.B., Vimalchand, P., and Donohue, M.D. 1989. High–pressure binary phase equilibria of aromatic hydrocarbons with carbon dioxide and ethane. *Journal of Chemical & Engineering Data* 34(4):391–395.

Kim, J.H., Paxton, T.E., and Tomasko, D.L. 1996. Microencapsulation of naproxen using rapid expansion of supercritical solutions. *Biotechnology Progress* 12(5):650–661.

Kim, M.S., Jin, S.J., Kim, J.S., Park, H.J., Song, H.S., Neubert, R.H.H., and Hwang, S.J. 2008b. Preparation, characterization and in vivo evaluation of amorphous atorvastatin calcium nanoparticles using supercritical antisolvent (SAS) process. *European Journal of Pharmaceutics and Biopharmaceutics* 69:454–465.

Kim, M.S., Lee, S., Park, J.S., Woo, J.S., and Hwang, S.J. 2007b. Micronization of cilostazol using supercritical antisolvent (SAS) process: effect of process parameters. *Powder Technology* 177:64–70.

Kim, S.R., Rhee, M.S., Kim, B.C., and Kim, K.H. 2007a. Modeling the inactivation of *Escherichia coli* O157:H7 and generic *Escherichia coli* by supercritical carbon dioxide. *International Journal of Food Microbiology* 118:52–61.

Kim, Y.H., and Shing, K.S. 2008a. Supercritical fluid-micronized ipratropium bromide for pulmonary drug delivery. *Powder Technology* 182:25–32.

Kim, Y.H., Sioutas, C., Fine, P., and Shing, K.S. 2008c. Effect of albumin on physical characteristics of drug particles produced by supercritical fluid technology. *Powder Technology* 182:354–363.

Kincal, D., Hill, W.S., Balaban, M.O., Portier, K.M., Sims, C.A., Wei, C.I., and Marshall, M.R. 2006. A continuous high pressure carbon dioxide system for cloud and quality retention in orange juice. *Journal of Food Science* 71(6):C338–C344.

Kincal, D., Hill, W.S., Balaban, M.O., Portier, K.M., Wei, C.I., and Marshall, M.R. 2005. A continuous high pressure CO_2 system for microbial reduction in orange juice. *Journal of Food Science* 70(5):M249–M254.

Kincal, D. 2000. Continuous cold pasteurization of orange juice with high pressure CO_2. MSc thesis, University of Florida, Gainesville, FL.

King, M.B., Mubarak, A., Kim, J.D., and Bott, T.R. 1992. The mutual solubilities of water with supercritical and liquid carbon dioxide. *Journal of Supercritical Fluids* 5(4):296–302.

King, J.S., and Mabbitt, L.A. 1982. Preservation of raw milk by the addition of carbon dioxide. *Journal of Dairy Research* 49:439–447.

Klotz, B., Pyle, D.L., and Mackey, B.M. 2007. A new mathematical modeling approach for predicting microbial inactivation by high hydrostatic pressure. *Applied and Environmental Microbiology* 73:2468–2478.

Knapman, K. 2000. Polymorphic predictions. *Modern Drug Discovery* 3:53.

Knapp, H., Yang, X., and Zhang, Z. 1990. Vapor–liquid equilibria in ternary mixtures containing nitrogen, methane, ethane and carbon dioxide at low temperatures and high pressures. *Fluid Phase Equilibria* 54:1–18.

Knez, Z. 2000. Micronization of pharmaceuticals using supercritical fluids. *Proceedings of the 7th Meeting on Supercritical Fluids*, Antibes, France, vol. 1, pp. 21–26. Nancy, France: International Society for the Advancement of Supercritical Fluids (ISASF).

Kobayashi, F., Yazama, F., Ikeura, H., Hayata, Y., Muto, N., and Osajima, Y. 2009. Inactivation of microorganisms in untreated water by a continuous flow system with supercritical CO_2 bubbling. *Journal of Water and Environment Technology* 7(4):241–250.

Kojima, K., and Tochigi, K. 1979. *Prediction of Vapour–Liquid Equilibria by the ASOG Method*. Amsterdam: Elsevier.

Kordikowski, A., Shekunov, T., and York, P. 2001. Polymorph control of sulfathiazole in supercritical CO_2. *Pharmaceutical Research* 18:682–688.

Korshid, N., Hossain, M.M., and Farid, M.M. 2007. Precipitation of food protein using high pressure carbon dioxide. *Journal of Food Engineering* 79:1214–1220.

Kosikowski, F.V., and Brown, D.P. 1973. Influence of carbon dioxide and nitrogen on microbial population and shelf-life of cottage cheese and sour cream. *Journal of Dairy Science* 56(1):12–18.

Koutchma, T., and Murakami, E. 2005. *Effect of Carbon Dioxide and Pressure Processing on Microbial and Enzyme Inactivation in Food and Beverages*. Chicago: National Center in Food Safety and Technology. http://www.foodtechinternational.com/papers/effectofcarbondioxide.htm

Kumagai, H., Hata, C., and Nakamura, K. 1997. CO_2 sorption by microbial cells and sterilization by high-pressure CO_2. *Bioscience, Biotechnology and Biochemistry* 61:931–935.

Kurz, F., Rumpf, B., and Maurer, G. 1996a. Simultaneous solubility of ammonia and carbon dioxide in aqueous solutions of ammonium sulfate and (ammonium sulfate+sodium sulfate) at temperatures from 202 K to 282 K and pressures to 10 MPa. *Journal of Chemical Thermodynamics* 28:497–520.

Kurz, F., Rumpf, B., Sing, R., and Maurer, G. 1996b. Vapor–liquid and vapor–liquid–solid equilibria in the system ammonia–carbon dioxide–sodium chloride–water at temperatures from 313 to 393 K and pressures up to 3 MPa, *Industrial Engineering Chemistry Research* 35:3795–3802.

Langelaan, H.C., Bartels, P.V., and Hulleman, S.H.D. 1999. Method for the extraction of a substance from a starting material and extraction apparatus for carrying out the method. WO Patent 99/26707.

Lauret, A., Richon, D., and Renon, H. 1994. Measurements of physical and thermodynamic properties of products obtained through direct coal liquefaction. *International Journal of Energy Research* 18(2):267–275.

Law, B.A., and Mabbitt, L.A. 1983. New methods for controlling the spoilage of milk and milk products. In: T.A. Roberts and F.A. Skinner (eds.), *Food Microbiology: Advances and Prospects*, pp. 131–150. London: Academic Press.

Lecky, M., and Balaban, M.O. 2004. Continuous high pressure carbon dioxide processing of watermelon juice. Paper presented at the Institute of Food Technologists annual meeting, Las Vegas, NV.

Lee, L.Y., Wang, C.H., and Smith, K.A. 2008b. Supercritical antisolvent production of biodegradable micro- and nanoparticles for controlled delivery of paclitaxel. *Journal of Controlled Release* 125:96–106.

Lee, R.J., and Chao, K.C. 1988. Extraction of 1–methylnaphthalene and m–cresol with supercritical carbon dioxide and ethane. *Fluid Phase Equilibria* 43(2–3):329–340.

Lee, S.Y., Jung, I.I., Kim, J.K., Lim, G.B., and Ryu, J.H. 2008a. Preparation of itraconazole/HP– β–CD inclusion complexes using supercritical aerosol solvent extraction system and their dissolution characteristics. *Journal of Supercritical Fluids* 44:400–408.

Lele, A.K., and Shine, A.D. 2002. Effect of RESS dynamics on polymer morphology. *Industrial & Engineering Chemistry Research* 33(6):1476–1485.

Lele, A.K., and Shine, A.D. 1992. Morphology of polymers precipitated from a supercritical solvent. *AIChE Journal* 38(5):742–752.

Lewis, M., and Heppell, N. 2000. *Continuous Thermal Processing of Foods: Pasteurization and UHT Sterilization*. Gaithersburg, MD: Aspen Publishers.

Li, D., and Duan, Z. 2007. The speciation equilibrium coupling with phase equilibrium in the H_2O–CO_2–NaCl system from 0 to 250°C, from 0 to 1000 bar, and from 0 to 5 molality of NaCl. *Chemical Geology* 244:730–751.

Li, G., Chu, J., Song, E. S., Row, K.H., Lee, K.H., and Lee, Y.W. 2006. Crystallization of acetaminophen micro-particle using supercritical carbon dioxide. *Korean Journal of Chemical Engineering* 23:482–487.

Li, Y., Yang, D.J., Chen, S.L., Chen, S.B., and Chan, A.S.C. 2008. Process parameters and morphology in puerarin, phospholipids and their complex microparticles generation by supercritical antisolvent precipitation. *International Journal of Pharmaceutics* 359:35–45.

Li, Z., Jiang, J., Liu, X., Tang, H., and Wei, W. 2009. Experimental investigation on the micronization of aqueous cefadroxil by supercritical fluid technology. *Journal of Supercritical Fluids* 48(3):247–252.

Li, Z., Dong, M., Li, S., and Dai, L. 2004. Densities and solubilities for binary systems of carbon dioxide + water and carbon dioxide + brine at 59°C and pressures to 29 MPa. *Journal of Chemical & Engineering Data* 49(4):1026–1031.

Liao, X., Zhang, Y., Bei, J., Hu, X., and Wu, J. 2009. Alterations of molecular properties of lipoxygenase induced by dense phase carbon dioxide. *Innovative Food Science & Emerging Technologies* 10:47–53.

Lim, S., Yagiz, Y., and Balaban, M.O. 2006. Continuous high pressure carbon dioxide processing of mandarin juice. *Food Science and Biotechnology* 15(1):13–18.

Lin, H-M., Cao, N.J., and Chen, L-F. 1994. Antimicrobial effect of pressurized carbon dioxide on *Listeria monocytogenes*. *Journal of Food Science* 59:657–659.

Lin, H-M., Yang, Z.Y., and Chen, L-F. 1993. Inactivation of *Leuconostoc dextranicum* with carbon dioxide under pressure. *Chemical Engineering Journal and the Biochemical Engineering Journal* 52:B29–B34.

Lin, H-M., Yang, Z.Y., and Chen L-F. 1992. Inactivation of *Saccharomyces cerevisiae* by supercritical and subcritical carbon dioxide. *Biotechnology Progress* 8:458–461.

Lin, H-M., Chan, E-C., Chen, C., and Chen, L-F. 1991. Disintegration of yeast cells by pressurized carbon dioxide. *Biotechnology Progress* 7:201–204.

Liu, Y., Hu, X.S., Zhao, X.Y., and Zhang, C. 2011.Inactivation of polyphenol oxidase from watermelon juice by high pressure carbon dioxide treatment. *Journal of Food Science and Technologies* doi:10.1007/s13197-011-0356-6.

Liu, Y., Zhang, C., Zhao, X.Y., Ma, Y., Li, W., Liao, X.J., and Hu, X.S. 2010. Inactivation of polyphenol oxidase from frozen red raspberry (*Rubus idaeus* L.) by high pressure carbon dioxide treatment. *International Journal of Food Science and Technologies* 45:800–806.

Liu, D.D., and Prausnitz, J.M. 1979. Thermodynamics of gas solubilities in molten polymers. *Journal of Applied Polymer Science* 24:725–733.

Liu, D.D., and Prausnitz, J.M. 1980. Calculation of phase equilibria for mixtures of ethylene and low-density polyethylene at high pressures. *Industrial and Engineering Chemistry Process Design and Development* 19:205–211.

Liu, G-T., and Nagahama, K. 1996. Application of rapid expansion of supercritical solutions in the crystallization separation. *Industrial & Engineering Chemistry Research* 35(12):4626–4634.

Liu, X., Gao, Y., Peng, X., Yang, B., Xu, H., and Zhao, J. 2008. Inactivation of peroxidase and polyphenol oxidase in red beet (*Beta vulgaris* L.) extract with high pressure carbon dioxide. *Innovative Food Science & Emerging Technologies* 9:24–31.

Loss, C.R. 2001. Effect of dissolved carbon dioxide on the thermal resistance of microorganisms in milk. MS dissertation, Cornell University, Ithaca, NY.

Loss, C.R., and Hotchkiss, J.H. 2002. Effect of dissolved carbon dioxide on thermal inactivation microorganisms in milk. *Journal of Food Protection* 65:1924–1929.

Loth, H., and Hemgesberg, E. 1986. Properties and dissolution of drugs micronized by crystallization from supercritical gases. *International Journal of Pharmaceutics* 32(2–3):265–267.

Lucien, F.P., and Foster, N.R. 1999. Phase behavior and solubility. In: P.G. Jessop and W. Leitner (eds.), *Chemical Synthesis Using Supercritical Fluids*, pp. 37–53. Weinheim: Wiley-VCH.

Ma, Y., Barbano, D.M., and Santos, M. 2003. Effect of CO_2 addition to raw milk on proteolysis and lipolysis at 4°C. *Journal of Dairy Science* 86(5):1616–1631.

Ma, Y., Barbano, D.M., Hotchkiss, J.H., Murphy, S., and Lynch, J.M. 2001. Impact of CO_2 addition to milk on selected analytical testing methods. *Journal of Dairy Science* 84(9):1959–1968.

Ma, Y., Ryan, C., Barbano, D.M., Galton, D.M., Rudan, M.A., and Boor, K.J. 2000. Effects of somatic cell count on quality and shelf-life of pasteurized fluid milk. *Journal of Dairy Science* 83(2):264–274.

Mabbitt, L.A. 1982. Preservation of refrigerated milk. *Kieler Milchwirtschaftliche Forschungsberichte* 34:28–31.

Macedo, E.A., Skovborg, P., and Rasmussen, P. 1990. Calculation of phase equilibria for solutions of strong electrolytes in solvent/water mixtures. *Chemical Engineering Science* 45:875–882.

Mader, U.K. 1991. H_2O–CO_2 mixtures: a review of P–V–T–X data and an assessment from a phase–equilibrium point of view. *Canadian Mineralogist* 29:767–790.

Madigan, M.T., Martinko, J.M., and Parker, J. 2002. *Brock Biology of Microrganisms*. Upper Saddle River, NJ: Prenctice Hall.

Malcolmson, R.J., and Embleton, J.K. 1998. Dry powder formulations for pulmonary delivery. *Pharmaceutical Science and Technology Today* 1:394–398.

Malinin, S.D., and Savelyeva, N.I. 1992. The solubility of CO_2 in NaCl and $CaCI_2$ solutions at 25, 50, and 75°C under elevated CO_2 pressures. *Geochemistry International* 9:410–418.

Malinin, S.D., and Kurovskaya, N.A. 1975. Investigation of CO_2 solubility in a solution of chlorides at elevated temperatures and pressures of CO_2. *Geochemistry International* 12:199–201.

Maniar, A.B., Marcy, J.E., Bishop, R.J., and Duncan, S.E. 1994. Modified atmosphere packaging to maintain direct-set cottage cheese quality. *Journal of Food Science* 59(6):1305–1308.

Marshall, C.E. 1902. The aeration of milk. *Michigan Agricicultural Experiment Station Special Bulletin* 16:25–30.

Martin, J.D., Werner, B.G., and Hotchkiss, J.H. 2003. Effects of carbon dioxide on bacterial growth parameters in milk as measured by conductivity. *Journal of Dairy Science* 86(6):1932–1940.

Martin, H.J., Schmidt, P.C., Wahl, M.A., Hils, P., Helfgen, B., Turk, M., and Schaber, K. 2000. Nanoscale particles for pharmaceutical purpose by rapid expansion of supercritical solutions (RESS). Part II: Characterization of the product and use. In: M. Perrut and E. Reverchon (eds.), *Proceedings of the Seventh Meeting on Supercritical Fluids Tome 1.*

Marty, A., Chulalaksananukul Willemot, R.W.M., and Condoret, J.S. 1992. Kinetics of lipase-catalyzed esterification in supercritical CO_2. *Biotechnology and Bioengineering* 39:273–280.

Marx, M., Stuparic, M., Schieber, A., and Carle, R. 2003. Effects of thermal processing on trans–cis-isomerization of β-carotene in carrot juices and carotene-containing preparations. *Food Chemistry* 4(83):609–617.

Matos, H.A., De Azevedo, E.G., Simoes, P.C., Carrondo, M.T., and Nunes da Ponte, M. 1989. Phase equilibria of natural flavours and supercritical solvents. *Fluid Phase Equilibria* 52:357–364.

Matous, J., Sobr, J., Novak, J. P., and Pick, J. 1969. Solubilities of carbon dioxide in water at pressure up to 40 atm. *Collection of Czechoslovak Chemical Communications* 3:3982–3985.

Matsuyama, K., Mishima, K. Hayashi, K-I., Ishikawa, H., Matsuyama, H., and Harada, T. 2003. Formation of microcapsules of medicines by the rapid expansion of a supercritical solution with a nonsolvent. *Journal of Applied Polymer Science* 89(3):742–752.

Mattick, K.L., Legan, J.D., Humphrey T.J., and Peleg, M. 2001. Calculating *Salmonella* inactivation in non-isothermal heat treatments from isothermal non linear survival curves. *Journal of Food Protection* 64:606–613.

Meyssami, B., Balaban, M.O., and Teixeira, A.A. 1992. Prediction of pH in model systems pressurized with carbon dioxide. *Biotechnology Progress* 8:149–154.

Michelsen, M.L. 1990. A method for incorporating excess Gibbs free energy models in equations of state. *Fluid Phase Equilibria* 60:47–58.

Miller, F.A., Brandão, T.R.S., Teixeira, P., and Silva, C.L.M. 2006. Recovery of heat-injured *Listeria innocua. International Journal of Food Microbiology* 112:261–265.

Mishima, K. 2008. Biodegradable particle formation for drug and gene delivery using supercritical fluid and dense gas. *Advanced Drug Delivery Reviews* 60:411–432.

Mishima, K., Matsuyama, K., Tanabe, D., Yamauchi, S., Young, T.J., and Johnston, K.P. 2000. Microencapsulation of proteins by rapid expansion of supercritical solution with a nonsolvent. *AIChE Journal* 46(4):857–865.

Moir, C.J., Eyles, M.J., and Davey, J.A. 1993. Inhibition of *Pseudomonads* in cottage cheese by packaging in atmospheres containing carbon dioxide. *Food Microbiology* 10(4):345–351.

Molin, G. 1983. The resistance to carbon dioxide of some food related bacteria. *European Applied Microbiology and Biotechnology* 18(4):214–217.

Molina-Gutierrez, A., Stippl, V., Delgado, A., Gänzle, M.G., and Vogel, R.F. 2002. In situ determination of the intracellular pH of *Lactobacillus plantarum* during pressure treatment. *Applied and Environmental Microbiology* 68:4399–4406.

Mollerup, J. 1986. A note on derivation of mixing rules from excess Gibbs free energy models. *Fluid Phase Equilibria* 25:323–327.

Moneghini, M., Kikic, I., Voinovich, D., Perissutti B., Alessi P., and Cortesi A. 2003. Study of the solid state of carbamazepine after processing with gas anti-solvent technique. *European Journal of Pharmaceutics and Biopharmaceutics* 56:281–289.

Moneghini, M., Kikic, I., Voinovich, D., Perissutti, B., and Filipovic-Grcic, J. 2001. Processing of carbamazepine-PEG 4000 solid dispersions with supercritical carbon dioxide: preparation, characterization, and in vitro dissolution. *International Journal of Pharmaceutics* 222:129–138.

Montgomery, R., and Swenson, C. 1969. *Quantitative Problems in the Biochemical Sciences*. San Francisco: W.H. Freeman.

Moribe, K., Tsutsumi, S., Morishita, S., Shinozaki, H., Tozuka, Y., Oguchi, T., and Yamamoto, K. 2005. Micronization of phyenylbutazone by rapid expansion of supercritical CO_2 solution. *Chemical & Pharmaceutical Bulletin* 53(8):1025–1028.

Morr, C.V. 1989. Whey proteins: manufacture. *Developments in Dairy Chemistry* 4(6):245–284.

Moshashaee, S., Bisrat, M., Forbes, R.T., Nyqvist, H., and York, P. 2000. Supercritical fluid processing of proteins. I: lysozyme precipitation from organic solution. *European Journal of Pharmaceutical Sciences* 11:239–245.

Muhrer, G., and Mazzotti, M. 2003. Precipitation of lysozyme nanoparticles from dimethyl sulfoxide using carbon dioxide as antisolvent. *Biotechnology Progress* 19:549.

Muller, G., Bender, E., and Maurer, G. 1988. Das Dampf-Flussigkeits gleichgewicht des ternaren Systems Ammoniak-Kohlendioxid-Wasser bei holen Wassergehalten im Bereich zwischen 373 und 473 Kelvin. *Berichte der Bunsengesellschaft für Physikalische Chemie* 92:148–160.

Muller, L.L., and Hayes, J.F. 1962. Improved equipment for continuous precipitation of acid casein. *Australian Journal of Dairy Technology* (October–December):189–193.

Mun, S., Hahn, J.S., Lee, Y.W., and Yoon, J. 2011. Inactivation behavior of *Pseudomonas aeruginosa* by supercritical N_2O compared to supercritical CO_2. *International Journal of Food Microbiology* 144:372–378.

Murakami, E., Reddy, N.R., Larkin, J.W., Sizer, C.E., Meneghel, R., Ting, E.Y., Takeuchi, K., and Chirtel, S.J. 2004. Effects of adding either CO_2 or air during high pressure processing of fluid foods. Paper presented at the Institute of Food Technologists annual meeting, Las Vegas, NV, 12–16 July.

Myers, J.A., Sandler, S.I., and Wood, R.H. 2002. An equation of state for electrolyte solutions covering wide ranges of temperature, pressure and composition. *Industrial & Engineering Chemistry Research* 41:3282–3297.

Nakamura, K., Enomoto, A., Fukushima, H., Nagai, K., and Hakoda, M. 1994. Disruption of microbial cells by the flash discharge of high-pressure carbon dioxide. *Bioscience, Biotechnology and Biochemistry* 58:1297–1301.

Neau, E., Garnier, S., Alessi, P., Cortesi, A., and Kikic, I. 1996. Modelling solubility of biological compounds in supercritical fluids. In: *Proceedings of the 3rd International Symposium on High Pressure Technology*, pp. 265–270, Zurich.

Ng, H-J., and Robinson, D.B. 1985. Hydrate formation in systems containing methane, ethane, propane, carbon dioxide or hydrogen sulfide in the presence of methanol. *Fluid Phase Equilibria* 21(1–2):145–155.

Nicolas, S., Karsten, P., and Jonathan, E.K. 2003. CO_2–H_2O mixtures in the mutual solubilities from 12 to 100°C and up to 600 bar. *Geochimica et Cosmochimica Acta* 67:3015–3031.

Nighswander, J.A., Kaiogerakis, N., and Mehrotra, A.K. 1989. Solubilities of carbon dioxide in water and 1 wt % NaCl solution at pressures up to 10 Mpa and temperatures from 80 to 200°C. *Journal of Chemical & Engineering Data* 34(3):355–360.

Niu, L., Hu, X., Wu, J., Liao, X., Chen, F., Zhao, G., and Wang, Z. 2010a. Effect of dense phase carbon dioxide process on physicochemical properties and flavor compounds of orange juice. *Journal of Food Processing and Preservation* 34:530–548.

Niu, S., Xu, Z., Fang, Y., Zhang, L., Yang, Y., Liao, X., and Hu, X. 2010b. Comparative study on cloudy apple juice qualities from apple slices treated by high pressure carbon dioxide and mild heat. *Innovative Food Science and Emerging Technologies* 11:91–97.

Noll, C.I., and Supplee, G.C. 1941. Factors affecting the gas content of milk. *Journal of Dairy Science* 24:993–998.

Norde, W. 1992. The behaviour of proteins at interfaces in relation to their structural stability. In: *Proceedings of an International Symposium*, pp. 3–11, Maastricht, the Netherlands.

Noriega, L., Gueimonde, M., Alonso, L., and de los Reyes-Gavilan, C.G. 2003. Inhibition of *Bacillus cereus* growth in carbonated fermented bifidus milk. *Food Microbiology* 20(5):519–526.

North, J.W., Blackwell, V.R., and Morgan, J.J. 1998. Studies of CO_2 hydrates formation and dissolution. *Environmental Science Technology* 32:676–681.

Ohgaki, K., Kobayashi, H., Katayama, T., and Hirokawa, N. 1990. Whisker formation from jet of supercritical fluid solution. *Journal of Supercritical Fluids* 3(3):103–107.

Oliveira, D., Feihrmann, A.C., Rubira, A.F., Kunita, M.H., Dariva, C., and Oliveira, J.V. 2006. Assessment of two immobilized lipases activity treated in compressed fluids. *Journal of Supercritical Fluids* 38:373–382.

Orbey, N., and Sandler, S.I. 1995. Reformulation of Wong–Sandler mixing rules for cubic equations of state. *AIChE Journal* 41:683–690.

Osajima, Y., Shimoda, M., and Miyake, M. 2004. Continuous processing method and continuous processing apparatus for liquid-form substance, and liquid-form substance processed thereby. US Patent 6,821,481 B1.

Osajima, Y., Shimoda, M., Takada, M., and Miyake, M. 2003. Method of and system for continuously processing liquid materials, and the product processed thereby. US Patent 6,616,849 B1.

Osajima, Y. 2000. Unfolding of poly-L-glutamic acid by microbubbling of supercritical carbon dioxide. *Bioscience, Biotechnology, and Biochemistry* 64(6):1295–1297.

Osajima, Y., Shimoda, M., Kawano, T., and Okubo, K. 1999a. System for processing liquid foodstuff or liquid medicine with a supercritical fluid of carbon dioxide. US Patent 5,869,123.

Osajima, Y., Shimoda, M., Kawano, T., and Okubo, K. 1999b. Method of and system for processing liquid foodstuff or liquid medicine with a supercritical fluid of carbon dioxide. European Patent 0,786,513 B1.

Osajima, Y., Shimoda, M., Kawano, T., and Okubo, K. 1998. System for processing liquid foodstuff or liquid medicine with a supercritical fluid of carbon dioxide. US Patent 5,704,276.

Osajima, Y., Shimoda, M., and Kawano, T. 1997. Method for inactivating enzymes, microorganisms and spores in a liquid foodstuff. US Patent 5,667,835.

Osajima, Y., Shimoda, M., and Kawano, T. 1996. Method for modifying the quality of liquid foodstuff. US Patent 5,520,943.

Oslon, J.D. 1989. Measurement of a vapor–liquid equilibria by ebulliometry. *Fluid Phase Equilibria* 52:209–218.

Oulé, M.K., Tano, K., Bernier, A-M., and Arul, J. 2006. *Escherichia coli* inactivation mechanism by pressurized CO_2. *Canadian Journal of Microbiology* 52:1208–1217.

Palakodaty, S., and York, P. 1999. Phase behavioral effects on particle formation processes using supercritical fluids. *Pharmaceutical Research* 16:976–985.

Panagiotopoulos, A.Z., and Reid, R.C. 1986. New mixing rules for cubic equations of state for highly polar, asymmetric mixtures, equation of state – theories and applications. *ACS Symposium Series* 300:571–582.

Papaiconomou, N., Simonin, J.P., Bernard, O., and Kunz, W. 2003. Description of vapor–liquid equilibria for CO_2 in electrolyte solutions using the mean spherical approximation. *The Journal of Physical Chemistry B* 107:5948–5957.

Park, S-J., Lee, J-I., and Park, J. 2002. Effects of a combined process of high-pressure carbon dioxide and high hydrostatic pressure on the quality of carrot juice. *Journal of Food Science* 67:1827–1834.

Park, S-B., Shim, C-S., Lee, H., and Lee, K-H. 1997. Solubilities of carbon dioxide in the aqueous potassium carbonate and potassium carbonate-poly(ethylene glycol) solutions. *Fluid Phase Equilibria* 134(1–2):141–149.

Parton, T., Bertucco, A., and Bertoloni, G. 2007. Pasteurisation of grape must and tomato paste by dense-phase CO_2. *Italian Journal of Food Science* 19:425–437.

Parton, T., Toniolo, C., Elvassore, N., and Bertucco, A. 2003. Preservation of nutritive properties of tomato sauce by high pressure CO_2 pasteurization. In: *Proceedings of the 6th International Symposium on Supercritical Fluids*, Versailles, France.

Parton, T., Spilimbergo, S., Elvassore, N., and Bertucco, A. 2002. UV-vis spectroscopy for the determination of diffusion coefficient and pH in aqueous solutions/SC–CO_2 systems. In: *Proceedings of the 4th International Symposium on High Pressure Technology and Chemical Engineering*, pp. 447–452, Lido di Venezia, Italy, September.

Pasquali, I., Bettini, R., and Giordano, F. 2008. Supercritical fluid technologies: an innovative approach for manipulating the solid-state of pharmaceuticals. *Advanced Drug Delivery Reviews* 60:399–410.

Patel, N.C., and Teja, A.S. 1982. A new cubic equation of state for fluids and fluid mixtures. *Chemical Engineering Science* 37:463–473.

Peleg, M. 2006. *Advanced Quantitative Microbiology for Food and Biosystems: Models for Predicting Growth and Inactivation*. Boca Raton FL: CRC Press.

Peleg, M., Normand, M.D., and Corradini, M.G. 2005. Generating microbial survival curves during thermal processing in real time. *Journal of Applied Microbiology* 98:406–417.

Peleg, M. 2003. Microbial survival curves: interpretation, mathematical modeling and utilization. *Comments on Theoretical Biology* 8:357–387.

Peleg, M. 2002. Simulation of *E. coli* inactivation by carbon dioxide under pressure. *Journal of Food Science* 67:896–901.

Peleg, M., and Cole, M.B. 1998. Reinterpretation of microbial survival curves. *Critical Reviews in Food Science and Nutrition* 38:353–380.

Peng, D.Y., and Robinson, D.B. 1976. A new two-constant equation of state. *Industrial and Engineering Chemistry Fundamentals* 15:59–64.

Perez de Diego, Y., Pellikaan, H.C., Wubbolts, F.E., Borchard, G., Witkamp, G.J., and Jansens, P.J. 2006. Opening new operating windows for polymer and protein micronisation using the PCA process. *Journal of Supercritical Fluids* 36:216–224.

Perrut, M., Jung, J., and Leboeuf, F. 2005a. Enhancement of dissolution rate of poorly-soluble active ingredients by supercritical fluid processes: part I: micronization of neat particles. *International Journal of Pharmaceutics* 288(1):3–10.

Perrut, M., Jung, J., and Leboeuf, F. 2005b. Enhancement of dissolution rate of poorly soluble active ingredients by supercritical fluid processes: part II: preparation of composite particles. *International Journal of Pharmaceutics* 288(1):11–16.

Perrut, M., and Clavier, J-Y. 2003. Supercritical fluid formulation: process choice and scale-up. *Industrial & Engineering Chemistry Research* 42(25): 6375–6383.

Pestov, D., Levit, N., and Leboeuf, F. 2003. Development of encapsulated microspheres from a multiple-component, rapidly expanding supercritical solution for sensing applications. In: *Proceedings of the 6th International Symposium on Supercritical Fluids*, Versailles, France.

Pfohl, O., Timm, J., Dohrn, R., and Brunner, G. 1996. Measurement and correlation of vapor–liquid–liquid equilibria in the glucose + acetone + water + carbon dioxide system. *Fluid Phase Equilibria* 124(1–2):221–233.

Pitzer, K.S. 1973. Thermodynamics of electrolytes. I. Theoretical basis and general equations. *Journal of Physical Chemistry* 77:268–277.

Plaza, M.L. 2010. Quality improvement of guava puree by dense phase carbon dioxide treatment. PhD dissertation, University of Florida, Gainesville.

Plaza, M.L., Ramirez-Rodrigues, M.M., Ferrentino, G., and Balaban, M.O. 2009. Optimization of dense phase carbon dioxide pasteurization of guava puree. In: *Proceedings of the IFT Annual Meeting*, Anaheim, CA, 5–10 June.

Prausnitz, J.M., Lichtenthaler, R.N., and de Azevedo, E.G. 1999. *Molecular Thermodynamics of Fluid Phase Equilibria*, 3rd ed. Englewood Cliffs, NJ: Prentice Hall.

Primo, M.S., Ceni, G.C., Marcon, N.S., Antunes, O.A.C., Oliveira, D., Oliveira, J.V., and Dariva, C. 2007. Effects of compressed carbon dioxide treatment on the specificity of oxidase enzymatic complexes from mate tea leaves. *Journal of Supercritical Fluids* 43:283–290.

Pruess, K., and Garcia, J. 2002. Multiphase flow dynamics during CO_2 disposal into saline aquifers. *Environmental Geology* 42(2–3):282–295.

Pyo, D., Lim, C., Cho, D., and Oh, D. 2007. Size prediction of recombinant human growth hormone nanoparticles produced by supercritical fluid precipitation. *Analytical and Bioanalytical Chemistry* 387:901–907.

Quan, C., Johan, C., and Charlotta, T. 2009. Carotenoids particle formation by supercritical fluid technologies. *Chinese Journal of Chemical Engineering* 17(2):344–349.

Raghuraman, B., Muhammad, M., Gao, J., Borman, C., Gustavson, G., and Rabbito, P. 2007. Spectroscopic pH measurement at high-temperature and/or high-pressure. US Patent Application 2007/0178595 A1.

Rajagopal, M., Werner, B.G., and Hotchkiss, J.H. 2005. Low pressure CO_2 storage of raw milk: microbiological effects. *Journal of Dairy Science* 88:3130–3138.

Ramirez-Rodrigues, M.M., Plaza, M.L., Ferrentino, G., and Balaban, M.O. 2009. Dense phase CO_2 process optimization of a *Hibiscus sabdariffa* beverage. In: *Proceedings of the IFT Annual Meeting*, Anaheim, CA, 5–10 June.

Raouche, S., Dobenesque, M., Bot, A., Lagaude, A., Cuq, J-L., and Marchesseau, S. 2007. Stability of casein micelle subjected to reversible CO_2 acidification: impact of holding time and chilled storage. *International Dairy Journal* 17(8):873–880.

Randolph, T.W., Blanch, H.W., and Clark, D.S. 1991. Biocatalysis in supercritical fluids. In: J.S. Dordick (ed.), *Biocatalysis for Industry*, pp. 219–237. New York: Plenum Press.

Rashed, M.A., Mehanna, N.M., and Mehanna, A.S. 1986. Effect of carbon dioxide on improving the keeping quality of raw milk. *International Journal of Dairy Technology* 39(2):62–64.

Raso, J., and Barbosa-Cánovas, G.V. 2003. Nonthermal preservation of foods using combined processing techniques. *Critical Reviews in Food Science and Nutrition* 43:265–285.

Redlich O., and Kwong J.N.S. 1949. On the thermodynamic of solutions. V. Fugacities of gaseous solutions. *Chemical Reviews* 44:233–244.

Rehman, M., Shekunov, B.Y., York, P., Lechuga-Ballesteros, D., Miller, D.P., Tan, T., and Colthorpe, P. 2004. Optimisation of powders for pulmonary delivery using supercritical fluid technology. *European Journal of Pharmaceutical Science* 22:1–17.

Rehman, M., Shekunov, B.Y., York, P., and Colthorpe, P. 2001. Solubility and precipitation of nicotinic acid in supercritical carbon dioxide. *Journal of Pharmaceutical Sciences* 90:1570–1582.

Renon, H., Laugier, S., Schwartzentruber, J., and Richon, D. 1989. New determinations of high pressure vapor–liquid equilibria in binary systems containing n–propylbenzene with nitrogen or carbon dioxide consistent with the Prausnitz–Keeler test. *Fluid Phase Equilibria* 51:285–298.

Reverchon, E., Adami, R., Cardea, S., and Porta, G.D. 2009. Supercritical fluids processing of polymers for pharmaceutical and medical applications. *Journal of Supercritical Fluids* 47:484–492.

Reverchon, E., Lamberti, G., and Antonacci, A. 2008. Supercritical fluid assisted production of HPMC composite microparticles. *Journal of Supercritical Fluids* 46(2):185–196.

Reverchon, E., De Marco, I., and Torino, E. 2007. Nanoparticles production by supercritical antisolvent precipitation: a general interpretation. *Journal of Supercritical Fluids* 43:126–138.

Reverchon, E., and Antonacci, A. 2007a. Drug–polymer microparticles produced by supercritical assisted atomization. *Biotechnology and Bioengineering* 97(6):1626–1637.

Reverchon, E., and Antonacci, A. 2007b. Polymer microparticles production by supercritical assisted atomization. *Journal of Supercritical Fluids* 39(3):444–452.

Reverchon, E., and Marco, I.D. 2006. Supercritical antisolvent precipitation of cephalosporins. *Powder Technology* 164:139–146.

Reverchon, E., Adami, R., and Caputo, G. 2006. Supercritical assisted atomization: performance comparison between laboratory and pilot scale. *Journal of Supercritical Fluids* 37(3):298–306.

Reverchon, E., and Antonacci, A. 2006. Cyclodextrins micrometric powders obtained by supercritical fluid processing. *Biotechnology and Bioengineering* 94(4):753–761.

Reverchon, E., and Spada, A. 2004. Erythromycin micro-particles produced by supercritical fluid atomization. *Powder Technology* 141(1–2):100–108.

Reverchon, E., Della Porta, G., Spada, A., and Antonacci, A. 2004. Griseofulvin micronization and dissolution rate improvement by supercritical assisted atomization. *Journal of Pharmacy and Pharmacology* 56:1379–1387.

Reverchon, E., and De Marco, I. 2004. Supercritical antisolvent micronization of Cefonicid: thermodynamic interpretation of results. *Journal of Supercritical Fluids* 31:207–215.

Reverchon, E., De Marco, I., Caputo, G., and Della Porta, G. 2003a. Pilot scale micronization of amoxicillin by supercritical antisolvent precipitation. *Journal of Supercritical Fluids* 26:1–7.

Reverchon, E., and Della Porta, G. 2003b. Micronization of antibiotics by supercritical assisted atomization. *Journal of Supercritical Fluids* 26(3): 243–252.

Reverchon, E., and Della Porta, G. 2003c. Terbutaline microparticles suitable for aerosol delivery produced by supercritical assisted atomization. *International Journal of Pharmaceutics* 258(1–2):1–9.

Reverchon, E., Della Porta, G., and Spada, A. 2003d. Ampicillin micronization by supercritical assisted atomization. *Journal of Pharmacy and Pharmacology* 55:1465–1471.

Reverchon, E. 2002. Supercritical-assisted atomization to produce micro- and/or nanoparticles of controlled size and distribution. *Industrial & Engineering Chemistry Research* 41(10):2405–2411.

Reverchon, E., De Marco, I., and Della Porta, G. 2002. Rifampicin microparticles production by supercritical antisolvent precipitation. *International Journal of Pharmaceutics* 243:83–91.

Reverchon, E., Della Porta, G., and Falivene, M. G. 2000. Process parameters and morphology in amoxicillin micro and submicro particles generation by supercritical antisolvent precipitation. *Journal of Supercritical Fluids* 17:239–248.

Rezaei, K., Temelli, F., and Jenab, E. 2007. Effects of pressure and temperature on enzymatic reactions in supercritical fluids. *Biotechnology Advances* 25:272–280.

Rezaei, K., and Temelli, F. 2000. Lipase-catalyzed hydrolysis of canola oil in supercritical carbon dioxide. *JAOCS* 77(8):903–909.

Richard, J. 2003. Microparticle engineering for drug delivery using a supercritical fluid-based process: application to injectable sustained-release formulations of biomolecules. In: *Abstracts of Papers*, MSE-409, 226th ACS national meeting, New York, 7–11 September.

Richard, J., Dulieu, C., Le Meurlay, D., and Benoit, J.P. 2001. Microparticles for pulmonary administration comprising at least a coating agent. In: *PCT International Patent Applications*, WO Patent 0112160.

Riondet, C., Cachon, R., Waché, Y., Alcaraz, G., and Diviès, C. 1997. Measurement of the intracelullar pH in *Escherichia coli* with the internally conjugated fluorescent probe 5-(and 6-)carboxyfluorescein succinimidyl ester. *Biotechnology Techniques* 11:735–738.

Roberts, R.F., and Torrey, G.S. 1988. Inhibition of psychrotrophic bacterial growth in refrigerated milk by addition of carbon dioxide. *Journal of Dairy Science* 71(1):52–60.

Rodrigues, M.M.R. 2010. Processing of a hibiscus beverage using dense phase carbon dioxide. PhD dissertation. University of Florida, Gainesville.

Rodrigues, M.A., Li, J., Padrela, L., Almeida, A., Matos, H.A., and de Azevedo, E.G. 2009. Anti-solvent effect in the production of lysozyme nanoparticles by supercritical fluid-assisted atomization processes. *Journal of Supercritical Fluids* 48(3):253–260.

Rosenbauer, R.J., Bischoff, J.L., and Koksalan, T. 2001. An experimental approach to CO_2 sequestration in saline aquifers: application to Paradox Valley. In: *Proceedings of the American Geophysical Union*, Abstract V32B-0974.

Rosenkranz, K., Kasper, M.M., Werther, J., and Brunner, G. 2008. Encapsulation of irregularly shaped solid forms of proteins in a high-pressure fluidized bed. *Journal of Supercritical Fluids* 46(3):351–357.

Rosenthal, L., Rosen, B., Bernstein, S., and Popel, G. 1991. Preservation of fresh cheeses in a CO_2-enriched atmosphere. *Milchwissenschaft* 46:706–708.

Ross, A.I.V., Griffiths, M.W., Mittal, G.S., and Deeth, H.C. 2003. Combining nonthermal technologies to control foodborne microorganisms. *International Journal of Food Microbiology* 89:125–138.

Ruas-Madiedo, P., Bascaran, V., Brana, A.F., Bada-Gancedo, J.C., and de los Reyes-Gavilan, C.G. 1998. Influence of carbon dioxide addition to raw milk on microbial levels and some fat-soluble vitamin contents of raw and pasteurized milk. *Journal of Agriculture and Food Chemistry* 46(4):1552–1555.

Ruas-Madiedo, P., Bada-Gancedo, J.C., Fernandez-Garcia, E., Gonzalez de Llano, D., and de los Reyes-Gavilan, C.G. 1996. Preservation of the microbiological and biochemical quality of raw milk by carbon dioxide addition: a pilot-scale study. *Journal of Food Protection* 59:502–508.

Rumpf, B., Xia, J., and Maurer, G. 1998. Solubility of carbon dioxide in aqueous solutions containing acetic acid or sodium hydroxide in the temperature range from 313 to 433 K and at total pressures up to 10 MPa. *Industrial & Engineering Chemistry Research* 37(5):2012–2019.

Rumpf, B., Xia, J., and Maurer, G. 1997. An experimental investigation of the solubility of carbon dioxide in aqueous solutions containing sodium nitrate or ammonium nitrate at temperatures from 313 K to 433 K and pressures up to 10 MPa. *Journal of Chemical Thermodynamics* 29(10):1101–1111.

Rumpf, B., Nicolaisen, H., Öcal, C., and Maurer, G. 1994. Solubility of carbon dioxide in aqueous solutions of sodium chloride: experimental results and correlation. *Journal of Solution Chemistry* 23:431–448.

Rumpf, B., and Maurer, G. 1993. An experimental and theoretical investigation on the solubility of carbon dioxide in aqueous solutions of strong electrolytes. *Berichte der Bunsengesellschaft für Physikalische Chemie* 97(1):85–97.

Sako, T., Sugeta, T., Nakazawa, N., Okubo, T., Sato, M., Taguchi, T., and Hiaki, T. 1991. Phase equilibrium study of extraction and concentration of furfural produced in reactor using supercritical carbon dioxide. *Journal of Chemical Engineering of Japan* 24(4):449–455.

Salinas-Hernández, R., Ruiz-Trevino, F.A., Ortiz-Estrada, C-H., Luna-Bárcenas, G., Prokhorov, Y., Alvarado, J.F.J., and Sanchez, I.C. 2008. Chitin microstructure formation by rapid expansion techniques with supercritical carbon dioxide. *Industrial & Engineering Chemistry Research*. 48(2):769–778.

Sander, B., Fredenslund, A., and Rasmussen, P. 1986. Calculation of vapour–liquid equilibria in mixed solvent/salt systems using an extended UNIQUAC equation. *Chemical Engineering Science* 41:1171–1183.

Sandler, S.I. 1989. *Chemical and Engineering Thermodynamics*, 2nd ed. New York: John Wiley & Sons, Inc.

Sane, A., and Limtrakul, J. 2009. Formation of retinyl palmitate-loaded poly (l-lactide) nanoparticles using rapid expansion of supercritical solutions into liquid solvents (RESOLV). *Journal of Supercritical Fluids* 51(2):230–237.

Sapru, V., Teixeira, A.A., Smeradge, G.H., and Lindsay, J.A. 1992. Predicting thermophilic spore population dynamics for UHT sterilization processes. *Journal of Food Science* 57:1248–52, 1257.

Sarkari, M., Darrat, I., and Knutson, B.L. 2003. CO_2 and fluorinated solvent-based technologies for protein microparticle precipitation from aqueous solutions. *Biotechnology Progress* 19:448–454.

Sattelmayer, T., and Wittig, S. 1986. Internal flow effects in prefilming airblast atomizers: mechanisms of atomization and droplet spectra. *Journal of Engineering for Gas Turbines and Power* 108:465–472.

Scharlin, P. 1996. Carbon dioxide in water and aqueous electrolyte solutions. In: J.W. Lorimer (ed.), *IUPAC Solubility Data Series*, vol. 62. Oxford: Oxford University Press.

Schmidt, C.F., Zimmermann, R., and Gaub, H.E. 1990. Multilayer adsorption of lysozyme on a hydrophobic substrate. *Biophysical Journal* 54:577–588.

Schneider, G.M. 1975. Phase equilibria of liquid and gaseous mixtures at high pressures. In: B. Le Neindre and B. Vodar (eds.), *Experimental Thermodynamics*, pp. 787–801. London: Butterworth Press.

Sellers, S.P., Clark, G.S., Sievers, R.E., and Carpenter, J.F. 2001. Dry powders of stable protein formulations from aqueous solutions prepared using supercritical CO_2 assisted aerosolization. *Journal of Pharmaceutical Sciences* 90(6): 785–797.

Sencar-Bozic, P., Srcic, S., Knez, Z., and Kerc, J. 1997. Improvement of nifedipine dissolution characteristics using supercritical CO_2. *International Journal of Pharmaceutics* 148:123–130.

Sethia, S., and Squillante, E. 2002. Physicochemical characterization of solid dispersions of carbamazepine formulated by supercritical carbon dioxide and conventional solvent evaporation method. *Journal of Pharmaceutical Science* 91:1948–1957.

Shimoda, M. 2004. Method of processing liquid food and processing apparatus. WO Patent 2004/045316 A1.

Shimoda, M., Kago, H., Kojima, N., Miyake, M., Osajima, Y., and Hayakawa I. 2002. Accelerated death kinetics of *Aspergillus niger* spores under high-pressure carbonation. *Applied and Environmental Microbiology* 68(8): 4162–4167.

Shimoda, M., Cocunubo-Castellanos, J., Kago, H., Miyake, M., Osajima, Y., and Hayakawa, I. 2001. The influence of dissolved CO_2 concentration on the death kinetics of *Saccharomyces cerevisiae*. *Journal of Applied Microbiology* 91:306–311.

Shimoda, M., Yamamoto, Y., Cocunubo-Castellanos, J., Tonoike, H., Kawano, T., Ishikawa, H., and Osajima, Y. 1998. Antimicrobial effects of pressured carbon dioxide in a continuous flow system. *Journal of Food Science* 63:709–712.

Shinozaki, H., Oguchi, T., Suzuki, S., Aoki, K., Sako, T., Morishita, S., Tozuka, Y., Moribe, K., and Yamamoto, K. 2006. Micronization and polymorphic conversion of tolbutamide and barbital by rapid expansion of supercritical solutions. *Drug Development and Industrial Pharmacy* 32(7):877–891.

Shyu, G-S., Hanif, N.S.M., Hall, K.R., and Eubank, P.T. 1997. Carbon dioxide–water phase equilibria results from the Wong–Sandler combining rules. *Fluid Phase Equilibria* 130:73–85.

Sievers, R.E., Quinn, B.P., Cape, S.P., Searles, J.A., Braun, C.S., Bhagwat, P., Rebits, L.G., McAdams, D.H., Burger, J.L., Best, J.A., Lindsay, L., Hernandez, M.T., Kisich, K.O., Iacovangelo, T., Kristensen, D., and Chen, D. 2007. Near-critical fluid micronization of stabilized vaccines, antibiotics and anti-virals. *Journal of Supercritical Fluids* 42(3):385–391.

Sievers, R.E., Huang, E.T.S., Villa, J.A., Engling, G., and Brauer, P.R. 2003. Micronization of water-soluble or alcohol-soluble pharmaceuticals and model compounds with a low-temperature Bubble Dryer®. *Journal of Supercritical Fluids* 26(1): 9–16.

Sievers, R.E., Huang, E.S., Villa, J.A., and Walsh, T.R. 2002. CAN-BD process for rapidly forming and drying fine particles. In: *Proceeding of the 8th Meeting on Supercritical Fluids: Chemical Reactivity and Material Processing in Supercritical Fluids*. Bordeaux, France: International Society for the Advancement of Supercritical Fluids.

Sievers, R.E., Milewski, P.D., Sellers, S.P., Miles, B.A., Korte, B.J., Kusek, K.D., Clark, G.S., Mioskowski, B., and Villa, J.A. 2000. Supercritical and near-critical carbon dioxide assisted low-temperature bubble drying. *Industrial & Engineering Chemistry Research* 39(12):4831–4836.

Sievers, R.E., Milewski, P.D., Sellers, S.P., Kusek, K.D., Kleutz, P.G., and Miles, B.A. 1998. Supercritical CO_2-assisted methods for the production and pulmonary administration of pharmaceutical aerosols. *Journal of Aerosol Science* 29(Suppl. 2):S1271–S1272.

Sih, R. 2008. New process development of dense gas technology for the processing of pharmaceuticals. PhD thesis, University of New South Wales, Australia.

Sims, M., and Estigarribia, E. 2003. Membrane carbon dioxide sterilization of liquid foods: scale up of a commercial continuous process. In: *Proceedings of the 6th International Symposium on Supercritical Fluids*, pp. 1457–1460, Versailles, France, 28–30 April.

Sims, M., and Estigarribia, E. 2002. Continuous sterilization of aqueous pumpable food using high pressure CO_2. In: A. Bertucco (ed.), *Proceedings of 4th International Symposium on High Pressure Process Technology and Chemical Engineering*, pp. 921–927. Venice, Italy: AIDIC Servizi S.r.I.

Sims M. 2000. Method and membrane system for sterilizing and preserving liquids using carbon dioxide. US Patent 6,331,272 B1 and WO Patent 00/41805.

Singh, A., Yeager, R., and McFeters, G.A. 1986. Assessment of in vivo revival, growth, and pathogenicity of *Escherichia coli* strains after copper- and chlorine-induced injury. *Applied and Environmental Microbiology* 52:832–837.

Sirisee, U., Hsieh, F., and Huff, H.E. 1998. Microbial safety of supercritical carbon dioxide processes. *Journal of Food Processing and Preservation* 22:387–403.

Smelt, J.P.P.M., and Rijke, G.G.F. 1992. High pressure treatment as a tool for pasteurization of foods. *High Press Biotechnology* 224:361–364.

Smith J.M., Van Ness H.C., and Abbott M.M. 2005. *Introduction to Chemical Engineering Thermodynamics*, 7th ed. New York: McGraw-Hill.

Smith, A.C. 1964. The carbon dioxide content of milk during handling, processing and storage and its effect upon the freezing point. *Journal of Milk Food Technology* 27:38–45.

Soave G. 1972. Equilibrium constants from a modified Redlich–Kwong equation of state. *Chemical Engineering Science* 39:357–369.

Song, K.Y., and Kobayashi, R. 1987. Water content of CO_2 in equilibrium with liquid water and/or hydrates. *SPE Formation Evaluation* 2(4):500–508.

Sousa, M.J., Ardo, Y., and McSweeney, P.L.H. 2001. Advances in the study of proteolysis during cheese ripening. *International Dairy Journal* 11(4–7):327–345.

Southward, C.R., and Walker, N J. 1980. The manufacture and industrial use of casein. *New Zealand Journal of Dairy Science and Technology* 15:201–217.

Span, R., and Wagner, W. 1996. A new equation of state for carbon dioxide covering the fluid region from the triple–point temperature to 1100 K at pressures up to 800 MPa. *Journal of Physical and Chemical Reference Data* 25(6):1509–1596.

Spilimbergo, S., and Ciola, L. 2010. Supercritical CO_2 and N_2O pasteurisation of peach and kiwi juice. *International Journal of Food Science and Technology* 45:1619–1625.

Spilimbergo, S., Quaranta, A., Garcia-Gonzalez, L., Contrini, C., Cinquemani, C., and Van Ginneken, L. 2010. Intracellular pH measurement during high-pressure CO_2 pasteurization evaluated by cell fluorescent staining. *Journal of Supercritical Fluids* 53:185–191.

Spilimbergo, S., Mantoan, D., Quaranta, A., and Della Mea, G. 2009. Real-time monitoring of cell membrane modification during supercritical CO_2 pasteurization. *Journal of Supercritical Fluids* 48:93–97.

Spilimbergo, S., Bertucco, A., Basso, G., and Bertoloni, G. 2005. Determination of extracellular and intracellular pH of *Bacillus subtilis* suspension under CO_2 treatment. *Biotechnology and Bioengineering* 92:447–451.

Spilimbergo, S., and Bertucco, A. 2003. Non-thermal bacteria inactivation with dense CO_2. *Biotechnology and Bioengineering* 84:627–638.

Spilimbergo, S., Dehghani, F., Bertucco, A., and Foster, N.R. 2003a. Inactivation of bacteria and spores by pulse electric field and high pressure CO_2 at low temperature. *Biotechnology and Bioengineering* 82(1):118–125.

Spilimbergo, S., Elvassore, N., and Bertucco, A. 2003b. Inactivation of microorganisms by supercritical CO_2 in a semi-continuous process. *Italian Journal of Food Science* 15:15–24.

Spilimbergo, S., Elvassore, N., and Bertucco, A. 2002. Microbial inactivation by high pressure. *Journal of Supercritical Fluids* 22:55–63.

Spilimbergo, S. 2002. A study about the effect of dense CO_2 on microorganisms. PhD dissertation, University of Padova, Italy.

Spycher, N., Pruess, K., and Ennis-King, J. 2003. CO_2–H_2O mixtures in the geological sequestration of CO_2. I. Assessment and calculation of mutual solubilities from 12 to 100°C and up to 600 bar. *Geochimica et Cosmochimica Acta* 67:3015–3031.

Staby, A., and Mollerup, J. 1991. Measurement of solubilities of 1 – pentanol in supercritical ethane. *Journal of Supercritical Fluids* 4(1):233–237.

Stahl, E., Quirin, K.W., and Gerard, D. 1986. Application of dense gassing to extraction and refining. In: *Dense Gases for Extraction and Refining*, pp. 220–221, Berlin: Springer-Verlag.

Steckel, H., Pichert, L., and Muller, B.W. 2004. Influence of process parameters in the ASES process on particle properties of budesonide for pulmonary delivery. *European Journal of Pharmaceutics and Biopharmaceutics* 57: 507–512.

Steckel, H., Thies, J., and Miller, B. W. 1997. Micronizing of steroids for pulmonary delivery by supercritical carbon dioxide. *International Journal of Pharmaceutics* 152:99–110.

Su, C-S., Tang, M., and Chen, Y-P. 2009. Micronization of nabumetone using the rapid expansion of supercritical solution (RESS) process. *Journal of Supercritical Fluids* 50(1):69–76.

Subramaniam, B., Lyon, C.J., and Arunajatesan, V. 2002. Environmentally benign multiphase catalysis with dense phase carbon dioxide. *Applied Catalysis B: Environmental* 37(4):279–292.

Suppes, G.J., and McHugh, M.A. 1989. Phase behavior of the carbon dioxide–styrene system. *Journal of Chemical & Engineering Data* 34(3):310–312.

Takenouchi, S., and Kennedy, G.C. 1965. The solubility of carbon dioxide in NaCl solutions at high temperatures and pressures. *American Journal of Science* 263:445–454.

Takenouchi, S., and Kennedy, G.C. 1964. The binary system H_2O–CO_2 at high temperatures and pressures. *American Journal of Science* 262:1055–1074.

Tamine, A.Y., and Deeth, H.C. 1980. Yogurt technology and biochemistry. *Journal Food Protection* 43:939–977.

Tanaka, H., Yamaki, Y., and Kato, M. 1993. Solubility of carbon dioxide in pentadecane, hexadecane and pentadecane + hexadecane. *Journal of Chemical & Engineering Data* 38(3):386–388.

Tandya, A., Mammucari, R., Dehghani, F., and Foster, N.R. 2007. Dense gas processing of polymeric controlled release formulations. *International Journal of Pharmaceutics* 328:1–11.

Tandya, A., Dehghani, R., and Foster, N.R. 2006. Micronization of cyclosporine using dense gas techniques. *Journal of Supercritical Fluids* 37(3):272–278.

Taniguchi, M., Kamihira, M., and Kobayashi, T. 1987. Effect of treatment with supercritical carbon dioxide on enzymatic activity. *Agricultural and Biological Chemistry* 51(2):593–594.

Taran, A.I., Averciyanova, O.A., and Podshivalenko, N.S. 2002. *Antimicrobial Properties of CO_2 Extracts*. Moscow: Pishchevaya Promyshlennostc.

Tedjo, W., Eshtiaghi, M.N., and Knorr, D. 2000. Impact of supercritical carbon dioxide and high pressure on lipoxygenase and peroxidase activity. *Journal of Food Science* 65(8):1284–1287.

Teixeira, A. 1992. Thermal process calculations. In: D.R. Heldman and D.B. Lund (eds.), *Handbook of Food Engineering*. New York: Marcel Dekker.

Teng, H., and Yamasaki, A. 1998. Solubility of liquid CO_2 in water at temperatures from 278 K to 293 K and pressures from 6.44 MPa to 29.49 MPa and densities of the corresponding aqueous solutions. *Journal of Chemical &. Engineering Data* 43(1):2–5.

Tenorio, A., Gordillo, M.D., Pereyra, C., and Martınez de la Ossa, E.J. 2007. Controlled submicro particle formation of ampicillin by supercritical antisolvent precipitation. *Journal of Supercritical Fluids* 40:308–316.

Tetra Pak Processing Systems AB. 1998. *The Orange Book*. Lund, Sweden: Tetra Pak Processing Systems AB.

Thakur, R., and Gupta, R.B. 2006. Formation of phenytoin nanoparticles using rapid expansion of supercritical solution with solid cosolvent (RESS-SC) process. *International Journal of Pharmaceutics* 308(1–2):190–199.

Thakur, R., and Gupta, R.B. 2005. Rapid expansion of supercritical solution with solid cosolvent (RESS-SC) process: formation of griseofulvin nanoparticles. *Industrial & Engineering Chemistry Research* 44:7380–7387.

Thiering, R., Dehghani, F., Dillow, A., and Foster, N.R. 2000a. Solvent effects on the controlled dense gas precipitation of model proteins. *Journal of Chemical Technology & Biotechnology* 75:42–53.

Thiering, R., Dehghani, F., Dillow, A., and Foster, N.R. 2000b. The influence of operating conditions on the dense gas precipitation of model proteins. *Journal of Chemical Technology & Biotechnology* 75:29–41.

Timsina, M.P., Martin, G.P., Marriott, C., Ganderton, D., and Yianneskis, M. 1994. Drug delivery to the respiratory tract using dry powder inhalers. *International Journal of Pharmaceutics* 101:1–13.

Tisi, D.A. 2004. Effect of dense phase carbon dioxide on enzyme activity and casein proteins in raw milk. PhD dissertation, Cornell University, Ithaca, NY.

Todheide, K., and Franck, E.U. 1963. Das Zweiphasen gebiet und die kritische Kurve im System Kohlendioxid-Wasser bis zu Drucken von 3500bar. *Zeitschrift für Physikalische Chemic* 37:387–401.

Toews, K.L., Shroll, R.M., and Wai, C.M. 1995. pH-defining equilibrium between water and supercritical CO_2: influence of SFE of organics and metal chelates. *Analytical Chemistry* 67:4040–4043.

Tom, J.W., Lim, G.B., Debenedetti, P.G., and Prudhomme, R.K. 1993. Application of supercritical fluids in the controlled release of drugs. In: E. Kiran and J.F. Brennecke (eds.), *Supercritical Fluid Engineering Science and Technology*, pp. 238–257. ACS Symposium Series 514. Washington, DC: American Chemical Society.

Tomasula, P.M., and Boswell, R.T. 1999. Measurement of the solubility of carbon dioxide in milk at high pressures. *Journal of Supercritical Fluids* 16(1): 21–26.

Tomasula, P.M., Boswell, R.T., and Dupre, N.C. 1999a. Buffer properties of milk treated with high pressure carbon dioxide. *Milchissenschaft* 54:667–670.

Tomasula, P.M., and Boswell, R.T. 1999b. Measurement of the solubility of carbon dioxide in milk at high pressure. *Journal of Supercritical Fluids* 16(1):21–26.

Tomasula, P.M., Craig, J.C., and Mcaloon, A. J. 1998. Economic analysis of a continuous casein process using carbon dioxide as precipitant. *Journal of Dairy Science* 81: 3331–3342.

Tomasula, P.M., Craig, J.C., and Boswell, T.R. 1997. A continuous process for casein production using high-pressure carbon dioxide. *Journal of Food Engineering* 33(3–4):405–419.

Tomasula, P.M., Craig, J.C., Boswell, T.R., Cook, R.D., Kurantz, M.J., and Maxwell, M. 1995. Preparation of casein using carbon-dioxide. *Journal of Dairy Science* 78(3):506–514.

Tong, H.H.Y., Shekunov, B.Y., York, P., and Chow, A.H.L. 2001. Characterization of two polymorphs of salmeterol xinafoate crystallized from supercritical fluids. *Pharmaceutical Research* 18:852–858.

Trebble M.A., and Bishnoi P.R. 1987. Development of a new four-parameter equation of state. *Fluid Phase Equilibria* 35:1–18.

Tsuji, M., Sato, Y., and Komiyama, Y. 2005. Inactivation of microorganisms and enzymes in juices by supercritical carbon dioxide method with continous flow system. *Nippon Shokuhin Kagaku Kogaku Kaishi* 52:528–531.

Türk, M. 2009. Manufacture of submicron drug particles with enhanced dissolution behaviour by rapid expansion processes. *Journal of Supercritical Fluids* 47(3):537–545.

Türk, M., Upper, G., and Hils, P. 2006. Formation of composite drug–polymer particles by co-precipitation during the rapid expansion of supercritical fluids. *Journal of Supercritical Fluids* 39(2):253–263.

Türk, M., Hils, P., Helfgen, B., Schaber, K., Martin, H.J., and Wahl, M.A. 2002. Micronization of pharmaceutical substances by the rapid expansion of supercritical solutions (RESS): a promising method to improve bioavailability of poorly soluble pharmaceutical agents. *Journal of Supercritical Fluids* 22(1):75–84.

US Department of Energy. 1999. *Carbon Sequestration Research and Development*. Report doe/sc/fe-1. Washington, DC: Office of Science, Office of Fossil Energy.

US Department of Health and Human Services. 1999. *Grade "A" Pasteurized Milk Ordinance*, rev. ed., vol. 229. Washington, DC: US Department of Health and Human Services, Public Health Service, Food and Drug Administration.

US Food and Drug Administration (FDA). 2000. Kinetics of microbial inactivation for alternative food processing technologies. http://www.cfsan.fda.gov/~comm/ift-over.html

Ulmer, H.M., Burger, D., Ganzle, M.G., Engelhardt, H., and Vogel, R.F. 2002. Effect of compressed gases on the high pressure inactivation of *Lactobacillus plantarum* TMW 1.460. *Progress Biotechnology* (*Trends in High Pressure Bioscience and Biotechnology*) 19:317–324.

Valtz, A., Chapoy, A., Coquelet, C., Paricaud, P., and Richon, D. 2004. Vapour–liquid equilibria in the carbon dioxide–water system, measurement and modelling from 278.2 to 318.2 K. *Fluid Phase Equilibria* 226:333–344.

Valverde, M.T., Marín-Iniesta, F., and Calvo, L. 2010. Inactivation of *Saccharomyces cerevisiae* in conference pear with high pressure carbon dioxide and effects on pear quality. *Journal of Food Engineering* 98:421–428.

van Boekel, M.A.J.S. 2002. On the use of the Weibull model to describe thermal inactivation of microbial vegetative cells. *International Journal of Food Microbiology* 74:139–159.

Van Ginneken, L., Weyten, H., Willems, L., and Lodewijckx, B. 2004. Method and apparatus for the inactivation of microorganisms in liquid food and feed products. WO Patent 2004/039180 A1.

Van Halsema, F.E.D., van der Wielen, L.A.M., and Luyben, K.C.A.M. 1998. The modeling of carbon dioxide–aided extraction of carboxylic acids from aqueous solutions. *Industrial and Engineering Chemistry Research* 37: 748–758.

Van Slyke, L.L., and Keeler, R.F. 1920. The carbon dioxide content as basis for distinguishing heated from unheated milk. *New York Agricultural Experiment Station Bulletin* 78:111–119.

Van Slyke, L.L., and Baker, J.C. 1919. Carbonic acid and carbonate in cow's milk. *New York Agricultural Experiment Station Bulletin* 69:10–25.

Vandenberg, M.G. 1979. Partial-pressure of carbon-dioxide in milk and its relation to the freezing-point. *Netherlands Milk and Dairy Journal* 33:91–111.

Varshosaz, J., Hassanzadeh, F., Mahmoudzadeh, M., and Sadeghi, A. 2009. Preparation of cefuroxime axetil nanoparticles by rapid expansion of supercritical fluid technology. *Powder Technology* 189(1):97–102.

Velaga, S.P., Ghaderi, R., and Carlfors, J. 2002a. Preparation and characterisation of hydrocortisone particles using a supercritical fluids extraction process. *International Journal of Pharmaceutics* 231:155–166.

Velaga, S.P., Berger, R., and Carlfors, J. 2002b. Supercritical fluids crystallization of budesonide and flunisolide. *Pharmaceutical Research* 19:1564–1571.

Ventosa, N., Sala, S., and Veciana, J. 2003. DELOS process: a crystallization technique using compressed fluids 1: comparison to the GAS crystallization method. *Journal of Supercritical Fluids* 26:33.

Ventosa, N., Sala, S., Torres, J., Llibre, J., and Veciana, J. 2001a. Depressurization of an expanded liquid organic solution (DELOS): a new procedure for obtaining submicron or micron sized crystalline particles. *Crystal Growth and Design* 1:299.

Ventosa, N., Sala, S., Veciana, J., and Llibre, J. T. 2001b. *Proceeding of the 6th Conference on Supercritical Fluids and Their Applications*, pp. 325–328, Maiori, Italy.

Ventosa, N., Veciana, J., Rovira, C., and Sala, S., S. 2000. Method for precipitating finely divided solid particles. PCT/ES01/00327. WO Patent 2002/016003.

Villa, J., Huang, E. S., Cape, S., and Sievers, R. 2005. Synthesis of composite microparticles with a mixing cross. *Aerosol Science and Technology* 39(6):473–484.

Vinderola, C.G., Gueimonde, M., Delgado, T., Reinheimer, J.A., and de los Reyes-Gavilan, C.G. 2000. Characteristics of carbonated fermented milk and survival of probiotic bacteria. *International Dairy Journal* 10(3):213–220.

Walstra, P., and Jenness, R. 1984. *Dairy Chemistry and Physics*. New York: Marcel Dekker.

Warwick, B., Dehghani, F., Foster, N.R., Biffin, J.R., and Regtop, H.L. 2002. Micronisation of copper indomethacin using gas anti-solvent processes. *Industrial & Engineering Chemistry Research* 41:1993–2004.

Warwick, B., Dehghani, F., and Foster, N.R. 2000. Synthesis, purification, and micronization of pharmaceuticals using the gas antisolvent technique. *Industrial & Engineering Chemistry Research* 39:4571–4579.

Watanabe, T., Furukawa, S., Hirata, J., Koyama, T., Ogihara, H., and Yamasaki M. 2003a. Inactivation of *Geobacillus stearothermophilus* spores by high-pressure carbon dioxide treatment. *Applied and Environmental Microbiology* 69(12):7124–7129.

Watanabe, T., Furukawa, S., Tai, T., Hirata, J., Narisawa, N., Ogihara, H., and Yamasaki M. 2003b. High pressure carbon dioxide decreases the heat tolerance of the bacterial spores. *Food Science Technology Research* 9:342–34.

Waterling, U., Zheng, D., and Knapp, H. 1991. Vapor–liquid equilibria at high temperatures and pressures in binary mixtures containing H_2, CH_4 and CO_2 with high boiling hydrocarbons: experimental equipment and results. *Chemical Engineering and Processing* 29(3):155–164.

Wei, C.I., Balaban, M.O., Fernando, S.Y., and Peplow, A.J. 1991. Bacterial effect of high-pressure CO_2 treatment on foods spiked with *Listeria* or *Salmonella. Journal of Food Protection* 54:189–193.

Weidner, E., Petermann, M., and Knez, Z. 2003. Multifunctional composites by high pressure spray processes. *Current Opinion in Solid State & Materials Science* 7:385–390.

Weidner, E., Knez, Z., and Novak, Z. 1995. Process for preparing particles or powders. WO Patent 95/21688.

Wendland, M., Hasse, H., and Maurer, G. 1999. Experimental pressure–temperature data on three- and four-phase equilibria of fluid, hydrate, and ice phases in the system carbon dioxide–water. *Journal of Chemical & Engineering Data* 44(5):901–906.

Wendland, M., Hasse, H., and Maurer, G. 1994. Multiphase high–pressure equilibria of carbon dioxide–water–acetone. *Journal of Supercritical Fluids* 7(4):245–250.

Wendland, M., Hasse, H., and Maurer, G. 1993. Multiphase high–pressure equilibria of carbon dioxide–water–isopropanol. *Journal of Supercritical Fluids* 6(4):211–222.

Wenli, Y., Fei, X., Heyang, J., Changchun, L., Yaping, Z., Siyuan, J., and Lin, H. 2008. Production of submicroparticles of β-sitosterol using an aerosol solvent extraction system. *Chinese Journal of Chemical Engineering* 16:956–960.

Werner, B.G., and Hotchkiss, J.H. 2006. Continuous flow nonthermal CO_2 processing: the lethal effects of subcritical and supercritical CO_2 on total microbial populations and bacteria spores in raw milk. *Journal of Dairy Science* 89(3):872–881.

Werner, B.G., and Hotchkiss, J.H. 2002. Effect of carbon dioxide on the growth of *Bacillus cereus* spores in milk during storage. *Journal of Dairy Science* 85(1):15–18.

Whitaker, M.J., Hao, J., Davies, O.R., Serhatkulu, G., Stolnik-Trenkic, S., Howdle, S.M., and Shakesheff, K.M. 2005. The production of protein-loaded microparticles by supercritical fluid enhanced mixing and spraying. *Journal of Controlled Release* 101(1–3):85–92.

White, A., Burns, D., and Christensen, T.W. 2006. Effective terminal sterilization using supercritical carbon dioxide. *Journal of Biotechnology* 123:504–515.

Wiebe, R., and Gaddy, V.L. 1941. Vapor phase composition of carbon dioxide–water mixtures at various temperatures and at pressures to 700 atmospheres. *Journal of the American Chemical Society* 63:475–477.

Wiebe, R., and Gaddy, V.L. 1940. The solubility of carbon dioxide in water at various temperatures from 12 to 40°C and at pressures to 500 atm. *Journal of the American Chemical Society* 62:815–817.

Wiebe, R., and Gaddy, V.L. 1939. The solubility in water of carbon dioxide at 50°, 75°, and 100°C at pressures to 700 atm. *Journal of the American Chemical Society* 61:315–318.

Wildasin, R.E., Forbes, J., Robey, R.J., and Paradis, A.J. 2002. Method and apparatus for continuous flow reduction of microbial and/or enzymatic activity in a liquid product using carbon dioxide. US Patent Application 2002/0122860 A1.

Wilkes, J.G., Conte, E.D., Kim, Y., Holcomb, M., Sutherland, J.B., and Miller, D.W. 2000. Sample preparation for the analysis of flavors and off-flavors in foods. *Journal of Chromatography* 880:3–33.

Wimmer, Z., and Zarevúcka, M. 2010. A review on the effects of supercritical carbon dioxide on enzyme activity. *International Journal of Molecular Sciences* 11:233–253.

Wisniewska, B., Gregorowicz, J., and Malanowski, S. 1993. Development of a vapor–liquid equilibrium apparatus to work at pressures up to 3 MPa. *Fluid Phase Equilibria* 86:173–186.

Witter, L.D., Berry, J.M., and Folinazzo, J.F. 1958. The viability of *Escherichia coli* and a spoilage yeast in carbonated beverages. *Food Research* 23:133–142.

Wong S.S.H., and Sandler S.I. 1992. A theoretically correct mixing rule for cubic equations of state. *AIChE Journal* 38:671–680.

Wroblewski, S.V. 1883. The solubility of carbon dioxide in water. *Annual Review of Physical Chemistry* 18:290–308.

Wu, J., and Prausnitz, J.M. 1998. Phase equilibria for systems containing hydrocarbons, water and salt. *Industrial & Engineering Chemistry Research* 37:1634–1643.

Xia, J., Rumpf, B., and Maurer, G. 1999. Solubility of carbon dioxide in aqueous solutions containing sodium acetate or ammonium acetate at temperatures from 313 to 433 K and pressures up to 10 MPa. *Fluid Phase Equilibria* 155(1):107–125.

Yang, Y., Liao, X., Hu, X., Zhang, Y., Chen, F., and Wu, J. 2011. The contribution of high pressure carbon dioxide in the inactivation kinetics and structural alteration of myrosinase. *International Journal of Food Science and Technologies*. doi:10.1111/j.1365-2621.2011.02614.x.

Yang, S.O., Yang, I.M., Kim, Y.S., and Lee, C.S. 2000. Measurement and prediction of phase equilibria for water + CO_2 in hydrate forming conditions. *Fluid Phase Equilibria* 175(1–2):75–89.

Yener, M.E. 2001. Supercritical fluid processing of food and biomaterials. II. Crystalization, extrusion, microbial inactivation. *Gida* 26:27–31.

Yeo, S., Lim, G., Debenedetti, P.G., and Bernstein, H. 1993. Formation of microparticulate protein powders using a supercritical fluid antisolvent. *Biotechnology and Bioengineering* 41:341–346.

Yıldız, N., Tuna, S., Döker, O., and Çalımlı, A. 2007. Micronization of salicylic acid and taxol (paclitaxel) by rapid expansion of supercritical fluids (RESS). *Journal of Supercritical Fluids* 41(3):440–451.

Yoshimura, T., Furutera, M., Shimoda M., Ishikawa H., Miyake M., Matsumoto K., Osajima Y., and Hayakawa, L. 2002. Inactivation efficiency of enzymes in buffered system by continuous method with microbubbles of supercritical carbon dioxide. *Journal of Food Science* 67(9):3227–3231.

Yu, Z.R., Rizvi, S.S.H., and Zollweg, J.A. 1992. Enzymatic esterification of fatty acid mixtures from milk fat and anhydrous milk fat with canola oil in supercritical carbon dioxide. *Biotech Prog*. 8:508–513.

Yuanhui, J., Xiaoyan, J., Xin, F., Chang, L., Linghong, L., and Xiaohua, L. 2007. Progress in the study on the phase equilibria of the CO_2–H_2O and CO_2–H_2O–NaCl systems. *Chinese Journal of Chemical Engineering* 15:439–448.

Yuk, H-G., and Geveke, D.J. 2010. Nonthermal inactivation and sublethal injury of *Lactobacillus plantarum* in apple cider by a pilot plant scale continuous supercritical carbon dioxide system. *Food Microbiology* 28(3):377–383.

Zawisza, A., and Malesinska, B. 1981. Solubility of carbon dioxide in liquid water and of water in gaseous carbon dioxide in the range 0.2–5 MPa and at temperatures up to 473 K. *Journal of Chemical and Engineering Data* 26(4):388–391.

Zemel, G.P., Sims, C.A., Marshall, M.R., and Balaban, M.O. 1990. Low pH inactivation of polyphenoloxidase in apple juice. *Journal of Food Science* 55(2):562–563.

Zhang, J., Dalal, N., Matthews, M.A., Waller, L.N., Saunders, C., Fox, K.F., and Fox A. 2007. Supercritical carbon dioxide and hydrogen peroxide cause mild changes in spore structures associated with high killing rate. *Journal of Microbiological Methods* 70:442–51.

Zhang, J., Burrows, S., Gleason, C., Matthews, M.A., Drews, M.J., LaBerge, M., and An, Y.H. 2006a. Sterilizing *Bacillus pumilus* spores using supercritical carbon dioxide. *Journal of Microbiological Methods* 66:479–485.

Zhang, J., Davis, T.A., Matthews, M.A., Drews, M.J., LaBerge, M., and An, Y.H. 2006b. Sterilization using high-pressure carbon dioxide. *Journal of Supercritical Fluids* 38:354–372.

Zhao, E.S., Yu, M., Sauvé, R.E., and Khoshkbarchi, M.K. 2000. Extension of Wilson model to electrolyte solutions. *Fluid Phase Equilibria* 173:161–175.

Zhao, X., Zu, Y., Li, Q., Wang, M., Zu, B., Zhang, X., Jiang, R., and Zu, C. 2010. Preparation and characterization of camptothecin powder micronized by a supercritical antisolvent (SAS) process. *Journal of Supercritical Fluids* 51:412–419.

Zhi, X., Zhang, Y., Hu, X., Wu, J., and Liao, X. 2008. Inactivation of apple pectin methyleasterase induced by dense phase carbon dioxide. *Journal of Agricultural and Food Chemistry* 56:5394–5400.

Zhou, L., Zhang, Y., Leng, X., Liao, X., and Hu, X. 2010. Acceleration of precipitation formation in peach juice induced by high-pressure carbon dioxide. *Journal of Agricultural and Food Chemistry* 58(17):9605–9610.

Zhou, L., Wu, J., Hu, X., Zhi, X., and Liao, X. 2009a. Alterations in the activity and structure of pectin methylesterase treated by high pressure carbon dioxide. *Journal of Agricultural and Food Chemistry* 57:1890–1895.

Zhou, L., Zhang, Y., Hu, X., Liao, X., and He, J. 2009b. Comparison of the inactivation kinetics of pectin methylesterases from carrot and peach by high-pressure carbon dioxide. *Food Chemistry* 115:449–455.

Zhou, L., Wang, Y., Hu, X., Wu, J., and Liao, X. 2009c. Effect of high pressure carbon dioxide on the quality of carrot juice. *Innovative Food Science and Emerging Technologies* 10:321–327.

Index

Dense Phase Carbon Dioxide: Food and Pharmaceutical Applications, First Edition. Edited by Murat O. Balaban and Giovanna Ferrentino.